Scaling in Biology

SCALING IN BIOLOGY

EDITORS

James H. Brown

University of New Mexico
Albuquerque, New Mexico .

Geoffrey B. West

Los Alamos National Laboratory
Los Alamos, New Mexico

Sante Fe Institute
Studies in the Science of Complexity

OXFORD
UNIVERSITY PRESS

Oxford University Press

Oxford   New York
Athens   Auckland   Bangkok   Bogotá   Buenos Aires   Calcutta
Cape Town   Chennai   Dar es Salaam   Delhi   Florence   Hong Kong   Istanbul
Karachi   Kuala Lumpur   Madrid   Melbourne   Mexico City   Mumbai
Nairobi   Paris   São Paulo   Singapore   Taipei   Tokyo   Toronto   Warsaw

and associated companies in
Berlin   Ibadan

Published by Oxford University Press, Inc.
198 Madison Avenue, New York, New York 10016

Oxford is a registered trademark of Oxford University Press

Library of Congress Cataloging-in-Publication Data
Scaling in biology / editors, James H. Brown, Geoffrey West.
p.   cm. — (Santa Fe Institute studies in the science of complexity)
Includes bibliographical references and index.
ISBN 0-19-513141-X; ISBN 0-19513142-8 (pbk.)
1. Body size.   2. Morphology (Animals).   3. Biomathematics.
I. Brown, James H., 1942, Sept. 25.   II. West, Geoffrey B.
III. Sante Fe Institute (Santa Fe, N.M.)
QL799.S25   2000
571.3 21—dc21   99-042515

9 8 7 6 5 4 3
Printed in the United States of America
on acid-free paper

About the Santa Fe Institute

The *Santa Fe Institute* (SFI) is a private, independent, multidisciplinary research and education center, founded in 1984. Since its founding, SFI has devoted itself to creating a new kind of scientific research community, pursuing emerging science. Operating as a small, visiting institution, SFI seeks to catalyze new collaborative, multidisciplinary projects that break down the barriers between the traditional disciplines, to spread its ideas and methodologies to other individuals, and to encourage the practical applications of its results.

All titles from the *Santa Fe Institute Studies in the Sciences of Complexity* series will carry this imprint which is based on a Mimbres pottery design (circa A.D. 950–1150), drawn by Betsy Jones. The design was selected because the radiating feathers are evocative of the out-reach of the Santa Fe Institute Program to many disciplines and institutions.

Contributors List

R. McNeill Alexander, *Department of Biology, The University of Leeds, Leeds LS2 9JT, United Kingdom; email: r.m.alexander@leeds.ac.uk*

Andrew A. Biewener, *Department of Organismal Biology & Anatomy, University of Chicago, 1027 East 57th Street, Chicago, IL 60637; e-mail: a-biewener@uchicago.edu*

John Tyler Bonner, *Department of Ecology & Evolutionary Biology, Princeton University, Princeton, NJ 08544; e-mail: jtbonner@phoenix.princeton.edu*

James H. Brown, *Department of Biology, University of New Mexico, Albuquerque, NM 87131 and Santa Fe Institute; e-mail: jhbrown@unm.edu*

William A. Calder, *Department of Ecology & Evolutionary Biology, University of Arizona, Tucson, AZ 85721; e-mail: calderwa@u.arizona.edu*

Hélène Cyr, *Department of Zoology, University of Toronto, 25 Harbord St., Toronto, Ontario M5S 3G5, Canada; e-mail: helene@zoo.utoronto.ca*

Brian J. Enquist, *Santa Fe Institute, 1399 Hyde Park Road, Santa Fe, NM 87501; e-mail: benquist@unm.edu*

John Harte, *Energy & Resources Group, University of California, 310 Barrows Hall, Berkeley, CA 94720; e-mail: jharte@socrates.berkeley.edu*

Paul H. Harvey, *Department of Zoology, University of Oxford, South Parks Road, Oxford OX1 3PS United Kingdom; e-mail: paul.harvey@zoo.ox.ac.uk*

Henry S. Horn, *Department of Ecology & Evolutionary Biology, Princeton University, Princeton, NJ 08544-1003; e-mail: hshorn@princeton.edu*

Rudolf Karch, *Department of Medical Computer Sciences, University of Vienna, Vienna, Austria*

Mimi Koehl, *Department of Integrative Biology, University of California, Berkeley, CA 94720-3140*

Jan Kozlowski, *Institute of Environmental Biology for Institute of Environmental Sciences, Jagiellonian University, OLeandry 2A 30-063 Krakow, Poland; e-mail: kozlo@eko.uj.edu.pl*

Richard E. Lenski, *Center for Microbial Ecology, Michigan State University, East Lansing, MI 48824; e-mail: lenski@pilot.msu.edu*

John K-J. Li, *Department of Biomedical Engineering, Rutgers University, Piscataway, NJ 08854-8014; e-mail: jli@biomed.rutgers.edu*

Judith A. Mongold, *Center for Microbial Ecology, Michigan State University, East Lansing, MI 48824*

Friederike Neumann, *Department of Medical Computer Sciences, University of Vienna, Vienna, Austria*

Martin Neumann, *Institute of Experimental Physics, University of Vienna, Vienna, Austria*

Wolfgang Schreiner, *Department of Medical Computer Sciences, Institut für Medizinische Computerwissenschaften, Spitalgasse 23 A-1090 Vienna, Austria; e-mail: wolfgang.Schreiner@akh-wien.ac.at*

Geoffrey B. West, *Los Alamos National Laboratory, T-8 Mail Stop B285, Los Alamos, NM 87545 and Santa Fe Institute; e-mail: gbw@lanl.gov*

Mair Zamir, *Department of Applied Math, University of Western Ontario, Ontario, Canada; e-mail: zamir@julian.uwo.ca*

Contents

Preface   xi

Scaling in Biology: Patterns and Processes, Causes and Consequences
    James H. Brown, Geoffrey B. West, and Brian J. Enquist   1

Allometry and Natural Selection
    John Tyler Bonner and Henry S. Horn   25

Hovering and Jumping: Contrasting Problems in Scaling
    R. McNeill Alexander   37

Scaling of Terrestrial Support: Differing Solutions to Mechanical
Constraints of Size
    Andrew A. Biewener   51

Consequences of Size Change During Ontogeny and Evolution
    Mimi A. R. Koehl   67

The Origin of Universal Scaling Laws in Biology
    Geoffrey B. West, James H. Brown, and Brian J. Enquist   87

Scaling and Invariants in Cardiovascular Biology
    John K-J. Li   113

Vascular System of the Human Heart: Some Branching and Scaling
Issues
    Mair Zamir   129

Constrained Constructive Optimization of Arterial Tree Models
    *Wolfgang Schreiner, Rudolf Karch, Friederike Neumann,
    and Martin Neumann*   145

Quarter-Power Allometric Scaling in Vascular Plants: Functional
Basis and Ecological Consequences
    Brian J. Enquist, Geoffrey B. West, and James H. Brown   167

Scaling in Biology, edited by J. H. Brown and G. B. West.
Oxford University Press, 2000.

Twigs, Trees, and the Dynamics of Carbon in the Landscape
   Henry S. Horn   199

Cell Size, Shape, and Fitness in Evolving Populations of Bacteria
   Richard E. Lenski and Judith A. Mongold   221

Does Body Size Optimization Alter the Allometries for Production
and Life History Traits?
   Jan Kozłowski   237

Why and How Phylogenetic Relationships Should be Incorporated
into Studies of Scaling
   Paul H. Harvey   253

Individual Energy Use and the Allometry of Population Density
   Hélène Cyr   267

Diversity and Convergence: Scaling for Conservation
   William A. Calder   297

Scaling and Self-Similarity in Species Distributions: Implications for
Extinction, Species Richness, Abundance, and Range
   John Harte   325

Index   343

Preface

On October 27–29, 1997, the Santa Fe Institute hosted a symposium on
"Scaling in Biology: From Organisms to Ecosystems." This brought together
27 scientists with diverse backgrounds and research specializations who, nevertheless, shared a common interest in the scaling of biological systems. These
individuals worked at all levels of biological organization from organ systems
within organisms to whole animals and plants to entire floras and faunas inhabiting large geographic areas. They included experimental laboratory biologists, field ecologists, mathematical modelers, and computer simulation modelers. Despite initial concerns that so much diversity would either prevent
effective communication or lead to divisive discussions, this did not occur.
What emerged after three days of excellent presentations and lively discussions was a consensus on two things. First, there is much in common among
the questions, approaches, and tools that different scientists use to address
issues in biological scaling; and there is much to be learned by escaping the
narrow confines of subdisciplinary specialization to seek a broader integration
and synthesis. Second, the diverse points of view represented at the workshop
were so interesting, that they were worth preserving and presenting to a much
wider audience.

Scaling in Biology, edited by J. H. Brown and G. B. West.
Oxford University Press, 2000.   **xi**

The result is this book. Unfortunately, not all of the individuals who were invited were able to attend the symposium, and not all of the individuals who attended were able to contribute a chapter. Nevertheless, we believe that the 17 chapters in this volume showcase some of the quality and diversity of current research on scaling in biology. The breadth of material and approaches is impressive. As editors, we appreciate the efforts of the authors to make their material accessible to a broad audience and to make connections between their own work and that of other authors. The result, we hope, is something more than the typical volume of symposium proceedings. It is a book that not only presents the enormous range of current research on problems of scale and scaling in biology, but also develops some of the common themes that encompass and unite this work.

We hope that the book will be of interest to a wide audience of biologists and other scientists. We have tried to make the chapters accessible to non-specialists, without compromising the presentation of the most recent, state-of-the-art research. We hope that the book may give some renewed unity and vigor to both broad and specialized work on scaling. This is an area of biology which has a long and distinguished history. Problems of allometry and other scaling relationships have attracted the interest of some of the greatest scientists of the twentieth century: including Cecil Murray, D'Arcy Thompson, Julian Huxley, J. B. S. Haldane, Robert MacArthur, Knut Schmidt-Nielsen, George Bartholomew, and E. O. Wilson. Several influential synthetic books on the topic were published in the early 1980s. We hope that the present volume will bring the picture up to date and revive interest. In particular, we hope that it will be read by students and younger scientists and used in seminar courses and discussion groups.

We note with sadness the untimely death of Thomas McMahon, whose seminal work on biological scaling is frequently cited in this book. Tom participated enthusiastically in the symposium. Failing health prevented him from completing his chapter. The book will suffer from its absence, and the entire field of biological allometry will miss his incisive mind and creative contributions.

Neither the symposium nor this book would have been possible without the help of many people. We thank the Santa Fe Institute (SFI) and the Eugene V. and Clare E.Thaw Charitable Trust for their generous support for the symposium, the preparation of this book, and our own research on scaling. We are especially grateful to Andi Sutherland, Ginger Richardson, other SFI staff, and University of New Mexico graduate students for help in organizing and hosting the symposium. We thank Della Ulibarri, Marylee McInnes, Ronda Butler-Villa, and other SFI staff for preparing the volume for publication and for making our jobs as editors easier and more pleasant. We thank Brian Enquist, our collaborator in scaling research, for all his help with both the symposium and the book. We are grateful to the authors for their efforts to write and rewrite chapters so as to improve the overall quality and integration of the book, and for their patience and cooperation with our efforts to present their work to a wide audience in a timely fashion. Finally, we thank our home institutions, the University of New Mexico and Los Alamos

National Laboratory, for their support and encouragement of our research on scaling and work on this book.

James H. Brown
University of New Mexico

Geoffrey B. West
Los Alamos National Laboratory

20 April 1999

Scaling in Biology: Patterns and Processes, Causes and Consequences

James H. Brown
Geoffrey B. West
Brian J. Enquist

> It is only a slight overestimate to say that the most important attribute of an animal, both physiologically and ecologically, is its size. Size constrains virtually every aspect of structure and function and strongly influences the nature of most inter- and intraspecific interactions. Body mass, which in any given taxon is a close correlate of size, is the most widely useful predictor of physiological rates.
>
> G. A. Bartholomew (1981, p. 46)

Life is amazing. Even the smallest bacterium is far more complex in its structure and function than any known physical system. The largest, most complex organisms, large mammals and giant trees, weigh more than 21 orders of magnitude more than the simplest microbes, yet they use basically the same molecular structures and biochemical pathways to sustain and reproduce themselves. From these incontrovertible observations, two fundamental features of life follow. The first is that biological diversity is largely a matter of size. The variety of sizes plays a central role in the ability of organisms to make their living in so many different ways that they have literally covered the earth, exploiting nearly all of its environments. The second consequence is that in order to achieve such diversity, organisms must adjust their struc-

Scaling in Biology, edited by J. H. Brown and G. B. West.
Oxford University Press, 2000.

1

ture and function to compensate for the geometric, physical, and biological consequences of being different sizes. The principles of mathematics and physics are universal, but their biological consequences depend on the size of the organism.

There is a long history of interest in the correlates and consequences of body size and other issues related to scaling in biology. Theoretical and quantitative studies date back almost a century to the work of such luminaries as D'Arcy Thompson [66], Cecil Murray [47], and Julian Huxley [28]. The more recent contributions of Brody [6], Kleiber [30], Yoda et al. [73], MacArthur and Wilson [37], Taylor et al. [60, 61], and McMahon [42, 44] and influential reviews and syntheses by Peters [50], McMahon and Bonner [45], Schmidt-Nielsen [56], and Calder [13] laid the foundation for the current state of the field. Since the early 1980s, however, the pace of research has seemingly slowed. Excited about our own work [70] and eager to rekindle discussion and research on biological scaling, we decided to host a symposium at the Santa Fe Institute on October 27–29, 1997.

This book contains the published versions of most of the papers presented at that symposium. We hope that the book conveys something of the diversity, promise, and excitement of current research on scaling. In recent years, with the well-publicized advances in many subdisciplines of molecular biology, there has been less attention to higher levels of biological organization. It is becoming increasingly clear, however, that molecular approaches and techniques are inadequate to answer many of the most challenging questions about life. Many of these questions concern the complexities of biological systems at levels of organization from organisms to ecosystems. It is at these levels that research on scaling has made major contributions in the past and has the potential to lead to even greater advances in the future.

1 BACKGROUND

1.1 SIZE AND SCALING

As the size of a physical or biological system changes, the relationships among its different components and processes must be adjusted so that the organism can continue to function. Many anatomical and physiological attributes of organisms change with size in such a way that they remain self-similar. Over a wide range of scales—typically many orders of magnitude—the same relationships among critical structural and functional variables are maintained. Such self-similarity is said to be fractal, and the relationships among the variables can be described by a fractal dimension or a power function. The concept of fractals is fairly new [39], but the use of power functions to characterize scaling laws has a venerable history.

The use of power laws in biology is so well established that they are called allometric equations [45, 56]. They have the form

$$Y = Y_0 M^b \tag{1}$$

where Y is some dependent variable, Y_0 is a normalization constant, M is some independent variable, typically body mass, and b is the scaling exponent.

Biological scaling relationships are called allometric, because the exponent, b, typically differs from unity. If $b = 1$, the relationship is called isometric, and it plots as a straight line on both linear and logarithmic axes (Figure 1). When $b \neq 1$, the relationship is called allometric, and it plots as a curve on linear axes. However, power functions have the nice property that they are linear when plotted on logarithmic axes. This is readily seen by taking the logarithms of both sides of Eq. (1)

$$\log Y = \log Y_0 + b \log M \,. \tag{2}$$

This is equivalent to the equation for a straight line, where the dependent variable, $\log Y$, is equal to an intercept, $\log Y_0$, plus the product of the slope, b, times the independent variable, $\log M$ (Figure 1). Therefore, the exponent of the power function is the slope of the linear plot on logarithmic axes. The allometric exponent is expressed either as a decimal or as a simple

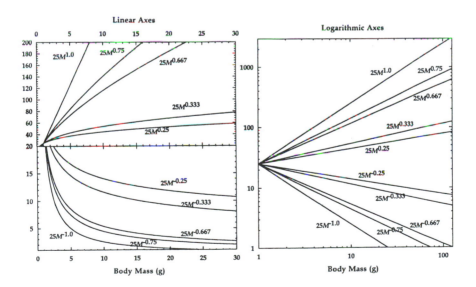

FIGURE 1   Plots of the same allometric relationships on linear and logarithmic axes. Note that the relationships are curvilinear when plotted on linear axes, but linear when plotted on logarithmic axes. Also, when the exponent, b, is > 1, the independent variable, Y, is an increasing function of body mass, M; when $b < 1$, it is a decreasing function.

fraction. The mathematical equivalence of Eqs. (1) and (2) means that it is fairly straightforward to derive empirical allometric relationships by using a least-squares regression technique to fit a linear regression to log-transformed data. Most of the allometric equations in this book were derived from data obtained in this fashion. Some care must be exercised, however, to choose the appropriate regression procedure [25].

The most familiar example of allometry is simple geometric scaling. If we have spheres or any objects of self-similar shapes, we can describe changes in surface area, A, or volume, V, as a function of a linear dimension, the radius, r, as follows: $A = \pi r^2$ and $V = 4/3\pi r^3$. And if the objects maintain a constant density as they vary in size, then $M \propto V$, and we can express their linear dimensions, l, or surface areas as functions of their mass

$$l = c_1 M^{1/3} \text{ and } A = c_2 M^{2/3},$$

where the values of the normalization constants c_1 and c_2 depend on the units of measurement. Since these same equations apply to any shape, if organisms preserve self-similar shapes as they vary in size, then their linear dimensions should vary as the $1/3$ and their surface areas as the $2/3$ powers of their mass. The naïve expectation would be that organisms of the same general body plan would scale geometrically. That is, they would resemble one of those sets of size-nested painted wooden dolls from the Ukraine.

It is an interesting fact of biology, however, that organisms do not usually exhibit such simple geometric scaling. This is because there are powerful constraints on structure and function that do not allow organisms to maintain the same geometric relationships among their components as size changes over several orders of magnitude. For example, as trees increase in size, the cross-sectional areas of their trunks and the total surface areas of their leaves increase more rapidly than expected from purely geometric considerations, as $M^{3/4}$ rather than $M^{2/3}$. The differential increase in trunk area provides for mechanical resistance to buckling due to gravity or wind, while the scaling of leaf area allows for increased gas exchange to support the increased phytomass. Similarly, as mammals increase in size, there is a differential increase in the thickness of their bones to provide mechanical support and in the surface area of the lungs to provide gas exchange for metabolism. So what are the biological scaling laws?

1.2 HISTORICAL PERSPECTIVE: THE RIDDLE OF QUARTER-POWER SCALING

Quantitative studies of biological allometry have a long and venerable history. Early in the twentieth century, two great biologists, D'Arcy Thompson [66] and Cecil Murray [47], produced large synthetic works on scaling. Both of these considered a wide range of size-related and scaling phenomena, and attempted to explain them largely in terms of geometric relationships and physical principles of mechanics. Their contemporary and another famous biologist of the early twentieth century, Julian Huxley [28], coined the term allometry. In interpreting scaling relationships as optimal solutions to problems of mechanical design, these early studies laid the foundation for the modern discipline of biomechanics. The current state of this field is illustrated by chapters in this book by Alexander, Biewener, Koehl, and Li.

In the 1930s Max Kleiber [31] and Samuel Brody [6, 7] showed that the metabolic rate of mammals ranging in size from mice to elephants scales as the 3/4 power of body mass. Prior to their work it had been assumed that the metabolic heat production of endothermic or warm-blooded birds and mammals scaled as $M^{2/3}$, reflecting simple geometric scaling of the body surface area available for heat dissipation. Subsequent to Kleiber's and Brody's work there was much excellent research, much of it stimulated by Knut Schmidt-Nielsen and his collaborators [56], on the scaling of anatomical and physiological characteristics of mammals. Others extended comparative research to other groups of animals and to plants. It was found that the metabolic rates of ectothermic or cold-blooded microbes, plants, and animals also scale as $M^{3/4}$ (Figure 2), and that many other biological scaling exponents were multiples of 1/4. Thus, for example, the radii of mammalian aortas and tree trunks scale as $M^{3/8}$, mammalian heart and respiratory rates scale as $M^{-1/4}$, circulation times for blood of mammals and sap of trees scale as $M^{1/4}$. In fact, most biological times, including those of respiratory and cardiac cycles, and gestation, postembryonic development, and life span scale as $M^{1/4}$, and consequently, most biological rate processes scale as the reciprocal of time, or as $M^{-1/4}$ (Figure 3). These scaling relationships were summarized and discussed in three influential books by McMahon and Bonner [45], Calder [13], and Schmidt-Nielsen [56] that appeared nearly simultaneously in the early 1980s. These syntheses called attention to the ubiquity of quarter-power scaling relationships in biology and discussed possible explanations for them.

It is easy to see why the structures and functions of organisms cannot simply be scaled geometrically, with exponents that are simple multiples of 1/3. It has not been clear, however, why biological scaling exponents characteristically take on values that are simple multiples of 1/4. The riddle of such quarter-power scaling has challenged biologists for more than half a century. Several theories have been proposed. One, the theory of elastic similarity [42, 43, 44, 46] is based on the biomechanical hypothesis of protection from buckling under the force of gravity or wind. This theory suggests that structures such as mammalian bones and tree trunks should exhibit a characteristic relationship between their length, l, and radius, r, such that $l^3 \propto r^2$ or $l \propto r^{2/3}$. This does not appear to explain, however, why metabolic rates of mammals or plants should scale as $M^{3/4}$, or why organisms, such as very small plants and animals or aquatic organisms, which are not significantly affected by gravitational forces, should also exhibit quarter-power scaling. Another theory [49] is based on rates of exchange of materials between aquatic organisms and their environment, and attributes 3/4-power scaling of metabolic rate to physical

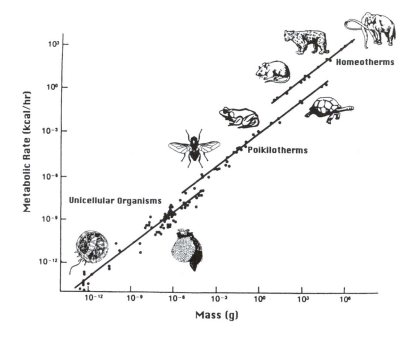

FIGURE 2 Hemmingsen's [26] classic plot of metabolic rate as a function of body size for three functional groups of organisms: endothermic birds and mammals, ectothermic vertebrates and invertebrates, and unicellular organisms. A line corresponding to $M^{3/4}$ has been fitted to each of the three data sets. Note, that the relationships are parallel, with identical allometric exponents, b, but different normalization constants, Y^0. Reprinted with Permission from Novo Nordisk of North America, Inc.

properties of diffusion across the boundary layer. Again, however, this idea seems too specific to account for the ubiquity of quarter-power scaling relationships in so many different kinds of organisms, including terrestrial ones.

In the absence of a theory of sufficient power and generality to explain biological scaling laws, there developed a tendency to treat scaling relationships and the allometric equations that describe them as purely empirical phenomena. Many researchers collected measurements of some biological structure or process from otherwise similar organisms spanning a wide range of body sizes. They plotted the dependent variable as a function of M (or some other measure of size: e.g., trunk diameter for trees), and used standard statistical procedures to derive a best-fit allometric equation. Although the exponents were often amazingly close to multiples of 1/4, there was often little if any discussion of the theoretical basis of these values. This approach is epitomized by R. H. Peters' [50] book, *The Ecological Consequences of Body Size*,

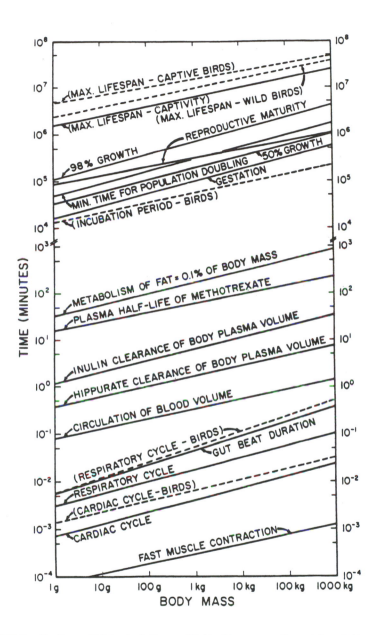

FIGURE 3   Scaling of biological times. Plotted here are allometric relationships for times of biological events, ranging from skeletal muscle contraction to life span, that have been derived for birds (dashed lines) and mammals (undashed lines) by various workers. Note that the slopes are all very similar, corresponding closely to a value of $b = 1/4$. (From Linstedt and Calder [35]. Reprinted with permission from The University of Chicago Press.)

which compiled from the literature hundreds of allometric equations for traits ranging from cellular structure and function to whole organism anatomy and physiology to life history and ecological attributes. The strictly empirical approach of Peters is characteristic of a large body of descriptive comparative research in the decades preceding and following his book. Typical of such work were questions such as whether passerine birds had higher metabolic rates (intercept or $\log Y_0$) than birds of other orders [32, 33], whether primates had different allometric relationships for brain size, post-embryonic development time, or life expectancy than other mammals (see Harvey and Harcourt [24] and Charnov [15], and the chapter by Harvey) or whether the relationship between population density had an exponent of $-3/4$, implying equal energy use independent of body size [17, 18], or some other value, implying no such intriguing relationship (see the chapter by Cyr).

The books of Peters [50], McMahon and Bonner [45], Schmidt-Nielsen [56], and Calder [13] summarized several decades of important work in allometry. Since these seminal publications in the early 1980s, with the notable exception of a few areas such as biomechanics [2, 34, 67], and life history [14, 15, 59], the pace of research has slowed and interest in allometry has seemingly waned. We hope that this book signals a new round of activity and progress. It illustrates the power of scaling not only to provide powerful insights into many specialized areas of biology, but also to make synthetic links between these specialties. It also illustrates the kinds of contributions that can come from different approaches to biological scaling, from mathematical models and computer simulations to compilations and analysis of new data sets.

1.3   GENERAL PRINCIPLES OF BIOLOGICAL SCALING

Recently, we [70] have proposed a mechanism and a general model to account for quarter-power scaling in biology. The idea is that biological rates and times are ultimately limited by the rates at which energy and materials can be distributed between surfaces where they are exchanged and the tissues where they are used (or produced in the case of wastes). Our model assumes that the distribution network: (i) branches to reach all parts of a three-dimensional organism, (ii) has terminal units (e.g., capillaries or terminal xylem) that do not vary with size, and (iii) minimizes the total resistance and hence the energy required to distribute resources. It predicts that the network will have a fractal-like architecture, that many anatomical and physiological features will scale as quarter powers of body mass, and, in particular, that whole-organism metabolic rate should scale as $M^{3/4}$. The model is presented in some detail in the chapter by West et al. Its specific application to vascular plants and its consequences for some ecological as well as anatomical and physiological characteristics of plants are explored in the chapter by Enquist et al.

It is too soon to assess whether our model will come to be accepted as a general explanation for the origin of scaling laws in biology. One thing that the model does do, however, is focus renewed attention on the theoretical basis for biological allometry. It suggests that there is a common mechanism that links the scaling of most anatomical, physiological, and even ecological characteristics of organisms so that their allometric exponents are predicted to be simple multiples of 1/4. There appear to be two reasons for this.

The first is because of the fundamental importance of metabolic rate. The metabolic rate is the rate at which organisms transform energy and materials. Because organisms are composed of similar elements and compounds, and these are formed and transformed by a common set of biochemical reactions, there are severe stoichiometric constraints. As a consequence, similar scaling exponents characterize the rates of uptake or release of all metabolites: uptake of photochemical energy, water, and carbon dioxide by photosynthetic plants, of organic molecules and oxygen for by aerobic organisms, and of all other metabolic requisites, such as nitrogen, phosphorus, and other inorganic nutrients by all organisms. And also because of the stoichiometry, rates of release of oxygen by plants, carbon dioxide and water by aerobic organisms, and heat and other waste metabolites by all organisms also scale identically.

Because the metabolic transformation of energy and materials provides both the physical power and the materials to build all biological structures and run all biological functions, metabolic rate limits all biological processes at all levels of organization from molecules and cells to individuals and populations. Thus at the cellular level, metabolic rate influences rates of membrane transport, biosynthesis, DNA replication, and cell division, while at the organismal level it governs rates of resource uptake (photosynthesis in plants or feeding in animals) and resource allocation to some combination of survival and maintenance, growth, and reproduction. Therefore, given that whole-organism metabolic rate scales as $M^{3/4}$, and cellular or mass-specific metabolic rate scales as $M^{-1/4}$, it should not be surprising that so many biological rate processes also scale as $M^{-1/4}$, and therefore most biological times scale as $M^{1/4}$.

The second reason for the pervasiveness of quarter-power scaling in biology is more abstract, but no less important. Even the simplest organism is an extremely complex system, which depends for its existence and reproduction on the integrated performance of its many component structures and functions. As organisms vary in size, these structures and functions, and the integration of all of them, must be preserved within narrow limits. The component systems cannot be optimized separately. Because they are all interdependent, all must be balanced with each other and scaled in some integrated way [13, 54, 56]. As a consequence, most biological scaling relationships are manifestations of a single underlying scaling process, which appears to be based on quarter powers and to be unique to living things.

Organisms have evolved a three-part solution to the problems of operating over a wide range of sizes: some of the most basic component structures and functions are held invariant, while the others are constrained to obey a common set of scaling laws, and natural selection imposes an economy of design that influences the coevolution of all traits. Charnov [15] calls attention to the importance of both invariant quantities and scaling relationships in the allometry of mammalian life histories. The dual roles of invariant and scaled components are elegantly illustrated in Li's chapter on mammalian cardiovascular system. Across all mammals, which span eight orders of magnitude from shrews to whales, some characteristics are invariant: radius of capillaries, velocity and pressure of blood in both the aorta and capillaries, and number of heartbeats per lifetime all vary as M^0. Other characteristics of the system scale with quarter-power exponents: radius of the aorta as $M^{3/8}$, power output of the heart as $M^{3/4}$, heartbeat frequency as $M^{-1/4}$, and blood circulation time as $M^{1/4}$. Note that most of these quarter-power exponents are geometrically related to each other, so they can be interconverted by taking reciprocals, squares or cubes, or square or cube roots. Such simple interconversion, and corresponding adjustment of different structures and functions would not be possible if, for example, some scaled as $M^{1/3}$ while others scaled as $M^{1/4}$.

Scaling relationships also appear to obey another principle called economy of design or symmorphosis [65, 68, 69]. The idea here is that biological structures and functions are designed so as to meet but not to exceed the maximal demand. Consequences of symmorphosis are seen in the normalization constants of allometric equations: i.e., in the values of Y_0, or the y-intercepts of linear regressions fitted to log-transformed data. An example is the relationship between height and trunk diameter in trees. The empirical values apparently reflect a safety factor to insure against breakage in the strongest winds normally encountered (see McMahon [43, 46] and also the chapters by Horn and Biewener.) The principle of symmorphosis also helps to explain why so many allometric relationships show relatively little variation around the regression lines: the biological designs which confer the highest fitness provide for the maximal demands, no more and no less (see the chapter by Bonner and Horn). This further implies that any substantial deviation of an individual or species from the value predicted from a general allometric relationship indicates either a suboptimal design, which should have a substantial fitness cost (e.g., excessively obese individuals), or a special adaptive response to selective pressures not operating on the other organisms used to derive the equation (e.g., the departures from predicted life history allometries discussed in the chapters by Calder and Harvey). In either case, these deviant phenotypes provide excellent opportunities to test the predictions of mechanistic models for allometric scaling.

These three principles of biological design—interrelated scaling exponents, invariant quantities, and symmorphosis—are fundamental to scaling the anatomical, physiological, and ecological characteristics of organisms. They probably also apply to many other kinds of complex nonliving systems where "fitness" is maximized or some performance criterion is optimized. For example, as the sizes of buildings vary, sizes of some basic components, such as furniture, floor and ceiling tiles, faucets, and electrical outlets remain invariant; others, such as walls and their support beams and the main water pipes and electrical cables scale in accordance with known engineering principles.

And these scaling relationships reflect solutions to just meet the maximal requirements.

2   LEVELS OF SCALING IN BIOLOGY

There are three levels at which biologists have traditionally studied scaling relationships: within individual organisms, among different individual organisms of varying size, and within assemblages of multiple individuals or species of organisms. Each of these levels of investigation has a history extending back well over a century. Each has attracted the interest of some of the greatest biologists of the time. Even advanced undergraduate and beginning graduate students will be able to recognize many of the names and identify some of the contributions cited in this and subsequent chapters of this book. These three levels of study of scaling will be introduced only briefly in this chapter. Examples of each level are developed in subsequent chapters.

2.1   WITHIN ORGANISMS

Many of the most obvious scaling relationships can be observed within a single individual organism. Organisms are composed of structural units that do not vary significantly with the size of the individual or among different parts of its body. Obvious examples include not only molecules and macromolecules, but also cells. For the organism to function as an integrated whole, these invariant structural units must be connected by systems that transport energy and nutrients, waste materials, and communication signals, and which provide protection and structural support. Natural selection for efficient design of such distribution and support (see the chapter by Bonner and Horn) has resulted in the evolution of networks with self-similar, hierarchically scaled architectures.

One example, well treated in this book, is the mammalian circulatory system. Every mammal, whether it is a 2 g shrew or a 200,000,000 g whale, has a network of vessels that branch in a regular architecture from the single aorta leaving the heart to the vast numbers of capillaries supplying the tissues and then back again in reverse order to the single vena cava returning blood to the heart. There are a few more generations of branching in the whale than in the shrew, but the mathematical scaling laws of how radii and lengths of branches change from parent to daughter branches are fundamentally the same, not only from shrew to shrew and whale to whale, but in shrews and whales and mammals of all sizes in between. West et al. discuss the theoretical derivation of this scaling law in their chapter. The chapter by Zamir documents the empirical scaling relationships for one part of this network, the system of carotid arteries and smaller vessels that supply the human heart. The chapter by Schreiner et al. shows that a simple computer algorithm that directs the self-assembly of such a network according to simple rules can produce a three-dimensional architecture which is amazingly similar to a real circulatory system.

Another example is afforded by the macroscopic architecture of higher land plants. Much of our information on the branching patterns and their functional basis comes from the forestry literature. Early efforts to provide a theoretical explanation for these observations were concerned largely or exclusively with biomechanics [43, 46]. This approach continues to provide valuable insights. For example, the chapter by Biewener (see also Niklas [48]) shows how the response of different-sized branches to the physical forces of gravity and wind can explain how scaling relationships change within a tree, from the main trunk to the finest branches. The chapter by Horn illustrates a very different approach: how different geometric rules of branching can give rise, within an individual plant over its ontogeny, to trees with different architectures (see also Horn [27]). An optimal branching architecture is derived theoretically in the chapter by Enquist et al. (see also West et al. [70]) based on the assumption that the system simultaneously delivers resources with minimal resistance through the multiple tiny tubes of xylem and phloem and provides mechanical support against buckling in response to the force of gravity or wind. The recent book by Niklas [48] brings together an enormous body of empirical information and theoretical considerations, much of it from the author's own work, to address many issues of scaling in plants.

2.2   AMONG INDIVIDUALS: TRADITIONAL ALLOMETRY

The majority of work on biological scaling has focused on variation among individuals. As indicated above, the real power of allometric theory and analysis comes when they can be applied over many orders of magnitude variation in body size. Often this involves comparisons of individuals belonging to different species. Usually such analyses are restricted to some taxonomic or functional group of species, to insure that the organisms are sufficiently similar so they have similar structures and functions but sufficiently different that there is enough variation in size to fit allometric equations. Equally valuable insights can come, however, from comparisons among individuals of the same species that differ substantially in size, often because they represent different stages of ontogenetic development. An example in this volume is the Chapter by Lenski and Mongold, which documents evolution of all size and shape in the bacterium, *Escherichia coli*, in less than 10,000 generations. Results of these comparative analyses indicate how different biological structures and functions scale as body size varies over phylogeny or ontogeny.

A number of methodological and conceptual issues are relevant to how data on size-related variation among individuals are compiled, analyzed, and interpreted. While empirical allometric equations are usually derived, as mentioned above, by fitting linear regressions to log-transformed data, this procedure is not so straightforward as it may sound. There are important statistical and biological issues that should be considered: (i) in deciding what data should be included in the analysis; (ii) which of several alternative regression models should be applied; (iii) whether and how phylogenetic relationships and other factors that affect the independence of the data points should be taken into account; (iv) how to interpret that statistical confidence derived values for the allometric exponents, b, and normalization constants, Y_0; and so on. These issues are too complicated and important to address in detail here. Some are raised in the chapter by Harvey, the excellent book by Harvey and Pagel [25], and in recent papers. The message that we would convey here is that these issues are complicated and many are still unresolved. Fitting allometric equations to empirical data is not an exercise that should be done casually, simply by plugging data into a computer and calculating a regression equation by following a recipe from a software package or from the literature.

There is an enormous body of literature in which size-related variation and allometric equations have been reported. For example, Peters [50] cites more than 500 references, and his appendices report the relevant statistics of about 1,000 allometric equations (sample sizes, values of b and Y_0, and often statistical confidence intervals). The dependent variables range from biochemical activity and structures at the molecular and cellular levels (e.g., mitochondrial density, enzyme and hormone activity, and rare element concentrations) to characteristics of whole-organism-level structure and function (e.g., brain size, metabolic rate, and blood circulation time), to aspects of life history and population dynamics (e.g., litter size, life span, territory size, and intrinsic rate of population increase). And Peters' survey represents only a fraction of the information available today. It does not include all of the data on plants in the extensive forestry and botanical literature (see Niklas [48]). Nor, obviously, can it include results of all the research since the early 1980s. We conclude, therefore, that there is an enormous body of information on the scaling of characteristics of organisms at the level of individuals and species.

Neither this introductory chapter nor this book can hope to analyze and synthesize all of this information. The best that it can do is to present some selected examples of biological systems and kinds of organisms that have been studied, and highlight some of the important results and remaining questions. Above and in our two other chapters on models for the origin of scaling laws in animals and plants, we try to point out some of the most general issues, especially the seeming universality of quarter-power scaling in biology. Other chapters consider particular examples in some detail: by Alexander, Koehl, and Biewener on biomechanics of structure, locomotion, and feeding in different kinds of animals; Horn on architecture and function of trees; Li on mammalian cardiovascular systems; Lenski and Mongold on the evolution of size in bacteria; Calder on life history traits of birds and mammals; and Cyr on population densities of different kinds of organisms in lakes.

The overall result, we believe, is a good representation of the current state of allometric research. Apparent is the tension between, on the one hand, intensive studies aimed at deriving and explaining allometric relationships for certain traits in specific kinds of organisms, and, on the other hand, efforts to relate the results to broader questions about biological scaling. Most of the chapters reveal how structural and functional constraints on the inter-relationships among the components of particular subsystems, such as the cardiovascular or musculo-skeletal locomotor system of mammals, the loco-motor and feeding appendages of arthropods, or the life history traits of birds and mammals, are reflected in particular allometric scaling laws that apply to those subsystems. Still largely missing, however, are treatments of how the structures and functions of the different subsystems are scaled and integrated so as to produce an entire organism with a body plan that functions effectively over many orders of magnitude variation in body size.

Some of the chapters point toward exciting areas for future research. Horn's chapter on trees raises interesting questions about the relationships among architecture, development, and ecological performance. Kozlowski's simulations of life history evolution address problems of how body size and size-related traits coevolve in response to natural selection. Calder's and Harvey's analyses of mammal and bird life histories suggest that complexes of multiple traits deviate in predictable ways from taxon-wide allometries, in ways that hint at underlying mechanisms. Despite all of the studies that have been performed and all of the data that have been compiled, therefore, there is still much to be done in traditional allometry. In fact, the wealth of excellent quantitative information on the scaling of so many biological attributes should stimulate a new round of integrative and synthetic studies.

2.3 SCALING OF POPULATIONS AND ASSEMBLAGES

As we have just seen, in traditional allometry the level of study is the individual organism and the independent variable in the allometric equation is body mass. This contrasts with scaling relationships within individuals, where the variables of interest reflect relationships among component structures and functions. There are also scaling relationships which are expressed at levels of biological organization greater than individual organisms, at levels of populations of multiple individuals and assemblages of multiple species. Compared to individual-level allometries, few higher-level relationships have been reported, but these hint at potentially important features of ecology and evolution. Here, we consider briefly some examples.

The first concerns the effect of body size on the abundance and resource use of organisms in ecological communities. Ecologists have long used allometric equations to characterize the influence of size on both the abundance of individuals and the diversity of species (e.g., Yoda et al. [73], Williams [71], May [40, 41], Peters [50], Damuth [17, 18], Dial and Marzluff [19], and Siemann et al. [58]). The so-called "thinning law" describing the relationship between the size of individual plants and their density in single- or mixed-species stands is perhaps the best example. The original and still most generally accepted theory is Yoda's $-3/2$ geometric model, which relates the spacing of competing plants to the volumes occupied by their canopies. We have developed an alternative model based on the resource use of individual plants (see Enquist et al. [21], and chapter in this book). Both empirical information and our model of plant vascular systems suggest that the metabolic rates of vascular plants should scale as $M^{3/4}$. This leads quite straightforwardly to the prediction that when the size of plants is limited by competition for resources, population density should scale as $M^{-3/4}$. This gives a $-4/3$ resource-based thinning law, which seems to fit the data better than Yoda's $-3/2$ geometric model. This relationship can then be used to derive additional testable predictions. For example, the model predicts and the data confirm that productivity of ecosystems should be independent of the size of the dominant plants. These results show how the scaling relationships at the level of individual organisms can affect higher levels of biological organization, such as populations and ecosystems. And note that these ecological phenomena appear to exhibit two general principles of biological scaling noted above: some attributes scale with exponents that are simple multiples of $1/4$, while others are invariant.

Another example of ecological scaling is the relationship between the size of a population and the variance in population size over either space or time. One pattern is often referred to as Taylor's power law, because it is due largely to data compiled and analyzed by L. R. Taylor [62, 63, 64]. Taylor showed that plots on a logarithmic scale of the variance in population density (S) recorded in repeated samples as a function of the mean density (N) for those same samples typically gives a straight line, such that $S = aN^b$ (Figure 4). Values of b typically range between 1 and 2 and sometimes even higher. Since $b = 1 = S/N$ indicates a randomly dispersion (e.g., a Poisson distribution), the common finding of $b > 1$ demonstrates a very general trend for populations to be aggregated or clumped in space and time. In contrast to size-related variation, however, the value of the exponent seems to vary widely with the kind of organism and environment. Taylor's analysis is typically used to assess patterns of spatial or temporal dynamics among different populations of a single species.

A related question concerns the frequency distribution of abundance, either within a single species over either space or time, or among the different species that coexist to form an ecological community (e.g., Fisher et al. [22], MacArthur [36], Preston [52, 53], and Williams [71]). There is an inverse relationship between the number of individuals counted and the frequency with which this number is observed. Most species are rare most of the time and in most of the places where they occur, and communities typically contain many rare species and only a few common ones. The mathematical forms and ecological causes of these distributions have long been debated. When the data are plotted on logarithmic axes, they are reasonably well fit by a power law and the slope is typically close to -1 [12]. Thus the spatial and temporal distribution of abundance within species and the distribution of relative abundances among coexisting species may be a special case of a much larger class of negative relationships between the magnitude (G) of events

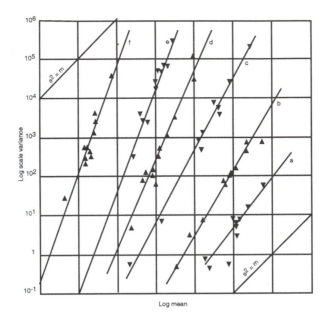

FIGURE 4   Relationship, plotted on logarithmic axes, between the variance and the mean for populations of six different species of aphids (a)–(f). These data illustrate "Taylor's power law" for the aggregated distribution of abundance, because $b > 1$. The Poisson random distribution, with a value of $b = 1$, is plotted for reference and labeled $S^2 = m$. (From Taylor et al. [64].) Reprinted with permission from Blackwell Science Ltd.

and their frequency (f). In many cases $G \propto 1/f$, so that $f = f_0 G^{-1}$. Such relationships are by no means confined to ecology; other examples include the frequency of earthquakes as a function of size (Gutenberg-Richter law), the frequency of words in written languages (Zipf's law), and the distribution of wealth among firms or nations (Pareto distribution). Whether such distributions of the form $G \propto 1/f$ reflect simply the mathematical consequences of a large class of stochastic processes or the more interesting common features of a special class of complex systems remains an open question [4, 23, 29, 57].

Another, and perhaps related, example of higher-level biological scaling relationships concerns the number of species in a genus or in some other higher taxonomic category. This, too, can be thought of as a relationship between magnitude (size of taxon) and frequency (number of species in that taxon). Recently, examples of such distributions have been compiled and analyzed by Burlando [8, 9]; see also Williams [71], and Dial and Marzluff [20]. Similar power-law relationships have been found in diverse kinds of organisms and over hundreds of millions of years of fossil history. When plotted on logarithmic axes, they appear to have slopes that vary somewhat but cluster in the range of -1 to -2. Again, while the pattern is fairly clear, its interpretation seems uncertain. Does this example of a $G \propto 1/f$ distribution reflect some fundamental processes of biological diversification, simple stochastic processes, or some feature of the human mind that influences how we place species—and perhaps other objects—into a hierarchical classification? At this point, we can only raise the questions, but not provide any easy answers.

A final example of an apparent biological power law at a level of biological organization higher than individuals is the species-area relationship, and the closely related species-time relationship. It has long been recognized that the number of species recorded increases with the size of a sample, where size is measured as either the number of individuals counted or the cumulative area of space or period of time surveyed [3, 16, 38, 52, 53, 55, 71, 72]. When number of species (S) is plotted as a function of either area of space (A) or period of time on logarithmic axes, the relationship is often quite linear (Figure 5). Not surprisingly, it has typically been fitted by a power function of the form $S = S_0 A^z$. Values of z vary considerably. Some authors have pointed out that the average value is often close to $1/4$, but others have called attention to patterns of variation. Samples of isolated environments such as islands, typically have steeper slopes (higher z) but smaller numbers of species (lower S_0) than nonisolated habitats [38, 55]. Furthermore, the relationship appears to change slope with spatial scale (see below).

Species-area relationships are considered in some detail in the chapter by Harte, so we will make only two general comments here. The same two concerns also apply to species-time relationships, which are discussed much less frequently in the literature (but see Rosenzweig [55]). First, as implied above by the fact that the slope of the relationship changes with spatial or temporal scale, there are grounds to question whether it should be thought of as a power law. Often the data encompass only a few orders of magnitude variation in area, which may be inadequate grounds to claim that the relationship reflects some fundamentally self-similar scaling process. This is not to deny the practical utility of using power functions to describe species-area relationships. There is no question that many empirical species-area relationships are well described by linear regressions fit to log-transformed data. In his chapter, Harte shows additional advantages of using the mathematical formalism of power functions to characterize the relationship between species diversity and area, at least over a limited range of spatial scales.

The second and related comment concerns the mechanistic processes that give rise to the species-area relationship. At the smallest spatial scales, the number of species encountered will necessarily depend on the number of individuals sampled, so S will typically increase with A, rapidly at first and then more gradually until the rarest species have been recorded. Once the rarest species have been sampled the slope would be expected to level off (approach 0) if the environment were homogeneous. If, however, the abiotic environment varies over space—and it does—and if different environmental conditions support distinctly different species—and they do—then number

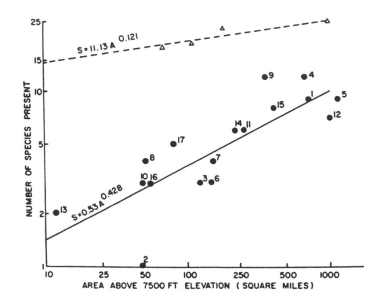

FIGURE 5   Species-area relationships for mammals inhabiting isolated mountain ranges (circles and undashed line) and nonisolated sample areas within large mountainous areas (triangles and dashed line). Note that a power function gives a good fit to both data sets, although the values of b and Y^0 are clearly different. (From Brown [10].) Reprinted with permission from The University of Chicago Press.

of species should increase as the increasing sample areas incorporate greater environmental heterogeneity. So the slope should steepen again because the larger areas encompass a greater diversity of environments and consequently support a greater diversity of species. This is just what is observed: a "collector's curve" with initially steep but decreasing slope as progressively rarer species are sampled, then very low z-values ($z \ll 0.25$) for relatively complete samples of ecological communities separated by small distances of space or time, and finally much higher values ($z \gg 0.25$) for biotas separated by geographical distances or geological time periods [11, 51, 55, 71]. This mechanistic interpretation of the species-area relationship implies that the processes which affect species diversity are not in fact self-similar over a wide range of spatial scales. Power functions fit to logarithmically transformed species-area data should probably be viewed only as approximate mathematical descriptions of more complicated relationships between the number of species and the size of the sample.

3 CONCLUDING REMARKS

The diversity of life is due in large part to the enormous diversity of sizes of organisms. Living organisms vary by more than 21 orders of magnitude in body size, from the tiniest microbes to the largest whales. On the one hand, the vital activities of all these organisms rely on a common set of molecules, biochemical reactions, and cellular structures and functions. On the other hand, the vital activities of all these organisms are strongly dependent on their body sizes. The sizes and shapes of structures and the rates of processes vary with body size in ways that can be described by mathematical scaling laws. These scaling laws reflect the both geometric, physical, and biological constraints that organisms have had to obey and the opportunities that organisms have been able to exploit as they have diversified in size.

The allometric equations used to characterize these scaling laws are power functions in which the magnitude of the trait of interest is expressed as a function to body mass, M, raised to some exponent, b. When organisms within a taxonomic or functional group of organisms vary in size over many orders of magnitude, these scaling laws typically can account for most but not all of the variation in their attributes. Research on allometry has had two longstanding and complementary themes. One has been to fit regression equations to data so as to describe empirical scaling relationships. The other has been to develop general theories to explain these patterns in terms of lawlike mechanistic processes. The deviant values responsible for the residual variation around fitted regression equations become particularly interesting once there is a general mechanistic model that predicts the parameters of the allometric equation. Without such a theory, there is little basis for interpreting the direction and magnitude of variation. With such a theory, the deviations from predicted values provide a powerful basis for testing and extending the theory.

A seemingly unique and pervasive feature of biological scaling relationships is that nearly all of them are fit by allometric equations which have exponents that are statistically indistinguishable from simple multiples of 1/4. Thus for example, metabolic rates of nearly all kinds of organisms scale as $M^{3/4}$, radii of tree trunks and mammalian aortas scale as $M^{3/8}$, and circulation times of sap in trees and blood in mammals, and embryonic development time and life span of mammals all scale as $M^{1/4}$. The fact that so many biological attributes scale as quarter powers of body mass has suggested a common basis for these biological scaling laws and stimulated the search for a general mechanistic model that can explain their origin. West et al. [70] have recently proposed such a model. Regardless of whether it is ultimately supported, this model focuses renewed attention on three seemingly universal principles of biological scaling: (i) an integrated set of scaling laws that have quarter-power exponents of body mass, (ii) some invariant quantities that do not scale with body size, so they go as M^0, and (iii) an economy of design imposed by natural selection so that the magnitudes of structures and functions tend to just meet maximum demands.

These three principles appear to apply to scaling in all kinds of organisms, from microbes to plants and animals, and at all levels of biological organization, from molecules and cells to organisms and ecosystems. Many of them are illustrated in detail in the different chapters of this book. We hope that the book, by providing examples of both the diversity and the common themes of current research on biological scaling, will stimulate renewed interest and further empirical and theoretical advances in allometry.

The variety of life is so overwhelming that many biologists seem to be content to describe the variation rather than to search for universal principles. Many seem to doubt that some general biological laws still remain to be discovered. We are more optimistic. We suggest that the common features and multiple consequences of biological scaling laws reflect the operation of universal principles. Scaling patterns and processes are so widely distributed across different kinds of organisms and so tightly integrated within organisms that they must be nearly universal. Distantly related organisms have independently evolved the same solutions to the problems of scaling structure and function as they vary in size. Simple mathematical models can be used to understand the causes and consequences of these scaling phenomena. This modeling effort has been underway for nearly a century, and considerable progress has been made. As the models have been tested and validated empirically, they have provided powerful insights into the geometric, physical, and biological processes that had to be solved as organisms diversified in size. But much theoretical and empirical research remains to be done on the pervasive influence of body size on the structure, function, and diversity of life.

REFERENCES

[1] Alexander, R. McN. *Animal Mechanics*, 2nd ed. Oxford: Blackwell, 1983.
[2] Alexander, R. McN. "Leg Design and Jumping Technique for Humans, Other Vertebrates, and Insects." *Phil. Trans. Roy. Soc. Lond.* B **347** (1995): 235–248.
[3] Arrhenius, O. "Species and Area." *J. Ecology* **9** (1921): 95–99.
[4] Bak, P. *How Nature Works*. New York: Springer-Verlag, 1996.
[5] Bartholomew, G. A. "A Matter of Size: An Examination of Endothermy in Insects and Terrestrial Vertebrates." In *Insect Thermoregulation*, edited by B. Heinrich, 45–78. New York: Wiley, 1981.
[6] Brody, S. *Bioenergetics and Growth*. New York: Reinhold, 1945. Reprint. Darien, CT: Haffner, 1964.
[7] Brody, S., R. C. Procter, and U. S. Ashworth. "Basal Metabolism, Endogenous Nitrogen, Creatinine and Neutral Sulphur Excretions as Functions of Body Weight." *Univ. Missouri Agr. Exp. Station* **220** (1934): 1–40.
[8] Burlando, B. "The Fractal Dimension of Taxonomic Systems." *J. Theor. Biol.* **146** (1990): 88–114.

[9] Burlando, B. "The Fractal Geometry of Evolution." *J. Theor. Biol.* **163** (1993): 161–172.

[10] Brown, J. H. "Mammals on Mountaintops: Nonequilibrium Insular Biography." *Amer. Natur.* **105** (1971): 467–478.

[11] Brown, J. H. *Macroecology.* Chicago, IL: The University of Chicago Press, 1995.

[12] Brown, J. H., D. H. Mehlman, and G. C. Stevens. "Spatial Variation in Abundance." *Ecology* **76** (1995): 586–604.

[13] Calder, W. A., III. *Size, Function, and Life History.* Cambridge, MA: Harvard University Press, 1984.

[14] Charnov, E. L "Evolutioin of Life History Variation Among Female Mammals." *Proc. Natl. Acad. Sci. USA* **88** (1991): 1134–1137.

[15] Charnov, E. L. *Life History Invariants.* Oxford: Oxford University Press, 1993.

[16] Connor, E. F., and E. D. McCoy. "The Statistics and Biology of the Species-Area Relationship." *Amer. Natur.* **113** (1979): 791–833.

[17] Damuth, J. "Population and Body Size in Mammals." *Nature* **290** (1981): 699–700.

[18] Damuth, J. "Of Size and Abundance." *Nature* **351** (1991): 268–269.

[19] Dial, K. P., and J. M. Marzluff. "Are the Smallest Organisms Most Diverse?" *Ecology* **69** (1988): 1620–1624.

[20] Dial, K. P., and J. M. Marzluff. "Nonrandom Diversification within Taxonomic Assemblages." *Syst. Zool.* **38** (1989): 26–37.

[21] Enquist, B. J., J. H. Brown, and G. B. West. "Allometric Scaling of Plant Energetics and Population Density." *Nature* **395** (1998): 163–165.

[22] Fisher, R. A., A. S. Corbet, and C. B. Williams. "The Relationship Between the Number of Species and the Number of Individuals in a Random Sample of an Animal Population." *J. Animal Ecol.* **12** (1943): 42–58.

[23] Gell-Mann, M. *The Quark and the Jaguar. Adventures in the Simple and the Complex.* New York: W. H. Freeman, 1994.

[24] Harvey, P. H., and A. H. Harcourt. "Sperm Competition, Testis Size, and Breeding System in Primates." In *Sperm Competition and Evolution of Animal Mating Systems*, edited by R. L. Smith, 589–600. New York: Academic Press, 1984.

[25] Harvey, P. H., and M. D. Pagel. *The Comparative Method in Evolutionary Biology.* Oxford: Oxford University Press, 1991.

[26] Hemmingsen, A. M. "Energy Metabolism as Related to Body Size and Respiratory Surfaces, and Its Evolution." Reports of the Steno Memorial Hospital and Nordisk Insulin Laboratorium **9** (1960): 6–110.

[27] Horn, H. S. *The Adaptive Geometry of Trees.* Princeton, NJ: Princeton University Press, 1971.

[28] Huxley, J. S. *Problems of Relative Growth.* London: Methuen, 1932.

[29] Keitt, T. H., and H. E. Stanley. "Dynamics of North American Bird Populations." *Nature* **393** (1998): 257–260.

[30] Kleiber, M. *The Fire of Life. An Introduction of Animal Energetics.* New York: Wiley, 1961.

[31] Kleiber, M. "Body Size and Metabolism." *Hilgardia* **6** (1932): 315–353.

[32] Lasiewski, R. C., and W. R. Dawson. "A Re-examination of the Relation Between Standard Metabolic Rate and Body Weight in Birds." *Condor* **69** (1967): 13–23.

[33] Lasiewski, R. C., and W. R. Dawson. "Calculation and Miscalculation of the Equations Relating Avian Standard Metabolism to Body Weight." *Condor* **71** (1969): 335–336.

[34] Li, J. K-J. *Comparative Cardiovascular Dynamics of Mammals.* Boca Raton, FL: CRC Press, 1996.

[35] Linstedt, S. L., and W. A. Calder. "Body Size, Physiological Time, and Longevity of Homeothermic Animals." *Quart. Rev. Biol.* **56** (1981): 1–16.

[36] MacArthur, R. H. "On the Relative Abundance of Bird Species. *Proc. Natl. Acad. Sci. USA* **43** (1957): 296–295.

[37] MacArthur, R. H., and E. O. Wilson. "An Equilibrium Theory of Insular Zoography." *Evolution* **17** (1963): 373–387.

[38] MacArthur, R. H., and E. O. Wilson. *The Theory of Island Biogeography.* Princeton, NJ: Princeton University Press, 1967.

[39] Mandelbrot, B. B. *The Fractal Geometry of Nature.* New York: W. H. Freeman, 1983.

[40] May, R. M. "Patterns of Species in Abundance and Diversity." In *Ecology and Evolution of Communities*, edited by M. L. Cody and J. M. Diamond, 81–120. Cambridge, MA: Harvard University Press, 1978.

[41] May, R. M. "How Many Species Are There on Earth?" *Science* **241** (1988): 1441–1449.

[42] McMahon, T. "Size and Shape in Biology." *Science* **179** (1973): 1201–1204.

[43] McMahon, T. "The Mechanical Design of Trees." *Sci. Am.* **233(1)** (1975): 92–102.

[44] McMahon, T. "Allometry and Biomechanics: Limb Bones in Adult Ungulates." *Amer. Natur.* **109** (1975): 547–563.

[45] McMahon, T., and J. T. Bonner. *On Size and Life.* New York: Scientific American Books, 1983.

[46] McMahon, T., and R. E. Kronauer. "Tree Structures: Deducing the Principle of Mechanical Design." *J. Theor. Biol.* **59** (1976): 443–466.

[47] Murray, C. D. "The Physiological Principle of Minimum Work. I. The Vascular System and the Cost of Blood Volume." *Proc. Natl. Acad. Sci. USA* **12** (1926): 207–214.

[48] Niklas, K. J. *Plant Allometry: The Scaling of Form and Process.* Chicago, IL: University of Chicago Press, 1994.

[49] Patterson, M. R. "A Mass-Transfer Explanation of Metabolic Scaling Relations in Some Aquatic Invertebrates and Algae." *Science* **255** (1992): 1421–1423.

[50] Peters, R. H. *The Ecological Implications of Body Size.* Cambridge, MA: Cambridge University Press, 1983.

[51] Preston, F. W. "Time and Space and the Variation of Species." *Ecology* **41** (1960): 785–790.

[52] Preston, F. W. "The Canonical Distribution of Commonness and Rarity: Part I." *Ecology* **43** (1962): 185–215.

[53] Preston, F. W. "The Canonical Distribution of Commonness and Rarity: Part II." *Ecology* **43** (1962): 410–432.

[54] Rashevky, N. *Mathematical Biophysics. Physico-Mathematical Foundations of Biology*, Vols. 1 and 2, 3rd ed. New York: Dover Publications, 1960.

[55] Rosenzweig, M. L. *Species Diversity in Space and Time.* New York: Cambridge University Press, 1995.

[56] Schmidt-Nielsen, K. *Scaling, Why Is Animal Size so Important?* Cambridge: Cambridge University Press, 1984.

[57] Schroeder, M. *Fractals, Chaos, and Power Laws.* New York: W. H. Freeman, 1991.

[58] Siemann, E., D. Tilman, and J. Haarstad. "Insect Species Diversity, Abundance, and Body Size Relationships." *Nature* **380** (1997): 704–706.

[59] Stearns, S. C. *The Evolution of Life Histories.* Oxford: Oxford University Press, 1992.

[60] Taylor, C. R., K. Schmidt-Nielsen, and J. L. Raab. "Scaling of Energetic Cost of Running to Body Size in Mammals." *Am. J. Physiol.* **219** (1970): 1104–1107.

[61] Taylor, C. R., N. C. Heglund, and G. M. O. Maloiy. "Energetics and Mechanics of Terrestrial Locomotion. I. Metabolic Energy Consumption as a Function of Speed and Body Size in Birds and Mammals." *J. Exp. Biol.* **97** (1982): 1–21.

[62] Taylor, L. R. "Aggregation, Variance, and the Mean." *Nature* **189** (1961): 732–735.

[63] Taylor, L. R. "Synoptic Dynamics, Migration, and the Rothamsted Insect Survey: Presidential Address to the British Ecological Society." *J. Animal Ecol.* **55** (1986): 1–38.

[64] Taylor, L. R., I. P. Woiwod, and J. N. Perry. "Variance and the Large-Scale Spatial Stability of Aphids, Moths, and Birds." *J. Animal Ecol.* **49** (1980): 831–854.

[65] Taylor, L. R., and E. R. Weibel. "Design of the Mammalian Respiratory System. I. Problem and Strategy." *Respir. Physiol.* **44** (1981): 1–10.

[66] Thompson, D. W. *On Growth and Form.* Cambridge, MA: Cambridge University Press, 1917.

[67] Vogel, S. *Life in Moving Fluids: The Physical Biology of Flow.* Boston, MA: Willard Grant Press, 1981.

[68] Weibel, E. R., C. R. Taylor, P. Gehr, H. Hoppeler, O. Mathiers, and G. M. O. Maloiy. "Design of the Mammalian Respiratory System. IX.

Functional and Structural Limits for Oxygen Flow." *Respir. Physiol.* **44** (1981): 151–164.

[69] Weibel, E. R., C. R. Taylor, and L. Bolis, eds. *Principles of Animal Design. The Optimization and Symmorphosis Debate.* Cambridge, MA: Cambridge University Press, 1998.

[70] West, G. B., J. H. Brown, and B. J. Enquist. "A General Model for the Origin of Allometric Scaling Laws in Biology." *Science* **276** (1997): 122–126.

[71] Williams, C. B. *Patterns in the Balance of Nature and Related Problems in Quantitative Ecology.* New York: Academic Press, 1964.

[72] Williamson, M. *Island Populations.* Oxford: Oxford University Press, 1981.

[73] Yoda, K., T. Kira, H. Ogawa, and K. Hozumi. "Self-Thinning in Over-crowded Pure Stands Under Cultivated and Natural Conditions." *J. Biol.* **14** (1963): 107–129.

Allometry and Natural Selection

John Tyler Bonner
Henry S. Horn

The whole point of scaling in general and allometry in particular is to understand the effect of changes in size on the construction of organisms. Going back to Galileo and D'Arcy Thompson, there are problems which are imposed by increased size when one considers that weight goes up as the cube of the linear dimensions and strength as the square. And there are all those interesting physiological matters which stem from the fact that, like weight, metabolism is related to the cube of the linear dimensions, while the diffusion of food and the exchange of gases, such as oxygen and carbon dioxide, is a square function dependent on the surface area of the individual.

It was D'Arcy Thompson's [14] position, and that of many of his followers, that the only consideration was a purely mechanical one; nature was bound by some physical law of maximum efficiency that automatically produced the perfect optimal construction for any particular size. Although he never said so, there is the definite implication that such physical forces per se drive evolution. One reason for thinking this is that he dismissed natural selection as irrelevant—in fact in one passage of his great book he scorns the idea that Darwinian selection could possibly play a constructive role in evolution. After discussing adaptive reasons often given for different colorations among animals he says, "To buttress the theory of natural selection the same instances of 'adaptation' (and many more) are used, as in an earlier but not too distant

Scaling in Biology, edited by J. H. Brown and G. B. West.
Oxford University Press, 2000.   **25**

age testified to the wisdom of the Creator and revealed to simple piety the immediate finger of God." Let us point out that this view was the norm for his generation and the idea that the reasons for efficiency were purely mechanical was a way of avoiding natural selection as an all encompassing explanation. It is of interest to note that Thompson also rejected genetics as being too mysterious a matter for consideration, and that in his big 1942 edition he dismissed allometry, which had been pushed by Julian Huxley [5], as being of little use. For him his mathematical-physical explanations were sufficient to enforce optima automatically with size changes. In his world a mathematical description was an explanation, not only of the pattern, but also of the prime cause. (See Ball [1], Pp. 6–8, for an excellent discussion of Thompson's views.)

Here we would like to take the opposite position: that natural selection is responsible for both the changes in size and the trend toward optimal efficiency in organisms. This means we are saying that both size and mechanical efficiency have a genetic basis, and it is in this way that selection can play that genetic instrument—the genome—which has led to the evolution of larger and smaller, more and less complex organisms, all with finely tuned, efficient constructions. Those constructions can be of optimal mechanical efficiency, but should that come to be it is because of selection for their biological properties, the mechanical properties follow. One place where purely mechanical considerations appear to take on a commanding role is in setting the limits of maximum or minimum size. For a particular body plan, be it that of an animal or a plant, there are physical limits as to how big or how small it can be. However, as those limits are approached, selection still decides what is possible—what can survive the culling of natural selection. Furthermore, those organisms near the upper and lower size limits can be selected for mutations that produce changes in the body plan to one with different physical constraints, an ever-present alternative that has often been followed.

1   SELECTION FOR SIZE

The idea that there is a natural selection for size certainly goes back to Charles Darwin, beginning with his emphasis on artificial selection in his "variation under domestication." He mentions in passing that horticulturists have selected for larger fruit, and those involved in animal husbandry for larger cattle. One need only think of our domestic dogs, which according to the latest report [15] all descended from a wolf ancestor. Selection over many years has produced giants such as the Irish wolfhound and midgets such as the Chihuahua, a size difference approaching two orders of magnitude.

Evidence for the evolution of size changes in nature is equally well known and the examples are abundant. For instance, there is a considerable literature showing that island races of animals are often larger or smaller than their close mainland relatives. A detailed and dramatic example of how this may come about is the work of Peter and Rosemary Grant [4] on beak size changes in the finches of the Galapagos archipelago. The selection for increased or decreased sizes in their bills depends much on the weather trends and the resulting size of the seeds on the island; large seeds need big bills to crush them, and if there is a relative abundance of small seeds, the birds with smaller beaks fare best. These shifts in beak size can be directly attributed to variation in the size of the available food. Even though the shifts involve relatively few generations and are driven by the prevailing weather, beak size is clearly to some degree an inherited character, and therefore the Grants are observing rapid natural selection.

This tells us that selection for the right size depends on the niches open in any one environment. Normally there is an almost continuous range of sizes of animals and plants when all the size niches are filled, but if through extinction there is a gap in this size spectrum, selection will see to it that it is soon filled, either by making a smaller organism larger, or a larger one smaller. There may be niches at the upper or the lower end of the size scale that are "open" in the sense that there are no larger or smaller competitors; but these niches may not be practical if they are ecologically or physiologically inefficient for allometric reasons.

We do not think there is any doubt in anyone's mind that the size of organisms is generally genetically determined, and that there are innumerable examples where selection has altered these size controlling genes to produce larger and smaller animals and plants.

2   THE EFFECTS OF SIZE CHANGE ON SHAPE

There are a number of different effects that size change can have on shape. In each case there is a selection for the maintenance, or the increase, of efficiency. All physiological and morphological aspects of an organism, to remain competitive, must approach optimal efficiency—those are the constructions that will succeed in the course of natural selection. And those constructions must have a genetic basis, for that is the only way natural selection can operate.

Simple Allometry.   The first and most familiar way in which size change affects shape is by simple allometry. That is, as an animal or a plant gets smaller or larger, its proportions will retain a constant allometric relation. To give an example among the almost limitless possibilities, if the log of the height of a tree is plotted against the log of the diameter of its trunk, the line will have a slope of 2/3 (Figure 1). This says that larger trees are disproportionately thick, which as McMahon [6] showed, is to be expected if the danger of buckling with increased size is to be averted. Furthermore, he pointed out that the buckling limit for different size trees, assuming elastic similarity, would also have a slope of 2/3 on an allometric plot. In other words, trees have, by the hit-and-miss of natural selection, evolved genes that produce a thickness that, for a given

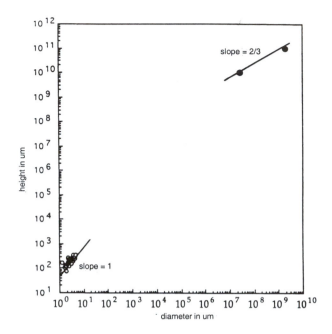

FIGURE 1   A log-log allometric plot comparing small cellular slime mold fruiting bodies (lower left) with large trees (upper right).

size, gives a rough but practical safety margin to avoid collapse by buckling in a wind.

A disproportionate increase in the thickness of the trunk of trees is not only evident at maturity, but the relationship holds during their early growth. Saplings are slender by comparison to the mature specimens of any one species, and as they grow they become relatively thicker, so that all points in their life cycle they retain a fixed allometric relation, and can support themselves by avoiding buckling. The genes that control thickness are acting continuously during the growth of the tree; selection produces a mechanism that provides the needed strength at all the stages of its development. And there is no wasteful, excessive increase in thickness before it is needed (see Horn this volume)—there has been a continuous selection at all stages for proportions with sufficient safety.

Isometry.   Another way in which the effect of size on shape can be shown is a special case of simple allometry, where the slope of the line is 1, and the proportions do not change with size. The small fruiting bodies of one species of cellular slime mold (*Dictyostelium discoideum*), over a large size range (from 100 to 350 μm in height) keep constant proportions, so that a plot of their stalk height against stalk width has a slope of 1 (Figure 1). The obvious interpretation of these observations is that compared to trees, where there has been a selection for a thicker trunk, in the minute slime mold the force of gravity is negligible, even in the largest fruiting bodies, and as a result there has been no selection for a relative increase in the thickness of the stalk.

This conclusion is misleading because other species of cellular slime molds, and many small fungi (molds) are also unaffected by gravity, yet they do not show geometric similarity. Rather, the stalk of their spore bearing structures will have a uniform thickness, but the length of the stalk will vary enormously, in some cases becoming very long (Figure 2). Clearly the stalk thickness is sufficient to carry apical spore masses regardless of stalk length—for them the weight-strength problem does not exist.

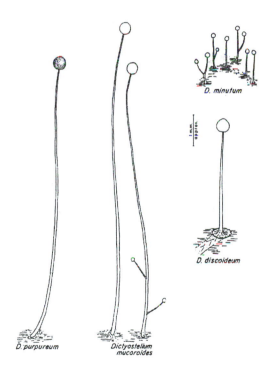

FIGURE 2   Different fruiting bodies of small cellular slime molds showing species that are geometrically similar (*D. discoideum*) and others that are clearly not (*D. mucoroide* and *D. purpureum*). (No measurements have been made on *D. minutum*.) Size in these slime molds is determined by how many amoebae enter an aggregate. Printed with permission Cornell University Press.

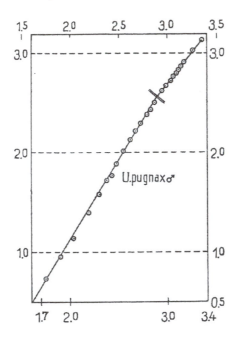

FIGURE 3   Increase during growth of the log of the chela weight plotted against the log of the body weight in fiddler crabs. Note the change in slope that coincides with the time of sexual maturity. (From Huxley, J. S. *Problems of Relative Growth*. London, UK: MacMillan, 1932, printed with permission.)

In other words the case of the geometric similarity found in *D. discoideum* is special and perhaps we must seek an explanation for its proportions in some particular condition in the soil which they inhabit, raising an interesting problem of soil ecology. Whatever the selective advantage of their isometry, the fact remains that it exists, something that would be impossible were it not for their small size.

Altered Allometry.   A subtle example of altered allometry is described by Niklas [8]. He examines plants of sizes that lie between McMahon's large trees and our slime molds (Figure 1) and finds that they adjust their allometries to favor height over diameter with a safety factor that declines with size.

Perhaps some of the best cases of altered allometry come from size relation of parts during growth, especially among arthropods. This was appreciated by Huxley [5] who has noted changes in crabs during growth of the allometric slope of body parts, such as carapace width plotted against interocular distance (Figure 3). More recently Nijhout and Wheeler [10] have shown, in a theoretical analysis of the growth of body parts of insects with complete metamorphosis (holometabolous), that if the size increase of an imaginal disc is plotted against body size, the allometric curve is not a straight line but a curved one, indicating an alteration of the allometric relations during development. In a further study, Nijhout and Emelin [9] have produced empirical evidence that there is competition between growing imaginal discs for substrates, and that during normal development it is the dynamics of these competitions that are responsible for the continuously changing allometries during growth.

Convergent Solutions to Size Increase.   There is yet another way in which size and shape can be shown to be selected for independently. Size changes will not always produce the same solutions in shape change, as can be seen in the many instances of convergent evolution of structures involved in the physiological well-being of an organism. For instance, both large vertebrates and large insects need special mechanisms to get oxygen to their internal tissues and to remove waste carbon dioxide to the exterior. Their ancestral construction has many radical differences. Vertebrates have perfected a lung (or gill) and a closed blood circulation system with efficient oxygen-carrying proteins, all of which permits gas exchange in the deepest tissues. Insects, on the other hand, use a tracheal system to bring the gases directly to and away from the tissues. Here are two totally different ways of solving the same physiological problem induced by size increase, ways which are both guided by genes in radically different fashions during their respective developments.

Another example of a separation of the selection for structure and for size, is to be found in the swimming of bacteria and protozoa. It is reasonable to assume that there is a selective advantage to speed, either as a means of pursuing food or escaping a predator. It is also apparent that regardless of the swimming mechanism, the larger the organism, the faster it can swim. This point is not only an empirical fact—from bacteria to tuna fish—but the basis of it is purely physical, and an eight-oared racing shell will go faster than one with four rowers, and two-oared shells and singles are progressively slower. With microbes we are in a world of low Reynolds numbers, and as a result there must be totally different means of locomotion depending on the size of the microbe (see Koehl this volume). Large ciliate protozoa are covered with many short motile cilia that serve essentially as oars, for they have a rigid, forward thrust stroke and a flexible, bending return. For bacteria, because of their even smaller size, water has become a highly viscous medium, so much so that they could not manage cilia. As Edward Purcell [12] has pointed out, it would be like a man trying to row a boat very slowly in thick molasses, without bringing the oars into the air. Bacteria have got around this problem by devising a totally different type of flagellum that is rotated at its base, like a miniature propeller. This is again a case of remarkable convergent evolution for speed—different mechanisms have been imposed by their size and

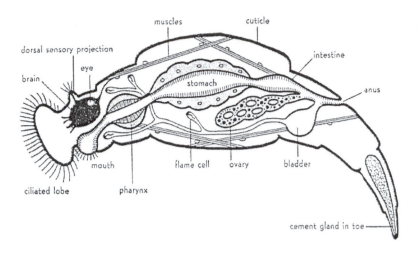

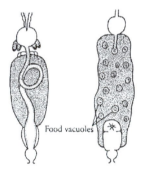

FIGURE 4   Above: A small rotifer (From Buchsbaum, R. *Animals Without Backbones*. Chicago, IL: University of Chicago Press, 1938.). Below: A comparison (not drawn to scale) of the conventional gut system of a large rotifer (left) with that of a small one (right) which shows food vacuoles passing through a continuous cytoplasm. (From McMahon and Bonner [7], after Pennak [11], based on literature cited in de Beachamps and Burger.)

the altered Reynolds numbers. The mechanism for traveling as fast as possible for a particular size is completely different for the larger and faster series of microorganisms. So again there has been a selection for efficiency using radically different motility mechanisms, each appropriate for the size of the organism.

New Anatomy to Accommodate Size Change.   There is an interesting example where the lower size limit is reached and the evolved, minute organism has undergone a radical change in its internal anatomy so that it can function on such a diminutive scale. It is commonly accepted that the ancestors of rotifers were large and that rotifers fill a niche in soil and in water that is close to that of ciliate protozoa. Most rotifers have a simple gut lined with cells running from the mouth to anus (Figure 4). In a number of very small species this gut is absent and replaced by a sac of continuous cytoplasm that forms food vacuoles at the mouth end, and after digestion ejects the contents of the vacuole at the anal end (Figure 4). This is a digestive system exactly like that found in the ciliate protozoa; it must be an efficient system of food digestion on a very small scale. Five such species have been described by de Beauchamps [3], and he mentions that a similar condition is found in small forms in other distantly related groups of rotifers, indicating that the phenomenon has been invented in small rotifers more than once.

In this curious instance, when the lower size limit is reached, the shape is not just modified to become a point off the allometric curve of close relatives, as was the case for the hearts of small mammals and birds, but there has been a change in the anatomy to produce another system of digestion. One cannot but wonder how such a complete change in the genes that control development might have arisen. Do the genes for an ancestral protozoan-like digestive system still exist in a dormant state in all rotifers, and can they be called forth at the lower size limit? Or for other reasons does the developmental switch between these two digestive systems involve only few mutational steps? In either case we have much to learn about the developmental genetics of these simple forms of digestion.

3 CONCLUSION

D'Arcy Thompson, and many who have followed him, see biological principles of engineering that govern the relation between size and shape and argue that these principles underlie why allometric relations are so constant and universal. We would agree in the sense that size imposes restrictions on the internal construction, and the construction imposes limits on the possible size; the two are totally bound up with one another by pure physical necessity. But we do not believe one need go beyond that simple and obvious statement. There is no mysterious bio-physical force that steers evolution; it is all to be explained by the trial-and-error of natural selection. Selection for size and selection for efficiency are central if we wish to understand the diversity of shapes and sizes of living organism.

This raises an important question: Why do we find that certain physical rules appear to correlate with allometric size changes? Sometimes the argument is that the predicted allometric relation correlates with the flow of materials involved in energy exchange and that this, therefore, is the fundamental basis for the relation. In other studies there is an excellent argument for elasticity being the common denominator. Sometimes the argument is for other mechanical properties involved in the generation of force needed for the movement of animals or the safe vertical stature of tall plants.

Our contention would be that most likely all these physical explanations of the allometric relations are correct. In some cases one may dominate, but in other cases more than one could be significant—they are not mutually exclusive. In each instance selection is culling for the best circulatory system for transport, for the maximum benefits that can be squeezed out of the elastic properties of the organism, for the most efficient way to resist compression and the other physical forces that impinge upon a living organism of a particular size, and for the optimum construction of those structures that depend upon diffusion. This selection for efficiency in each instance (with a few interesting exceptions) will result in organisms that follow clear and consistent allometric relations with different sizes. In these respects selection is doing no more than the engineer when he designs bridges, or automobiles, or any other ingenious mechanical device that human beings have created—in every instance, often by choosing among different possibilities—efficiency is maximized. But selection for size and efficiency come first, and because of the mechanical rules which cannot be divorced from function, the allometric relations follow. In this way the horse is appropriately put in front of the cart.

REFERENCES

[1] Ball, P. *The Self-Made Tapestry: Pattern Formation in Nature*. Oxford and New York: Oxford University Press, 1999.

[2] Buchsbaum, R. *Animals Without Backbones*. Chicago, IL: University of Chicago Press, 1938.

[3] de Beauchamps, P. "Contribution á l'tude du Genre Ascomorpha et des Processus Digestifs Chez les Rotiféres." *Societé Zoologique de France, Bulletin* **57** (1932): 428–449.

[4] Grant, B. R., and P. Grant. *Evolutionary Dynamics of a Natural Population*. Chicago: Chicago University Press, 1989.

[5] Huxley, J. S. *Problems of Relative Growth*. London: Methuen, 1932.

[6] McMahon, T. A. "Size and Shape in Biology." *Science* **179** (1973): 1201–1204.

[7] McMahon, T. A., and J. T. Bonner. *On Size and Life*. New York: Scientific American Books, 1983.

[8] Niklas, K. J. "The Allometry of Safety Factors for Plant Height." *Am. J. Bot.* **81** (1994): 345–351.

[9] Nijhout, H. F., and D. J. Emelin. "Competition Among Body Parts in the Development and Evolution of Insect Morphology." *Proc. Natl. Acad. Sci. USA* **95** (1998): 3685–3689.

[10] Nijhout, H. F., and D. E. Wheeler. "Growth Models of Complex Allometries in Holometabolous Insects." *Amer. Natur.* **148** (1996): 40–56.

[11] Pennak, R. W. *Freshwater Invertebrates of the United States*, 3rd ed. New York: John Wiley, 1983.

[12] Purcell, E. M. "Life at Low Reynolds Number." *Am. J. Physiol.* **45** (1977): 3–11.

[13] Raper, K. B. "The Communal Nature of the Fruiting Process in the Acrasieae." *Am. J. Bot.* **24** (1940): 436–448.

[14] Thompson, D.'A. W. *On Growth and Form.* Cambridge, MA: Cambridge University Press, 1942.

[15] Vila, C., P. Savolainen, J. E. Maldonado, I. R. Amorim, J. E. Rice, R. L. Honeycutt, K. A. Crandall, J. Lundeberg, and R. K. Wayne. "Multiple and Ancient Origins of the Domestic Dog." *Science* **276** (1997): 1687–1689.

Hovering and Jumping: Contrasting Problems in Scaling

R. McNeill Alexander

1 INTRODUCTION

Evolution is a powerful, optimizing process that tends to form animals and plants to designs that are, for them, close to the best possible. When similar organisms evolve a range of sizes, evolution will tend to scale them so that each has a near-optimal design for its particular size. Scaling problems are problems of optimization over a range of sizes.

This obvious and (I believe) uncontroversial point often seems to be overlooked in studies of scaling. Instead of asking what are the optimum designs for different sizes of animal, we tend to ask, In what way can we expect different-sized animals to be similar to each other? For example, should we expect different-sized mammals to be geometrically similar to each other, so that the small ones are exact scale models of the big ones? Alternatively, should we expect them to be elastically similar, so that their bodies sag to the same degree under gravity [13]? Or, another possibility, should we expect them to be designed so that equal stresses act in different bones or in different muscles (see Biewener [5] and Enquist this volume). This approach may assume optimization: it may be assumed that if one animal is designed optimally, a different-sized animal will also be optimal if it is similar to it in

Scaling in Biology, edited by J. H. Brown and G. B. West.
Oxford University Press, 2000. **37**

shape, in degree of sagging, in stresses, or in some other specified way. This may be assumed, but it is seldom stated.

The similarity approach can be useful, as I will show by discussing two groups of hovering animals, bees and hummingbirds. But for some scaling problems it seems necessary to introduce optimization explicitly. My examples of this are jumping animals, locusts and frogs. There is a basic difference between hovering and jumping which seems to demand different approaches to their study. The requirements for hovering are clear cut; the animal needs to be able to keep itself airborne, and it would be an advantage to be able to do so at the lowest possible energy cost. The wings must be able to produce enough lift to support the animal's weight, but there is no need for them to supply more. In contrast, for animals that jump to escape from enemies there is no preset level of performance that is "enough." It would be an advantage to the animal to be able to jump as fast and as far as possible, to have the best chance of escaping. To find the optimum design we have to realize that superb jumping legs imply costs as well as benefits. We must find out what the costs are and balance them against the benefits.

Hovering and jumping are good examples for my purpose. Both are amenable to analysis by simple mechanics. Hummingbirds and bees (which hover) and frogs and locusts (which jump) each come in a wide range of sizes. There are good published data about the allometry of these four groups of animals.

2   HUMMINGBIRDS AND BEES

Hovering like helicopters is an important skill for hummingbirds and bees. They keep themselves airborne in one place by using their wings to drive air downward. How can they be designed so as to be able to hover at least energy cost? We will think about just two major design features, wing length and wing beat frequency.

Think first about the aerodynamics of hovering. Suppose that we know the optimum design for one size of animal. Then we can predict that, for another size of animal, the optimum design will give the same pattern of airflow. In more technical terms, the patterns of air flow produced by the two animals, as they hover, should be dynamically similar.

Each wing beat produces a puff of downward-moving air, and a vortex ring forms around each puff. Thus a stack of vortex rings develops under the hovering animal (Figure 1). The rings will be spaced a distance ν/f apart, where ν is the speed of the downward-moving air and f is the wing beat frequency. Bigger animals will produce bigger vortex rings, but for the air flows below different-sized animals to be similar, ring spacing must be proportional to ring radius and so to wing length r. The different-sized animals must have equal values of fr/ν (a quantity that aerodynamicists will recognize as a Strouhal number). Helicopter theory tells us that ν must be proportional to

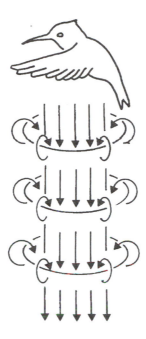

FIGURE 1   A diagram of a hovering hummingbird, showing how the air flow below it forms a stack of vortex rings. For the wakes of different-sized hummingbirds to be dynamically similar, the spacing of the rings must be proportional to their radius.

$(m/r^2)^{0.5}$ [1] so, to have equal values of fr/ν, animals of different masses m must have equal values of $fr^2 m^{-0.5}$.

The wing muscles of a hovering animal have to exert forces for two functions: they must overcome inertial forces to accelerate and decelerate the wings in every beat, and aerodynamic forces to move the wings through the air. It may seem reasonable to suggest as a similarity principle that might govern scaling, that different-sized hovering animals should be designed so as to have equal ratios of inertial forces to aerodynamic forces. Equations derived by Weis-Fogh [16] show that this would imply different-sized animals having equal values of r^2/m. (The argument assumes that they have wings of the same shape, operating at the same lift coefficient.)

We now have two suggested scaling rules: different-sized hovering animals should have equal values of $fr^2 m^{-0.5}$ and equal values of r^2/m. For them to have both, f must be proportional to $m^{-0.5}$ and r must be proportional to $m^{0.5}$.

Wing lengths r and hovering wing beat frequencies f have been measured for hummingbirds and bees, in each case over wide ranges of body mass. Table 1 shows exponents of allometric equations calculated from these data. For

TABLE 1   A comparison of the theoretical allometric exponents for hovering animals, derived in this chapter, with empirical exponents reported by Rayner [15] and Casey, May, and Morgan [7]. These are exponents b of equations of the form $y = a(\text{body mass})^b$.

	Predicted Exponent	Observed Exponent: Hummingbirds	Observed Exponent: Euglossine Bees
Wing length	0.5	0.53	0.42
Wing frequency	−0.5	−0.60	−0.35
$fr^2/m^{0.5}$	$0^{a\,a}$	−0.04	−0.01
r^2/m	$0^{b\,b}$	0.06	−0.16

[a]Condition for similar wakes.
[b]Condition for a constant ratio of inertial to hydrodynamic forces.

hummingbirds, f is proportional to $m^{-0.60}$ and r is proportional to $m^{0.53}$, reasonably close to our predictions. For bees, the exponents are a little different and the agreement is less good. This suggests that at least one of our similarity principles may be mistaken. A row lower down in the Table shows that for both groups of animals, $fr^2m^{-0.5}$ is very nearly proportional to m^0. In other words, it is almost the same for all sizes of animals. For different-sized hummingbirds and also for different-sized bees, the spacing of the vortex rings is about proportional to their diameter. However, the Table goes on to show that the exponents of r^2/m are less close to zero; the ratio of inertial to aerodynamic forces does not remain so constant, over the size ranges of these groups of animals.

Perhaps that is what we should have expected. There seems to be a good reason for expecting different-sized hovering animals to have similar patterns of airflow in their wakes. If the pattern is optimal for one animal, similar patterns are expected to be optimal for others of different sizes. However, there is less reason for expecting a particular ratio of inertial to aerodynamic forces to be optimal. The ratio has a big effect on the pattern of forces that the muscles have to exert, and it might be thought that one pattern of forces would be more favorable than another, but this is not necessarily the case. A theoretical analysis by Alexander [3] seems to show that the ratio will not affect the energy cost of flapping the wings, provided the physiological properties of the muscles and the elastic properties of their tendons are well adapted to the particular ratio.

3 LOCUSTS

Design of jumping animals presents a different kind of problem. A hovering animal has simply to produce enough lift force to keep itself airborne. In contrast, for an animal that jumps to escape enemies, the faster and farther it can jump the better. To explain the jumping ability of different-sized animals we have to ask, why have they not evolved the ability to jump farther? We must look for some cost that makes excessive jumping ability unduly expensive.

Locusts and other jumping insects use catapult mechanisms [4]. They prepare for a jump by contracting their muscles relatively slowly to store up strain energy in tendons or other elastic structures, while the legs are locked in the position of readiness for the jump by a catch mechanism. When the animal is ready to jump, it releases the catch and the stored energy is released in a rapid elastic recoil. Typically, this extends the legs much faster than could be done by muscle contraction, in the absence of elastic storage.

The muscles and the energy-storing springs should plainly be well matched; the spring must be able to store all the work that the muscle can do. Assuming that they are well matched, the bigger the muscles (in an animal of given mass) the better it will jump. However, there is a penalty for having excessively large jumping muscles; they use metabolic energy even when resting.

If muscles of different sizes exert equal stresses while shortening by equal fractions of their lengths, they do work proportional to their volumes, and so proportional also to their masses [8]. Let an insect of mass m have a fraction μ of jumping muscle in its body, and let this muscle be capable of doing work α per unit mass, in a single contraction. We can estimate the speed ν at which the insect can take off for a jump by equating the work done by the muscle to the kinetic energy given to the body

$$\begin{aligned}
\tfrac{1}{2}m\nu^2 &= \alpha\mu m\,, \\
\nu &= \sqrt{(2\alpha\mu)}\,.
\end{aligned} \tag{1}$$

Assume that the benefit conferred by the ability to jump is proportional to this speed; the faster the insect can take off the harder, in general, it will be for a predator to catch it. There are two reasons why this is likely to be true; it is harder to grab a prey animal that is moving fast than one which is moving slowly, and the faster the prey animal takes off the longer the jump can be. Assume also that the benefit conferred by a given takeoff speed is the same, irrespective of the size of the insect.

Now consider the insect's metabolic energy budget. Its jumping muscles may be important for escape, but their metabolism (even when the animal is resting) is a drain on its energy resources. In contrast, other parts of the body (feeding apparatus, digestive organs, blood system, etc.) enable it to replenish its energy resources and even to grow. Enlarging the leg muscles not only increases the burden of leg muscle metabolism, but also reduces the mass of organs available for taking in and processing food, if body mass is kept constant. An animal with bigger leg muscles will not only need more food to sustain them, but it will also be unable to take in and assimilate food as fast. Let the insect consist of a mass μm of leg muscle which metabolizes at a rate per unit mass that is a fraction R of the body's resting metabolic rate; and a mass $(1 - \mu)m$ of other tissues which accumulate energy at a net rate per unit mass (food intake minus metabolism) that is a multiple P of the body's metabolic rate. The net effect of the leg muscles on the animal's energy balance is a fraction $-\mu(P+R)$ of the metabolic rate. This is the cost to the animal of having the muscles. Assume that energy losses which are equal fractions of the resting metabolic rate are equally costly to animals of different sizes.

These arguments suggest the hypothesis that the function natural selection tends to maximize is

$$\begin{aligned} \Phi &= \nu - k_1\mu(P+R) \\ &= \sqrt{(2\alpha\mu)} - k_1\mu(P+R) \end{aligned} \tag{2}$$

(using Eq. (1)). The factor k_1 is the (unknown) exchange rate between the currencies of speed and metabolism. As the fraction m of muscle in the body increases, the benefit term and the cost term on the right-hand side of Eq. (2) both increase, and there is an optimum value of μ, that maximizes Φ (Figure 2). This optimum value will be the same for locusts of all sizes, because body mass m does not appear in the equation; similar jumping insects of different sizes are expected to have equal proportions of muscle in their bodies. You might argue that the benefit of jumping may not be proportional to take-off speed, but to some other function of speed, perhaps the square root. This

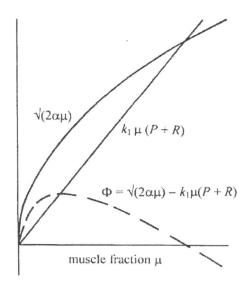

FIGURE 2   A schematic graph showing the benefit term ($\sqrt{(2\alpha\mu)}$) and the cost term ($k_1\mu(P+R)$) from Eq. (2). The broken line shows the difference between them, the function Φ, which is greatest when μ has its optimal value.

would not alter our conclusion. If there is an optimum proportion of jumping muscle, our argument says it should be the same for all sizes of locusts.

Now consider how long the insects' legs should be. A catapult that stores a given quantity of energy is expected to be able to accelerate a body of given mass to the same takeoff speed, irrespective of the length of the legs. However, the legs must be large enough to be able to accommodate the muscles and catapult. If insects of different sizes are geometrically similar, they will contain equal proportions of muscle. Also, equal stresses in geometrically similar structures, with the same elastic modulus, cause geometrically similar distortions. Therefore, geometrically similar catapults function equally well.

These arguments point to the conclusion that jumping insects of different sizes should be geometrically similar to each other in all features that affect jumping, should have equal takeoff speeds, and, consequently, should be able to jump equal distances. Observations by Katz and Gosline [9] on juvenile locusts enable us to test these predictions (see Table 2). Katz and Gosline found that the lengths of juvenile locust legs were proportional to (body mass)$^{0.38}$ and their diameters to (body mass)$^{0.31}$, not too different from the expectation of geometric similarity which would have made both proportional to (body mass)$^{0.33}$. They found that takeoff speed increased only slightly with increasing body size, in proportion to (mass)$^{0.05}$.

4 FROGS

Now we consider frogs whose jumps are powered, not by catapults, but directly by muscle contraction. Prediction of scaling rules for them is a more complex problem than it was for insects. We have to take account of the force-velocity properties of the muscles. In addition, we take account of the mass of the legs, including the mass of the skeleton as well as the muscles.

TABLE 2   A comparison of the theoretical allometric exponents for jumping animals, derived in this chapter, with empirical exponents reported by Katz and Gosline [9] and Marsh [11]. These are exponents b of equations of the form $y = a(\text{body mass})^b$.

	Juvenile Locusts		Frogs	
	Predicted	Observed	Predicted	Observed
Muscle mass	1.00		1.06+	1.12, 1.03, 1.08
Leg length	0.33	0.38	0.32	thigh 0.29
				lower leg 0.31
Muscle speed	no prediction		−0.23	−0.09
Takeoff speed	0.00	0.05	0.04+	
Jump length or	0.00	0.00	0.13+	0.20
height			(height)	(length)

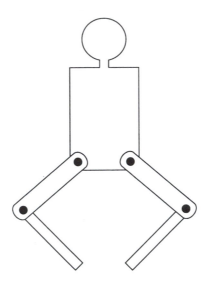

FIGURE 3   A diagram of the model of a jumping animal used by Alexander [2] and in this chapter.

Consider a frog of mass m which has leg muscles of mass μm and other leg tissues (principally bone and skin) of mass σm. As before, muscle has a resting metabolic rate per unit mass that is R times the body's resting metabolic rate, and the parts of the body other than the legs accumulate energy at a net rate that is P times the body's metabolic rate. We will ignore the metabolic rate of the nonmuscle parts of the legs; bone and skin have very low metabolic rates [12]. Thus the energy cost per unit body mass of having legs is $[\mu(R + P) + \sigma P]$, and the function to be maximized is

$$\Phi = \nu - k_1[\mu(R + P) + \sigma P].   \tag{3}$$

To calculate the takeoff speed ν, we will use Alexander's [2] mathematical model of a jumping animal (Figure 3). This two-dimensional model jumps vertically, powered by extensor muscles of the knees. The muscles have force-velocity properties governed by Hill's equation. The model is designed to take account of elastic compliance in the muscles' tendons, but in this chapter we will take this compliance to be zero.

We assume as in Alexander [2] that the animal will initially have its knees bent at an angle of 60°, so extension of the knee at takeoff (straightening the leg) will involve rotation through about 120°. Further, we assume that the muscles are arranged in such a way that extending the knee through this angle requires the muscle fibres to shorten by one quarter of their length. This assumption agrees reasonably well with Lutz and Rome's [10] observations of sarcomere length changes in the semimembranosus. Also, we take the maximum stress that the muscle can exert when contracting isometrically to be 300 kPa, a typical value for vertebrate striated muscle, including frog muscle [6]. These assumptions, together with a selected value for the maximum shortening speed (V_{max}) of the muscle, enable us to calculate the moments that the muscle exerts at the knee joint, and the speed of takeoff.

The stronger the muscles and the longer the legs, the heavier the leg skeleton must be and the more skin is needed to enclose the legs. The assumptions we have already made lead to the conclusion that, to be strong enough, the leg bones must have cross-sectional areas proportional to $(\mu m)^{0.67}$, and masses proportional to $s(\mu m)^{0.67}$, where s is leg segment length. Thus the total mass of each leg is $0.5[\mu m + k_2 s(\mu m)^{0.67}]$, where k_2 is a constant. As in Alexander [2], each leg consists of two segments of equal length, with two thirds of the mass in the thigh and one third in the lower leg.

Muscle mass is represented by the parameter μ, muscle mass as a fraction of body mass. It will also be convenient to define parameters λ (describing leg length) and ω (describing the maximum shortening speed of the muscles expressed as an angular velocity of knee extension):

$$\lambda = s\left(\frac{\rho}{m}\right)^{1/3}, \tag{4}$$

$$\omega = k_3 m^{0.25} V_{max}. \tag{5}$$

In Eq. (4), ρ is the density of the body (taken to be 1000 kg/m^3), so λ is leg segment length expressed as a multiple of the side of a cube of volume equal to the body. Geometrically similar animals of different sizes will have equal values of λ. In Eq. (5), V_{max} is the maximum shortening speed of the muscle (lengths per unit time) and the constant k_3 is 1.0 kg$^{-0.25}$s. Thus, for a muscle with a maximum shortening speed of one length per second in a 1 kg frog, $\omega = 1$. The factor $m^{0.25}$ has been introduced because the resting metabolic rates of muscles seem likely to be proportional to maximum shortening speeds, and resting metabolic rates per unit mass of similar animals are generally about proportional to (body mass)$^{-0.25}$ (Brown et al. this volume and Peters [14]). Thus frogs of different sizes with equal values of ω are expected to have equal values of R (the metabolic rate of muscle as a multiple of whole-body resting metabolic rate). We are now in a position to rewrite Eq. (3), which gives the quantity Φ that is to be maximized:

$$\Phi = \nu(\mu, \lambda, \omega, m) - k_1 \left[\mu(k_4\omega + P) + (\lambda\mu^{2/3}k_2 P/\rho^{1/3})\right]. \tag{6}$$

$$\underbrace{\qquad\qquad}_{\text{benefit term}} \qquad \underbrace{\qquad\qquad}_{\text{cost term}}$$

The function $\nu(\mu, \lambda, \omega, m)$ cannot be shown as an algebraic expression, but is computed for particular sets of parameter values using the model of Alexander [2].

Figure 4 shows graphs of ν against μ, λ, and ω for various values of m. In each graph, the parameters not being varied have been held at values

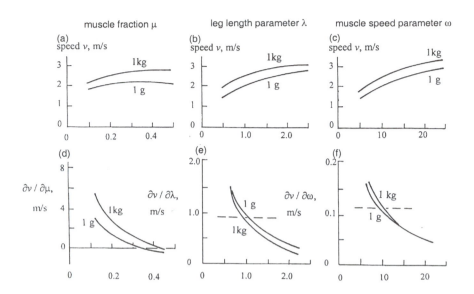

FIGURE 4   Graphs used in the discussion of jumping. In the upper row, predicted takeoff speed is plotted against (a) relative leg muscle mass, μ; (b) the leg length parameter, λ; and (c) the muscle speed parameter ω, for frogs of masses 1 g and 1 kg. In the lower row, (d), (e), and (f) show the gradients of (a), (b), and (c), respectively; these are the partial derivatives referred to in the text. The broken lines are explained in the text.

that seem to be realistic for frogs. The selected value for relative leg muscle mass, $\mu = 0.2$, lies in the range of observed values given by Marsh [11]. The selected leg length parameter, 1.0, gives leg segment lengths close to those predicted for the thigh and lower leg of *Rana catesbiana*, by Marsh's allometric equations. The selected muscle speed parameter, $\omega = 10$, gives $V_{\max} = 10$ lengths per second for a 1 kg frog and 25 lengths per second for a 25 g frog. These speeds are higher than have been recorded in experiments with excised frog muscles, but they give realistic jumping performance. (The highest shortening speed I know for frog leg muscle is 10 lengths per second for *Rana pipiens* at 25°C [10], but we need faster muscles to get realistic jumps because we have ignored tendon compliance.) The constant k_2 was given the value $3.3 \text{ kg}^{0.33}\text{m}^{-1}$, which makes the leg muscles 65% of total leg mass and the skeleton, skin etc. 35%, when μ and λ have the selected values of 0.2 and 1.0. This possibly underestimates the mass of skeleton and skin.

Figure 4(a)–(c) predicts, as expected, that takeoff speed will generally increase as the size of the jumping muscles increase, as leg length increases, and as the maximum shortening speed of the muscles increases. However, there is a point beyond which further increase in leg muscle mass would make takeoff speed fall again. This is predicted only for unrealistically large leg muscles, comprising 35% or more of body mass, and seems to be due to larger muscles increasing the mass of the legs. Figure 4(a)–(c) also predicts that larger frogs will take off at higher speeds than smaller ones with the same values of the parameters μ, λ, and ω. Figure 4(d)–(f) shows in all cases that the partial derivatives $\partial v/\partial \mu$, $\partial v/\partial \lambda$, and $\partial v/\partial \omega$, decrease as the values of the parameters increase.

Rates of energy intake and growth of similar animals tend to be proportional to (body mass)$^{3/4}$, like resting metabolic rate [14]. Therefore P, the rate of energy accumulation of nonleg parts of the body expressed as a multiple of resting metabolic rate, is expected to be independent of body mass. Thus the cost term in Eq. (6) is independent of body mass. Parameters μ, λ, and ω were defined with this end in view.

The cost term includes constants whose values we do not know, but the equation will nevertheless be useful. Note that the term increases linearly as λ increases and as ω increases; $\partial(\text{cost})/\partial\lambda$ and $\partial(\text{cost})/\partial\omega$ are independent of λ and ω, as well as being independent of body mass. The cost term also increases as μ increases, but not precisely linearly; $\partial(\text{cost})/\partial\mu$ must fall slowly as μ increases.

At the optimum, where the function Φ has its maximum value, $[\partial v/\partial\mu - \partial(\text{cost})/\partial\mu]$ must be zero, and similarly for λ and ω. This enables us to use Figure 4(d)–(f) to make predictions about scaling. It will be easiest to look first at Figure 4(e) and (f), representing the two cases in which the partial derivative of the cost function is constant. The broken line in Figure 4(e) shows a possible position for the line representing $\partial(\text{cost})/\partial\lambda$. If this were the correct position, the optimum value of λ for a ' g frog would be 1.05, where the appropriate graph of $\partial v/\partial\lambda$ intersects the broken line. Similarly, the optimum value for a 1 kg frog would be 0.95. If λ had these values, it would be proportional to (body mass)$^{-0.014}$. If λ were constant, leg length would be proportional to (body mass)$^{1/3}$. Thus the graph suggests that leg length should be proportional to (body mass)$^{0.32}$. If the broken line were placed higher or lower, the predicted exponent would be slightly different, but a considerable change would be needed to make the prediction much different from isometry. The line has been deliberately placed at a level which gives predicted optimum values for λ close to 1.0, as found in real frogs. It seems likely that evolution has optimized leg lengths.

In Figure 4(f), the broken line shows a suggested value for $\partial(\text{cost})/\partial\omega$. The continuous lines showing $\partial v/\partial\omega$ for 1 g and 1 kg frogs intersect it when ω is 0.95 and 1.05, respectively, suggesting that ω should be proportional to (body mass)$^{0.015}$. Constant values of ω make the maximum shortening speed of the leg muscles proportional to (body mass)$^{-0.25}$, so the graph suggests that maximum shortening speeds should be proportional to (body mass)$^{-0.23}$. The level of the broken line in Figure 4(f) has been chosen to predict optimum values of ω close to 10, the value selected for use when the other parameters were being varied. It is far from certain that this value is realistic, but the broken line could have been placed much higher or lower, with very little effect on the predicted exponent.

If a horizontal line were drawn on Figure 4(d) at the level required to predict realistic values of relative leg muscle mass μ, it would indicate optimal values of about 0.16 for a 1 g frog and 0.24 for a 1 kg frog, making relative muscle mass proportional to (body mass)$^{0.06}$ and muscle mass proportional to (body mass)$^{1.06}$. However, the line should not be precisely horizontal, because, as we have seen, $\partial(\text{cost})/\partial\mu$ must fall slowly as μ increases. This implies that the predicted exponent should be somewhat higher than 1.06; hence the entry 1.06+ in Table 2. Unfortunately, we do not have enough information to be more precise.

Table 2 compares the allometric exponents predicted by our theory with the observed exponents reported by Marsh [11]. Agreement is reasonably good, except in the case of muscle speed ($V_{\max}$), for which the theory predicts too strongly negative an exponent. One possible reason for the discrepancy is that our assumption, that the resting metabolic rate of a muscle is proportional to its $V_{\max}$, may be false.

5 DISCUSSION

Hovering and jumping have been convenient subjects, because they are amenable to simple mechanical analysis. There was nothing unconventional in our discussion of hovering; we postulated that optimal designs, for hoverers of different sizes, would make their movements similar in particular respects, and we worked out the consequences. Our discussion of jumping was more original. The benefits to jumping performance, that different designs offered, had to be balanced against their costs. The benefit and the cost were expressed in different currencies, takeoff speed and metabolic energy. There was no apparent way of converting these currencies to fitness, the fundamental currency that drives evolution. Our analysis showed that, even so, it was possible to make some progress in explaining how jumping animals scale.

We started with the hypothesis that jumping animals have evolved in a way that balances the benefits of high performance against the metabolic cost of possessing legs capable of giving that high performance. If I had been able to create animals to any desired specification, I might have tested the hypothesis by constructing frogs with a range of different designs, varying leg length and the properties and dimensions of the muscles. I would then have measured the jumping performance and energy balance of each of my creations, making comparisons between them and the natural products of evolution. Plainly, that approach was not an option, so I created a mathematical model which could be modified easily. It would not have been feasible to have made a mathematical model that matched or even approached the complexity of a real frog. I might have chosen instead to imitate as much of a frog's complexity as time and computing power allowed, but that would probably not have been a sensible strategy. I might have wasted a great deal of time and effort, and we would still have been left wondering which features of the model were essential to the result. Instead, I adopted a strategy that is often very productive, creating a model that is exceedingly simple but seems to capture the essentials of the problem (Figure 3). We used it to find out how the calculated costs and benefits are affected by changes in leg length, muscle size, and muscle speed.

REFERENCES

[1] Alexander, R. McN. *Animal Mechanics*, 2nd ed. Oxford: Blackwell, 1983.
[2] Alexander, R. McN. "Leg Design and Jumping Technique for Humans, Other Vertebrates, and Insects." *Phil. Trans. Roy. Soc. Lond. B* **347** (1995): 235–248.
[3] Alexander, R. McN. "Optimum Muscle Design for Oscillatory Movement." *J. Theor. Biol.* **184** (1997): 253–259.
[4] Bennet-Clark, H. C. "The Energetics of the Jump of the Locust *Schistocerca gregaria*." *J. Exp. Biol.* **63** (1975): 53–83.
[5] Biewener, A. A. "Scaling Body Support in Mammals: Limb Posture and Muscle Mechanics." *Science* **245** (1989): 45–48.
[6] Calow, L. J., and R. McN. Alexander. "A Mechanical Analysis of a Hind Leg of a Frog (*Rana temporaria*)." *J. Zool.* **171** (1973): 293–321.
[7] Casey, T. M., M. L. May, and K. R. Morgan. "Flight Energetics of Euglossine Bees in Relation to Morphology and Wing Stroke Frequency." *J. Exp. Biol.* **116** (1985): 271–289.
[8] Hill, A. V. "The Dimensions of Animals and Their Muscular Dynamics." *Science Progress* **38** (1950): 209–230.
[9] Katz, S. L., and J. M. Gosline. "Ontogenetic Scaling of Jump Performance in the African Desert Locust (*Schistocerca gregaria*)." *J. Exp. Biol.* **177** (1993): 81–111.
[10] Lutz, G. J., and L. C. Rome. "Built for Jumping: The Design of the Frog Muscular System." *Science* **263** (1994): 370–372.
[11] Marsh, R. L. "Jumping Ability of Anuran Amphibians." *Adv. Vet. Sci.* **38B** (1994): 51–111.
[12] Martin, A. W., and F. A. Fuhrman. "The Relationship Between Summated Tissue Respiration and Metabolic Rate in the Mouse and Dog." *Physiol. Zool.* **28** (1955): 18–34.
[13] McMahon, T. A. "Size and Shape in Biology." *Science* **179** (1973): 1201–1204.
[14] Peters, R. H. *The Ecological Implications of Body Size*. Cambridge, MA: Cambridge University Press, 1983.
[15] Rayner, J. M. V. "Form and Function in Avian Flight." *Curr. Ornithol.* **5** (1988): 1–66.

[16] Weis-Fogh, T. "Quick Estimates of Flight Fitness in Hovering Animals, Including Novel Mechanisms for Lift Production." *J. Exp. Biol.* **59** (1973): 169–230.

Scaling of Terrestrial Support: Differing Solutions to Mechanical Constraints of Size

Andrew A. Biewener

Terrestrial animals and plants span an enormous size range, and yet even distantly related groups are constructed of similar materials (e.g., bone, wood, muscle, and tendon). As with many physiological processes, evolutionary and ontogenetic changes in size impose constraints of scale on the mechanical design and function of skeletal support systems that are built of materials having similar properties. Adequate design requires that the capacity of skeletal elements (and muscles) for force transmission safely exceeds the levels required for biological support and movement. This is the case when the force transmitted per unit cross-sectional area of the material, defined as a mechanical stress ($= F/A$, e.g., N/mm^2), does not exceed the material's strength (the maximum stress that the material can withstand before failure). Clearly, larger structures can support larger forces more safely. The important design consideration, however, is whether changes in force requirements are matched by comparable changes in tissue cross-sectional area in order to keep maximal stresses and, thus, safety factors (defined as failure stress/peak functional stress) constant as size changes. Scale-invariant features (bone strength, timber strength, and peak muscle stress), therefore, require size-dependent changes in other features if the functional integrity of support systems is to be maintained over a broad size range (see also Li this volume). What are the features of terrestrial skeletal support systems that vary in a regular way with changes in size,

Scaling in Biology, edited by J. H. Brown and G. B. West.
Oxford University Press, 2000.

and are these scale-dependent features ones that operate across different size ranges and obey general biological scaling laws?

Because the forces acting on a structure are likely to vary in proportion to the organism's weight and, hence, are proportional to its volume (αV), stress is predicted to increase with size due to the disproportionate scaling of volume versus area (V/A). This also implies a scale-dependent increase in tissue strain (defined as the deformation of the tissue under mechanical load divided by its unloaded length, or dl/L). Stress (σ) and strain (ε) are related by the elastic modulus (E) of the material, so that for linearly elastic materials, $\sigma = E\varepsilon$. Isometric, or geometrically similar, scaling predicts that larger animals and plants are subjected to greater stresses and strains, which should increase $\alpha M^{1/3}$. For organisms built of similar materials this suggests a drastic reduction in safety factor at larger sizes. Consequently, geometrically similar organisms likely face major constraints for meeting the force requirements of support and movement on land with the evolution of large body size. In order to avoid an increase in tissue stress and strain, and an increased probability of mechanical failure (which results directly from excessive tissue strain), larger organisms must either scale with strong allometry, restrict their size range or functional capacity, or evolve a means for reducing weight-specific forces to match the mass-specific decrease in tissue cross-sectional area.

In this chapter, I explore how different-sized terrestrial mammals achieve generally similar safety factors. The scaling of muscle mass and area, which limits locomotor stress capacity and underlies musculoskeletal design, may be linked to the 3/4 power scaling of metabolic energy supply for force generation. However, in contrast to arguments that a single set of scaling laws may explain the design architecture of respiratory and vascular supply networks (see Brown et al. this volume) associated with the 3/4 power scaling of metabolism, I argue that similar safety factors are achieved in the mammalian musculoskeletal system by differing solutions to mechanical constraints of size. Further, I examine whether similar constraints operate over different size ranges within these animals. In particular, is peak stress relative to tissue strength the limiting constraint at all sizes? I also discuss how differing design constraints may also operate at different scales of size within a single individual, by examining the branching architecture of a tree. Finally, I consider how musculoskeletal scaling to maintain similar stress matches observed scaling patterns of whole-body and leg spring stiffness (Farley [12]).

1   SIZE-RELATED CHANGES IN LOCOMOTOR POSTURE AND MUSCLE MECHANICAL ADVANTAGE

Despite the constancy of their material properties, the skeletal and muscular systems of terrestrial mammals scale near isometry, or with only slight positive allometry over much of their size range ($> 10^4$) [1, 2, 7]. Because of this, adjustments of bone and muscle architecture are insufficient to maintain similar locomotor stress. In order to achieve comparable safety factors, terrestrial mammals, ranging in size from 0.1 to 300 kg body mass, have evolved size-dependent changes in limb posture [8]. A shift from crouched postures at small sizes to more upright postures at large sizes enables a reduction in the ratio of muscular force to ground reaction force (α body weight, W) by increasing the effective mechanical advantage (EMA = r/R) of limb muscles (Figure 1). This size-related shift in limb posture during terrestrial locomotion means that muscle forces do not scale proportionally to body weight but, rather, scale $\alpha W^{0.73}$. For example, whereas muscular forces (F) developed in a chipmunk at a gallop are 10 times the ground reaction force (G), they are nearly equal to the

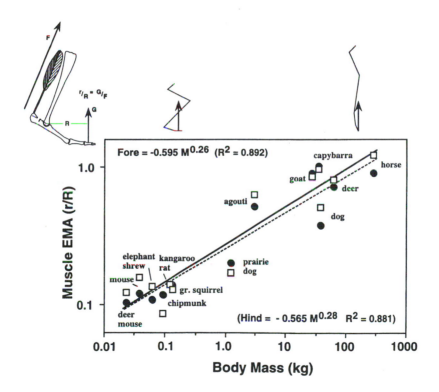

FIGURE 1 Scaling of posture-related muscle mechanical advantage (EMA = r/R, depicted in the upper left inset) in terrestrial mammals plotted against body mass on logarithmic coordinates. Least-squares (L-S) regression equations for forelimb and hindlimb show a similar pattern with the combined scaling of muscle EMAαM$^{0.27}$ (which implies muscle force $F\alpha$M$^{0.73}$). Changes from crouched locomotor postures in small mammals to more upright postures in larger mammals explains the similarity of peak bone and muscle stress in different-sized species.

TABLE 1   Allometric and theoretical scaling relationships of bone and muscle stress in terrestrial mammals.

	Area	Force	Stress	Reference
Bone	$\alpha M^{0.72}$	$\alpha M^{0.73}$	$\alpha M^{0.01}$	Biewener [7, 8]
Muscle	$\alpha M^{0.79}$	$\alpha M^{0.73}$	$\alpha M^{-0.06}$	Alexander et al. [1] and Biewener [8]
Theoretical	$\alpha M^{3/4}$	$\alpha M^{3/4}$	αM^0	

ground force exerted by a horse. By increasing limb mechanical advantage and reducing mass-specific muscle forces over this size range, terrestrial mammals are able to maintain nearly equivalent bone and muscle stresses (Table 1). This is supported by the observation of similar stresses (and strains) within the long bones of different-sized mammals [9, 26].

2   IS THERE A THEORETICAL BASIS FOR A LINK BETWEEN THE BIOMECHANICS OF MUSCLE SCALING AND ENERGY SUPPLY?

These empirical observations suggest the possibility that the scaling of muscle forces ($M^{3/4}$) over a considerable portion of the size range of terrestrial mammals may be related to the more general 3/4 power scaling of metabolic and transport processes (see Table 1 and Brown et al. this volume). Because the mechanical properties of vertebrate skeletal muscle are generally scale invariant (i.e., constant stress and strain), muscle force generating requirements must scale proportional to scale-dependent changes in muscle fiber cross-sectional area, which is achieved through the scaling of limb posture and muscle mechanical advantage. Evidence that the cost of force generation by skeletal muscles during locomotion largely determines the scaling of energy cost in different-sized mammals [18, 29] indicates a metabolic link to the scaling of muscle force. Because the muscles of different-sized mammals generate similar forces per unit volume (the decrease in mass-specific force is offset by the longer fibers of larger animals), the cost of generating muscle force appears to depend mainly on the rates of force development and muscle shortening. With their slower stride frequencies ($\alpha M^{-0.15}$) [15] and longer periods of limb support, larger animals expend less energy than small animals to support a given weight of their body while running. Consequently, the scaling of metabolic cost of transport and maximum aerobic capacity [28] scales close to $M^{3/4}$ (empirical range: $\alpha M^{0.70 \text{ to } 0.080}$). This suggests that the aerobic rate of ATP supply and the amount of ATP consumed by the muscles to move a given distance are matched to the biomechanical requirements of muscle force generation, at least for mammals ranging from 0.1 to 300 kg in size. Do similar mechanisms apply more generally across the full size range of terrestrial mammals?

3 STRESS-SIMILARITY SCALING AT GIANT SIZE: POSITIVE SKELETAL ALLOMETRY

Economos [11] suggests that different scaling relationships and, by implication, mechanical constraints may apply to large versus small terrestrial mammals (which he has estimated to occur at about 20 kg body mass). Consistent with this, but at a larger size, posture-related changes in limb mechanical advantage at sizes above 300 kg body mass appear to be constrained, such that positive allometric changes in skeletal shape are required to maintain adequate safety factors in extremely large terrestrial species. In a study comparing the scaling of different-sized carnivoran families with previous studies of bovids [20] and ceratomorphs (rhinos, tapirs, and their fossil relatives) [24], John Bertram and I [6] found evidence that, over different size ranges, these groups of terrestrial mammals exhibit differing allometric scaling (Figure 2), with ursids closely matching the elastically similar scaling ($L \alpha D^{0.67}$) [19] observed within bovids. As a larger size group, ceratomorphs scale with stronger positive allometry, close to static stress similarity ($L \alpha D^{0.5}$) [19]. At smaller sizes, carnivorans and small bovids scale closer to geometric similarity ($L \alpha D$). The increasingly more robust scaling within larger sized groups of terrestrial mammals, particularly within ceratomorphs, is also likely associated with reductions in locomotor performance (maximum speed and maneuverability); however, other than for largely anecdotal evidence [14] constraints on locomotor performance at extremely large sizes remain to be demonstrated. It is the case that elephants are unable to trot or gallop [13, 16] and their maximum running speed is considerably less than fleet bovids and cursorial carnivores, but measurements of limb mechanical advantage have not yet been made for these terrestrial giants.

4 GEOMETRIC SCALING AT SMALL SIZE: A CONSTRAINT OF STIFFNESS VERSUS STRENGTH?

In most cases, limits on physiological and mechanical performance are generally analyzed in terms of the effect of a size *increase*, rather than a consideration of the implications of an (evolutionary) size *decrease*. Choosing the particular size from which scale-dependent changes are considered is equally important. In an evolutionary context, the reference size for considering scale effects should be the size of the basal ancestral group from which the group evolved. For eutherian mammals this would be in the range of 0.1 to 1.0 kg. What are the consequences when a lineage of animals or plants evolve to

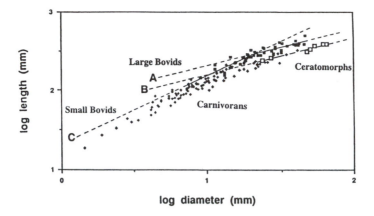

FIGURE 2 Differential scaling of tibial dimensions (length versus anteroposterior diameter on logarithmic coordinates) in bovids (small closed squares [20]), carnivorans (small crosses [6]) and ceratomorphs (open squares [24]). Small bovids (L-S slope = 0.89, $r = 0.90$) and small carnivorans (mustelids, procyonids, and viverrids: L-S slope = 0.85, $r = 0.98$) scale similarly, with only slight allometry; whereas, large bovids (L-S slope = 0.48, $r = 0.83$) and large carnivorans (ursids and felids: L-S slope = 0.70, $r = 0.95$) scale with strong allometry, approaching the extremely robust scaling observed for ceratomorphs (L-S slope = 0.47, $r = 0.98$). All regressions are significant at $p < 0.01$. L-S regression was used to compare the data from the three studies. The large-bovid line (A) parallels the ceratomorph line (B) but with a higher intercept, indicating their longer tibiae. The largest carnivoran species closely overlap the ceratomorph regression. From Bertram and Biewener [6]. Printed with permission from Wiley-Liss, Inc., a division of John Wiley & Sons. Inc.

smaller size? In terms of mechanical stress, geometric scaling as well as elastically similar scaling predict a decrease in stress with decreasing size (Figure 3). This results from decreases in weight-related forces that exceed reductions in bone, muscle, and tendon cross-sectional area. Consequently, smaller geometrically similar animals can also be expected to have relatively greater limb stiffness. That is to say, the structural elements of the limbs of smaller animals are likely to undergo smaller deflections for their size compared with the limb elements of larger animals. The nearly geometric scaling of small to medium-sized mammalian taxa (weighing 0.03 to 30 kg) suggests, therefore, that stiffness of support elements, and their effect on overall limb stiffness, may be the limiting design constraint rather than strength (i.e., peak stress).

Why is stiffness important? One property of muscles that is strongly affected by stiffness is their force-length relationship. All skeletal muscles exhibit an optimal range of length (L_{opt}) over which they can exert maximal force. Due to actin-myosin filament overlap, a muscle's ability to generate force is

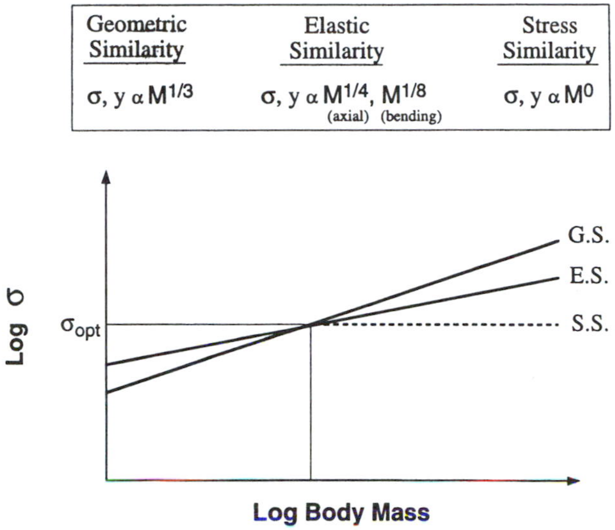

Geometric Similarity	Elastic Similarity	Stress Similarity
$\sigma, y \propto M^{1/3}$	$\sigma, y \propto M^{1/4}, M^{1/8}$ (axial) (bending)	$\sigma, y \propto M^{0}$

FIGURE 3 Theoretical logarithmic scaling of musculoskeletal stress (σ) versus body mass for the three similarity models: geometric (G.S.), elastic (E.S.), and static stress (S.S.). Only for S.S. does stress remain constant with change of size. For both G.S. and E.S., stress (and deflection of bone elements, y) scales with body mass, according to the above relationships (E.S. and S.S. relationships are based on McMahon [19]). For a given optimum stress (σ_{opt}) at a given mass, evolutionary decreases in body size according to G.S. or E.S., suggest reduced stress, increased stiffness, and increased safety factor. Evolutionary or ontogenetic size increases, on the other hand, predict increased stress, reduced stiffness, and reduced safety factor. These scaling trends suggest that, whereas peak stress and safety factor (i.e., strength) are important at large size, stiffness may be the key design constraint at small size.

greatly reduced at long lengths ($> 20\%$ of L_{opt}) and at short lengths ($< 20\%$ of L_{opt}). Consequently, the operating length of a muscle must be matched to the length of its fibers. If the tendons and bones of the limb become too slender relative to the forces that they must transmit, their resulting deflections might require excessive length change of the muscles' fibers, placing the muscles at a disadvantage for effective control of limb motion [25]. In most cases, the thickness of mammalian tendons seems to be disproportionately large relative to the forces that the tendon's muscle can exert, such that many tendons operate with safety factors in the range of 8–10 [17]. This suggests that stiffness can be as important a design constraint as strength. Geometric scaling to smaller body size is consistent with this observation.

While evolutionary decreases in size, at least within mammalian taxa, appear to be generally rare (the notable exception being insular island populations), interpretations of the importance of stiffness versus strength as constraints on mechanical design depend critically on what body size stress is considered to be "optimal" for the dimensions of the organism (σ_{opt}, Figure 3). Although no attempt is made here to define σ_{opt}, Alexander et al. [3] have defined an optimal bone stiffness in relation to bone stress, as that which would minimize the combined weight of bone and muscle in the limb at a peak stress of 70 MPa. Their analysis, however, does not consider scale effects of size. In the case of an evolutionary increase in size within a lineage (Cope's rule; see Stanley [27] and Alroy [4]), therefore, stiffness may be the limiting constraint at small size, with a shift to strength (and safety factor) as the lineage evolves to larger size along a geometrically similar trajectory. In order to distinguish this, it would be necessary to demonstrate a reduction in safety factor from what would be considered an "excessive" value at small size, which is not an easy task.

5   DIFFERENTIAL SCALE CONSTRAINTS WITHIN A TREE

As with the scaling of the mammalian musculoskeletal system, differential constraints on mechanical design also appear to apply to the scaling of branch architecture within trees. By sampling the branching architecture of a single tree, Bertram [5] has shown that two distinct size classes and scale patterns exist within a silver maple (*Acer saccharinum*). Nonperipheral branches (trunk and main supporting branches) scaled with strong positive allometry, closely matching McMahon's [19] elastic similarity model (Figure 4). Reduced major axis (RMA) and least-squares (L-S) regression give exponents that bracket the predicted 2/3 value for elastically similar deflections that McMahon and Kronauer [22] previously have found for white oak (*Quercus alba*). On the other hand, the peripheral (leaf bearing) branches scaled with substantial negative allometry: $L\alpha D^{1.39}$ (RMA regression), such that these smaller branches become relatively more slender as they grow. The divergence in scale pattern is clearly seen when the slenderness ratio (L/D) of different size classes of the tree's peripheral and nonperipheral branches is plotted versus branch diameter (Figure 4(c)).

These distinct scaling patterns suggest that, whereas strength and stiffness are important to the design of the nonperipheral weight-support branches, in which elastic similarity applies, flexibility is key to the function of the peripheral branches. Being flexible enables the tree's peripheral branches and leaves to reorient in the wind to reduce drag. The small diameter of these branches also means that they are difficult to break by bending: the smaller a beam's diameter, the less strain can be developed for a given bending curvature (an analogous but more extreme example of this is glass optical fibers, which have considerable flexibility and rarely break, despite the high stiffness and brittleness of glass as a material). An advantage of studying scaling patterns within a single individual, such as the sugar maple, is the absence of

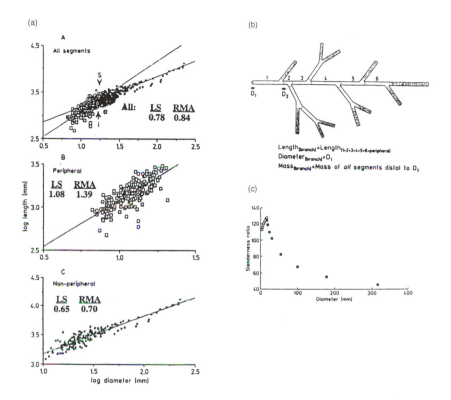

FIGURE 4 (a) Differential scaling of branch dimensions (length versus diameter) within a silver maple (*Acer saccharinum*). Branches were divided into two size classes: nonperipheral supporting branches (small solid squares) and peripheral leaf-bearing branches (larger open squares); branch length and diameter being determined as shown in (b). Nonperipheral and peripheral branches separated statistically into two size classes (large and small) and showed significantly different scale relations. Whereas nonperipheral branches scaled with strong positive allometry (slope < 1), close to McMahon's [19]) elastic similarity model and similar to the branching scaling observed for a white oak [22], the peripheral branches scaled with strong negative allometry (slope > 1), making them increasingly more slender as they grew in length. The change in scale pattern between the two branch size classes is clearly observed in (c), which shows the slenderness ratio (L/D) plotted against diameter (smaller peripheral branches, open squares; large nonperipheral branches solid squares).

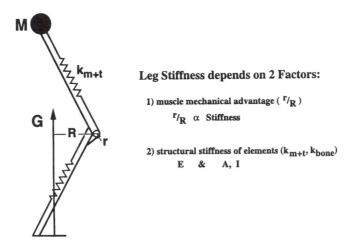

SPRING STIFFNESS OF ANIMAL LIMBS

FIGURE 5   Whole limb spring stiffness (elastic deformation of the limb during ground support) depends on two factors: (1) muscle mechanical advantage (r/R) and (2) the structural stiffness of muscle tendon (k_{m+t}) and bone (k_b) support elements (depicted here for the "knee" joint). The latter depend on the elastic modulus of the tissues and their shape (cross-sectional area, A, and second moment of area, I, for bending).

genetic variation; however, the disadvantage is that generalizations to other individuals within the species and across species is more limited. Additional studies of other trees are needed to test the generality of these intriguing results.

6   SCALING OF LIMB MECHANICAL ADVANTAGE AND LEG STIFFNESS

Changes in limb mechanical advantage that allow terrestrial mammals of vastly differing size to maintain similar levels of peak bone and muscle stress were discussed above. In addition to affecting the magnitude of muscle force required to support an animal's weight, changes in limb posture also likely affect the scaling of limb stiffness during running (Figure 5). Whole leg stiffness (k_{leg}, after McMahon and Cheng [21] and Farley et al. [12]) can be defined as the ratio of the displacement (ΔL) of the leg during the stance phase of a step to the peak ground reaction force (G): $k_{leg} = \Delta L/G$ (Figure 6). In running, trotting, galloping, and hopping gaits, when an animal lands on a limb, its CM falls during the first half of stance, compressing its leg spring. During the second half of stance its leg spring recoils, as its CM rises, until the limb leaves the ground. Displacements of the leg spring (ΔL) are a function of the cumulative angular excursions of the limb's joints ($\Delta\theta$) during its contact with the ground (Figure 6). The notion that the leg functions like a spring derives from the fact that these "bouncing" gaits all utilize a similar energy conserving mechanism [10], in which the potential and kinetic energy that is lost as the body's CM falls (Δy) and decelerates during the first half of stance is converted into and stored as elastic strain energy in the tendons, ligaments, and muscles of the limb. This elastic energy is subsequently recovered during the second half of stance, allowing the animal to regain its lost potential and kinetic energy as it leaves the ground. Consequently, compression of the "leg spring" actually reflects the stretching of muscle-tendon and ligamentous spring elements in the limb. (Displacements due to compression of articular cartilage and bone flexure, in comparison, are likely to be quite small: i.e., $k_b \gg k_{m+t}$.)

k_{leg} depends on two general factors (Figure 5): (1) the structural stiffness of limb support elements: the muscle-tendon units and bones (k_{m+t} and k_b), and (2) the limb's muscle mechanical advantage (r/R, defined in Figure 1). The former depends on the material moduli (E) of the elements and, as noted above, their shape (length relative to cross-sectional area and second moment of area). The latter reflects the relative magnitude of force that the muscle-tendon (and bone) elements must generate (or support) for a given magnitude of ground reaction force at the foot. In a recent study of seven bipedal and quadrupedal mammals, Farley et al. [12] found that k_{leg} scaled proportionally to $M^{0.67}$, with larger animals having stiffer leg springs. This resulted from the fact that peak ground reaction forces scaled directly with the animal's body weight ($G\alpha M^{0.97}$) and displacements of the their leg spring, ΔL, scaled proportional to $M^{0.30}$ (all exponents being close to simple fractional values predicted by geometric similarity: $M^{2/3}$, M^1, and $M^{1/3}$).

This pattern of leg spring stiffness matches the predicted changes in whole limb displacement associated with postural adjustments in muscle mechanical advantage. As discussed above, these changes in muscle mechanical advantage are required to maintain bone and muscle stress constant in different-sized mammals (0.03 and 300 kg) [8], by matching muscle force to the scaling of bone area and muscle area (Table 1). The stiffness of the muscle-tendon spring ($k_{m+t} = F/\Delta x$) reflects the ratio of force transmission relative to its stretch, which can be related to the joint moment arm (r) and joint angular displacement ($\Delta\theta$) according to

$$\Delta x = r \, \Delta\theta. \tag{1}$$

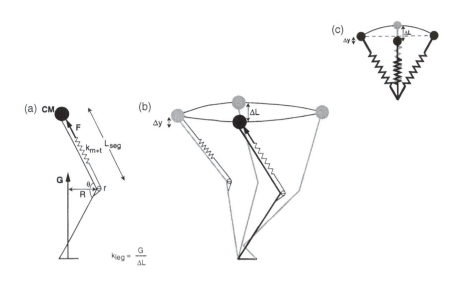

FIGURE 6   (a) Following McMahon and Cheng [21] and Farley et al. [12] the stiffness of the whole limb (k_{leg}) can be defined as the maximum displacement of the
limb (ΔL) divided by the ground reaction force (G) acting on the limb. ΔL represents the displacement of the limb, which occurs due to flexion of limb joints, as
the limb contacts the ground (initially ground force and muscle-tendon force are
zero), becoming compressed at mid-support when maximum G and muscle force
(F) are developed (dark lines in (b)). Subsequently, the limb rebounds as the "leg
spring" is unloaded, causing the CM to rise during the second half of stance. The
muscle-tendon model of leg spring stiffness depicted in (a) and (b), for the purpose
of analyzing how limb mechanical advantage affects leg spring stiffness, is analogous
to the simple mass-spring model (c) used by McMahon and Cheng [21] and Farley
et al. [12]. Reprinted with permission from Elsevier Science.

Correspondingly, the vertical displacement of the whole leg (ΔL) will be a
function of leg segment lengths and joint angular displacements. Given

$$L = 2L_{\text{seg}} \sin \theta \tag{2}$$

we can write

$$\Delta L = 2L_{\text{seg}} \cos \theta \Delta \theta . \tag{3}$$

Following a similar analysis by McMahon et al. [23], we can assess how
much changes in limb posture (reduced mechanical advantage, r/R) versus
muscle-tendon stiffness (k_{m+t}) contribute to L_{leg}. Given

$$k_{\text{leg}} = \frac{G}{\Delta L} \tag{4}$$

and

$$G = \frac{Fr}{R} \tag{5}$$

it follows that

$$k_{\text{leg}} = \frac{Fr}{\Delta LR}. \tag{6}$$

Substituting for F in terms of k_{m+t} and Δx, gives

$$k_{\text{leg}} = k_{m+t}\frac{\Delta xr}{\Delta LR}. \tag{7}$$

Using Eqs. (1) and (3), we can rewrite Eq. (7) as

$$k_{\text{leg}} = k_{m+t}\frac{r^2}{2L_{\text{seg}}\cos\theta R} \tag{8}$$

and recognizing that $R = L_{\text{seg}}\cos\theta$, Eq. (8) simplifies to

$$k_{\text{leg}} = k_{m+t}\frac{r^2}{2R^2}. \tag{9}$$

Hence, the ratio of k_{leg} to k_{m+t} scales as

$$\frac{k_{\text{leg}}}{k_{m+t}}\alpha\frac{r^2}{R^2}\alpha M^{0.54} \tag{10}$$

suggesting that k_{m+t} scales $\alpha M^{0.13}(M^{0.67}/M^{0.54})$. A theoretical basis for why the muscle-tendon stiffness scales in this manner $(\alpha M^{2/3})$ is unclear because it reflects a geometric change in leg stiffness $(\alpha M^{2/3})$ divided by an allometric change in limb mechanical advantage squared (approximately $\alpha M^{1/2}$). This result may also depend on modeling k_{leg} based on the displacement of a single joint. When other joints are taken into account, the scaling of leg stiffness relative to limb mechanical advantage, and hence, muscle-tendon stiffness, may differ from the analysis shown above. It will also be important to explore the function of muscles and muscle-tendon components in different-sized animals in order to determine whether their active force-length properties match the overall scaling predicted by whole leg stiffness and postural shifts in mechanical advantage.

The scaling of k_{m+t} suggested by the forgoing analysis indicates, therefore, that other size-related changes in muscle-tendon architecture are likely to contribute to the enhanced stiffness of larger animal limbs. Given that muscle stiffness is less than tendon stiffness, it seems likely that the relatively shorter muscle fibers $(\alpha M^{0.28})$ [1] of larger mammals and their relatively longer tendons may both contribute to the suggested overall increase in muscle-tendon stiffness. At present, these observations, and the predictions derived from them, require further study.

7 CONCLUSIONS

The observed scaling patterns of musculoskeletal design within terrestrial mammals suggests that differential mechanical constraints operate over different ranges of size and taxa. No one pattern, or general scaling model, explains the range of observed solutions. Whereas peak stress and safety factors are probably limiting constraints over much of the terrestrial size range, stiffness may be the key design constraint at smaller size. Further, differing mechanisms for maintaining uniform safety factors also appear to operate. Over much of their size range (0.1 to 300 kg), posture-related changes in limb mechanical advantage occur that enable terrestrial mammals to match locomotor forces to the scaling of bone and muscle areas. Intriguingly, within this size range there appears to be a link between the scaling of transport processes for energy supply and use, which obey a 3/4 power law, and the scaling of muscle force requirements for terrestrial locomotion. However, at larger sizes (> 300 kg), more extreme positive allometry (robust scaling) and/or reductions in locomotor performance appear to be required in order to keep peak stresses within a safe range. How these adjustments in locomotor support affect muscle-force generating requirements in relation to the metabolic supply of energy as yet remains unknown. Posture-related changes in limb-muscle mechanical advantage are also shown to be consistent with recently observed changes in the stiffness of animal limbs (k_{leg}) [12], in which the shift to a more upright posture to reduce musculoskeletal loading for constant safety factor also results in an increased limb stiffness, counter to the decrease in stiffness predicted by geometric or elastic similarity scaling at larger size (Figure 3). Consequently, changes in limb stiffness that would be predicted by the structural scaling of limb support elements (Figure 3) must also take into account size-related changes in limb mechanical advantage. Finally, as for the limbs of terrestrial mammals, differential scaling patterns and design constraints are also to be found within the branching architecture of a single tree: larger structural support branches scale with positive allometry to avoid excessive deflection, whereas small leaf-bearing branches scale with negative allometry precisely in order to achieve flexibility, reduce drag, and avoid bending failure.

Although common design principles and a single scaling model may help to explain general features of vascular and respiratory supply networks within biology, no one scaling model appears sufficient for explaining how mammals and trees have evolved to meet the mechanical demands of life on land. Nevertheless, it is the case that certain features, largely those that reflect the strength of the materials of which even distantly related organisms are constructed, are scale invariant. As a result, regular size-related changes in other features are required either to maintain a constant mechanical safety factor or to achieve an appropriate limb (or branch) stiffness.

REFERENCES

[1] Alexander, R. M., G. Goldspink, A. S. Jayes, G. M. O. Maloiy, and E. M. Wathuta. "Allometry of the Limb Bones of Mammals from Shrew (*Sorex*) to Elephant (*Loxodonta*)." *J. Zool., Lond.* **189** (1979): 305–314.

[2] Alexander, R. M., A. S. Jayes, G. M. O. Maloiy, and E. M. Wathuta. "Allometry of the Leg Muscles of Mammals." *J. Zool., London* **194** (1981): 539–552.

[3] Alexander, R. M., R. F. Ker, and M. B. Bennett. "Optimum Stiffness of Leg Bones." *J. Zool., London* **222** (1990): 471–478.

[4] Alroy, J. "Cope's Rule and the Dynamics of Body Mass Evolution in North American Fossil Mammals." *Science* **280** (1988): 731–734.

[5] Bertram, J. E. A. "Size-Dependent Differential Scaling in Branches: The Mechanical Design of Trees Revisited." *Trees* **4** (1989): 241–253.

[6] Bertram, J. E. A., and A. A. Biewener. "Differential Scaling of the Long Bones in the Terrestrial Carnivora and Other Mammals." *J. Morph.* **204** (1990): 157–169.

[7] Biewener, A. A. "Bone Strength in Small Mammals and Bipedal Birds: Do Safety Factors Change with Body Size?" *J. Exp. Biol.* **98** (1982): 289–301.

[8] Biewener, A. A. "Scaling Body Support in Mammals: Limb Posture and Muscle Mechanics." *Science* **245** (1989): 45–48.

[9] Biewener, A. A. "Biomechanics of Mammalian Terrestrial Locomotion." *Science* **250** (1990): 1097–1103.

[10] Cavagna, G. A., N. C. Heglund, and C. R. Taylor. "Mechanical Work in Terrestrial Locomotion: Two Basic Mechanisms for Minimizing Energy Expenditures." *Am. J. Physiol.* **233** (1977): R243–261.

[11] Economos, A. C. "Elastic and/or Geometric Similarity in Mammalian Design?" *J. Theor. Biol.* **103** (1983): 167–172.

[12] Farley, C. T., J. Glasheen, and T. A. McMahon. "Running Springs: Speed and Animal Size." *J. Exp. Biol.* **185** (1993): 71–86.

[13] Gambaryan, P. *How Mammals Run: Anatomical Adaptations.* New York: Wiley, 1974.

[14] Garland, T. J. "The Relation between Maximal Running Speed and Body Mass in Terrestrial Mammals." *J. Zool., London* **199** (1983): 157–170.

[15] Heglund, N. C., and C. R. Taylor. "Speed, Stride Frequency, and Energy Cost per Stride: How Do They Change with Body Size and Gait?" *J. Exp. Biol.* **138** (1988): 301–318.

[16] Hildebrand, M. B. *Analysis of Vertebrate Structure,* 3rd ed. New York: Wiley, 1988.

[17] Ker, R. F., R. M. Alexander, and M. B. Bennett. "Why Are Mammalian Tendons So Thick?" *J. Zool., London* **216** (1988): 309–324.

[18] Kram, R., and C. R. Taylor. "Energetics of Running: A New Perspective." *Nature* **346** (1990): 265–267.

[19] McMahon, T. A. "Using Body Size to Understand the Structural Design of Animals: Quadrupedal Locomotion." *J. Appl. Physiol.* **39** (1975): 619–627.

[20] McMahon, T. A. "Allometry and Biomechanics: Limb Bones in Adult Ungulates." *Amer. Natur.* **109** (1975): 547–563.

[21] McMahon, T. A., and G. C. Cheng. "The Mechanics of Running: How Does Stiffness Couple with Speed?" *J. Biomech.* **23** (1990): 65–78.

[22] McMahon, T. A., and R. E. Kronauer. "Tree Structures: Deducing the Principle of Mechanical Design." *J. Theor. Biol.* **59** (1976): 443–466.

[23] McMahon, T. A., G. Valiant, and E. C. Frederick. "Groucho Running." *J. Appl. Physiol.* **62** (1987): 2326–2337.

[24] Prothero, D. R., and P. C. Sereno. "Allometry and Paleoecology of Medial Miocene Dwarf Rhinoceroses from the Texas Gulf Coastal Plain." *Paleobiol.* **8** (1982): 16–30.

[25] Rack, P. M. H. "Stretch Reflexes in Man: The Significance of Tendon Compliance." In *Feedback and Motor Control in Invertebrates and Vertebrates*, edited by W. J. P. Barnes and M. H. Gladden, 217–229. London: Croom Helm, 1985.

[26] Rubin, C. T., and L. E. Lanyon. "Dynamic Strain Similarity in Vertebrates; an Alternative to Allometric Limb Bone Scaling." *J. Theor. Biol.* **107** (1984): 321–327.

[27] Stanley, S. M. "An Explanation for Cope's Rule." *Evolution* **27** (1973): 1–26.

[28] Taylor, C. R., G. M. O. Maloiy, E. R. Weibel, V. A. Langman, J. M. Z. Kamau, H. J. Seeherman, and N. C. Heglund. "Design of the Mammalian Respiratory System: III. Scaling Maximum Aerobic Capacity to Body Mass: Wild and Domestic Mammals." *Resp. Physiol.* **44** (1981): 25–37.

[29] Taylor, C. R. "Relating Mechanics and Energetics During Exercise." In *Comparative Vertebrate Exercise Physiology: Unifying Physiological Principles*, edited by James H. Jones, 181–215. Advances in Veterinary Science and Comparative Medicine, Vol. 38A. New York: Academic Press, 1994.

Consequences of Size Change During Ontogeny and Evolution

M. A. R. Koehl

Changes in body size can occur during the lifetime of an individual organism as it grows, or can occur over many generations during the evolution of a lineage. Most studies of body size either explore how function is maintained at different sizes, or seek mechanistic explanations for the patterns we see in how features such as shape, metabolic rate, or life history vary with body size [7, 40, 52] (e.g., Brown this volume; Biewener this volume). My purpose here is to complement this literature about the consequences of *being* one size or another with an exploration of some of the functional consequences of *changing* size.

Function can shift as size changes, but the particular consequences of a size change can depend on factors such as the environment and morphology of the organism. This chapter focuses first on functional consequences of changing size, and then considers how the effects of morphological features on performance can be altered as size changes. This chapter concludes by considering how the consequences of a size change might be affected by: (1) the size range in which the change occurs, (2) the habitat, and (3) the structural design of the organism. I introduce each of these topics with a few examples, mostly biomechanical, and then discuss in more general terms their ecological or evolutionary consequences.

Scaling in Biology, edited by J. H. Brown and G. B. West.
Oxford University Press, 2000.

1   TRANSITIONS IN FUNCTION AS SIZE CHANGES

1.1   EXAMPLES OF FUNCTIONAL SHIFTS ACCOMPANYING SIZE CHANGE

1.1.1   Shifts in Hydrodynamic Function.   Many important biological processes (such as respiration, and locomotion) depend on how organisms interact with the fluid medium, water or air, around them. The Reynolds number ($\Re$) of a structure moving through a fluid represents the relative importance of inertial to viscous forces determining how the fluid moves ($\Re = LU/\nu$, where L is a linear dimension of the structure, U is fluid velocity relative to it, and ν is kinematic viscosity of the fluid) [57]. At high $\Re$'s (e.g., large, rapidly moving structures), inertial forces predominate and flow is messy and turbulent, whereas at low $\Re$'s (e.g., small, slowly moving structures), viscosity damps out disturbances in the fluid and flow is smooth and orderly. Thus, we should expect changes in biological processes that involve fluids as body size (and hence $\Re$) changes.

1.1.2   Hairy Little Legs.   Many animals from different phyla use appendages bearing arrays of hairs (Figure 1(a)–(e)) to capture molecules from the surrounding fluid (e.g., feathery gills or olfactory antennae), to capture particles (e.g., hairy suspension-feeding appendages), or to move the fluid around them (e.g., setulose appendages used to fly, swim, or create ventilatory currents). When a particular type of structure is so ubiquitous among organisms and serves such critical biological functions, it is important to figure out how it works, and how its function is affected by its size.

The performance of all the functions mentioned above (e.g., capturing molecules or particles, moving water or air) depend on how the arrays of hairs interact with the fluid around them. Therefore, the first step in analyzing how these structures work is to figure out how fluid moves around and through them. The $\Re$'s at which the hairs on the types of structures listed above operate [8, 9, 30, 35, 38, 51] (using hair diameter for L) range between 10^{-5} and 10. In this $\Re$ range viscosity is very important in determining flow patterns (although we cannot ignore the effects of inertia at the upper end of this $\Re$ range). When fluid flows past a solid surface, the fluid in contact with the surface does not slip relative to the surface and a velocity gradient develops between the surface and the free-stream flow. The layer of fluid along a solid surface in which this velocity gradient exists is called the boundary layer. Similarly, when a structure moves through water or air, the fluid contacting its surface is carried along with it and a boundary layer of sheared fluid develops along the structure's surface. The lower the $\Re$, the thicker this layer of sheared fluid can be relative to the dimensions of the structure. Thus, in the case of viscous flow around a hair, the layer of fluid moving along with the hair can be quite thick relative to the hair's diameter, and little fluid may move through the gaps between adjacent hairs in an array [8, 9, 30]. Furthermore,

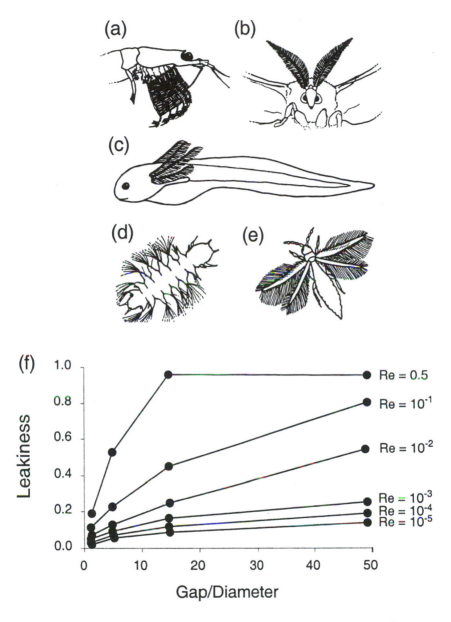

FIGURE 1   (a) Suspension-feeding appendages of a euphausid ("krill"), Phylum Arthropoda. (b) Olfactory antennae of a male moth, Phylum Arthropoda. (c) External gills of a larval African lungfish, Phylum Chordata. (d) Swimming parapodia of a nereid larva, Phylum Annelida. (e) Wings of a thrips, Phylum Arthropoda. (f) Leakiness to fluid flow of the gap between neighboring hairs, plotted as a function of the ratio of the width of that gap to the diameter of a hair (redrawn from Koehl [32]). Each line represents a different Reynolds number.

flow is laminar (i.e., there is no turbulent mixing) in the $\Re$ range in which these hairs operate; hence, the only way that molecules can be spread across streamlines is via molecular diffusion. Since humans operate at high $\Re$, we cannot trust our intuitions when considering the viscous flow around arrays of little hairs.

In order to understand how arrays of hairs capture molecules and particles, or push fluids around, the first thing that we need to figure out is whether fluid flows through the gaps between the hairs in an array, or flows around the array rather than through it. The leakiness (the proportion of the fluid encountering a gap between two hairs that flows through the gap rather than around the perimeter of the array of hairs [8]) of a hair-bearing structure determines, for example, whether or not the structure can function as a filter, the flux of molecules to hair surfaces, and the ability of the appendage to generate thrust or lift [30, 31]. Therefore, size ($\Re$) should affect all of these leakiness-dependent functions.

A general model of flow between neighboring hairs permits us to examine how the size, spacing, and speed of an array of hairs affects its leakiness (Figure 1(f)). Each line in Figure 1 represents a different $\Re$ (i.e., a different size). At small hair sizes ($\Re = 10^{-5}$ to 10^{-3}), arrays of hairs have very low leakiness (i.e., only a small proportion of the water or air encountered actually goes through the gaps between hairs, while most flows around the array) and function like nonporous paddles. In contrast, from $\Re$'s of 10^{-2} to 1, a transition in leakiness occurs: a structure that functioned like a paddle at small size becomes a leaky sieve at larger size and filtering becomes possible. Thus, as hair-bearing appendages change size across this critical $\Re$ range, their function switches.

A biological example of this transition in leakiness is provided by calanoid copepods, abundant planktonic crustaceans that play a critical link in many marine food webs between single-celled algae and higher trophic levels such as fish. Copepods capture single-celled algae using a pair of hairy feeding appendages, the second maxillae (M2's), which they fling apart from each other and then sweep back together [35]. Some species perform this capture motion with their setae (hairs) operating at $\Re$'s of order 10^{-2}; their nonleaky, paddlelike M2's capture food by drawing a parcel of water containing an alga toward the mouth during the fling. In contrast, other species operate their M2's at setal $\Re$'s of one and filter their food from the water as they sweep toward each other. Thus, even though their M2 feeding motions look qualitatively similar, the physical mechanisms by which these two species of copepods capture food are different because they operate at $\Re$'s above and below the transition from paddle to sieve [30]. These results suggest that we might expect similar functional transitions during the ontogeny of the many aquatic larvae with setulose appendages that grow across this hair $\Re$ range where the transition in leakiness occurs.

1.1.3 Swimming.   Other examples of transition in hydrodynamic function as size ($\Re$) changes are provided by studies of the ontogeny of swimming. As brine shrimp larvae get bigger, their propulsive mechanism switches from drag-based rowing at low $\Re$ to inertial swimming at higher $\Re$, even though the flapping motion of their appendages does not change [60, 61]. Similarly, larval fish switch from drag-based swimming at low $\Re$ to inertial propulsion when they grow to higher $\Re$ [4, 46]. Furthermore, as larval fish increase in size and the importance of viscous force declines at higher $\Re$, intermittent swimming becomes more energetically advantageous [59]. Scallops provide another example of transitions in swimming performance with changes in body size [11, 39]. Scallops swim by jet propulsion by squirting water out of the mantle cavity while clapping their shells together. Very small juvenile scallops cannot use this inertial mode of locomotion effectively and are sedentary, whereas larger scallops can jet; once $\Re > 3000$, scallops are big enough to use lift to get up off the substratum. However, when scallops grow even larger, they become poor swimmers again as their shells grow too heavy relative to the thrust they can generate.

1.1.4 Walking on Water.   Whether or not animals can walk on water depends on their size. Some animals, such as water striders, are held up by the surface tension acting along the perimeters of their feet. Since the force holding the animal up is proportional to length, while the weight of the animal is proportional to volume, there is a body size above which animals cannot use the surface-tension mechanism to walk on water [1].

Some larger animals, such as basilisk lizards, can run on water using a different physical mechanism: the force to support the lizard's body is provided by an upward impulse as the foot slaps onto the water surface, followed by an upward impulse as the foot strokes down into the water [17, 18]. There is also an upper limit to the body size for which this mechanism of locomotion on a water surface can work, since the weight that must be supported increases at a greater rate with body size than does the upward force that can be generated by the feet. These lizards also provide an example of another functional shift that can sometimes accompany increases in body size: at small size there can be permission for diversity in the ways in which the animals move their appendages without serious performance consequences, whereas at larger size limb movements can have a critical effect on performance [31]. Small basilisks, which have the capacity to generate a large force surplus relative to their body weight, can vary their limb movements considerably and still remain atop the water, whereas larger animals, which can generate barely enough force to support their weight, are constrained to a narrow range of leg and foot motions to run successfully on water [17, 18]. In the field, juveniles often run on water simply to move to another sunning spot, whereas adults venture onto the water only under duress.

1.1.5 Solar Panels and Wings.   An example of how an isometric change in body size has the potential to generate novel function is provided by wind-tunnel experiments using models of fossil insects of a range of sizes [26, 27]. The physical models were used to test various hypotheses about the aerodynamic and thermoregulatory consequences of changes in the length of protowings on early insects. At small body size, short thoracic protowings improved thermoregulatory performance, but had negligible effect on aerodynamic gliding, parachuting, or turning performance; in contrast, protowings of the same relative length on larger models improved aerodynamic performance. This illustrates that it is physically possible for a simple increase in body size to cause a novel function (i.e., a solar panel can become a wing) without requiring the invention of a novel structure. (Of course, whether or not protowings served thermoregulatory or aerodynamic roles in early insects remains speculative, as do other feasible hypothesized functions, such as sexual signaling, gas exchange, or skimming along the surface of a body of water.)

1.1.6 Trophic Role.   Examples of functional shifts that accompany size change can be found in ecological studies of the trophic roles played by certain species of animals as they grow. For instance, some benthic marine worms that feed on sediment particles as adults have juvenile stages that are herbivorous or carnivorous. This size-dependent switch in feeding mode is thought to occur because the guts of little juveniles are too small to permit adequate digestion of nutrient-poor sediment particles [22, 24, 50]. Similarly, the type of prey that can be caught and ingested by certain species of predators can shift as body size increases (e.g., snakes [19], fish [45]).

Some species of prey grow large enough to become invulnerable to particular species of predators. The classic example of this type of switch in function that occurs as organisms grow is the size refuge attained by large mussels, *Mytilus californianus*, from predation by starfish (*Pisaster ocraceus*) [48]. Since *Mytilus californianus* are important competitors for space on rocky shores along the Pacific coast of North America, their ability to undergo a transition from being the preferred prey of *Pisaster ocraceus* to being not eaten by this keystone predator can have profound effects on the structure of the communities of organisms in these habitats [48, 49].

1.2 WHERE TO EXPECT FUNCTIONAL SHIFTS TO OCCUR AS SIZE CHANGES

There are bound to be many yet to be studied examples of functional shifts that occur as size changes. Some hints of where to expect such changes in function can be gleaned from the dimensionless numbers, worked out by engineers, to describe the relative importance of different physical factors involved in a process. In some of the examples cited above, the relative importance of inertia to viscosity ($\Re$) depends on size (L). For biological functions involving momentum exchange between organisms and the water or air around them (e.g., swimming, flying, ventilation, circulation), we can expect to find functional shifts as $\Re$ changes. Similarly, for biological functions involving transport of molecules, such as gas exchange, nutrient uptake, or smelling, the importance of fluid motion relative to molecular diffusion is given by the Péclet number (Pé $= LU/D$), where L is a linear dimension, U is velocity of the fluid relative to the structure, and D is the diffusion coefficient of the molecule of interest in the fluid) [57]. Like $\Re$, Pé depends on size (L). Pedestrian locomotion (walking, running) and swimming at the air-water interface depend on the importance of gravity relative to inertia (Froude number $= U^2/[gL]$, where g is the acceleration due to gravity), which also depends on size (L).

1.3 ECOLOGICAL AND EVOLUTIONARY SIGNIFICANCE OF FUNCTIONAL SHIFTS THAT OCCUR AS SIZE CHANGES

The examples cited above are but a few of the transitions in function that must accompany the size changes that occur during the ontogeny of an individual or the evolution of lineage. An important ecological consequence of such transitions is that a single species can play several different roles in a community if, for example, their feeding mode or their vulnerability to predation or physical disturbance changes as they grow [45, 48, 49]. An important evolutionary consequence of functional transitions that accompany size change is that new selective pressures on morphology can occur if a novel function is acquired as a lineage changes size over evolutionary time [26, 30].

2 TRANSITIONS IN THE EFFECTS OF MORPHOLOGY AND BEHAVIOR AS SIZE CHANGES

2.1 EXAMPLES OF SIZE-DEPENDENT CONSEQUENCES OF MORPHOLOGY AND BEHAVIOR

2.1.1 Hairy Little Legs.   The hairy legs mentioned above provide examples of how the effects of a particular morphological characteristic or type of behavior can be altered as size changes. We have been using mathematical [8, 9, 30] and physical models [30, 38] to quantify how various structural or kinematic features of a row of hairs might affect its leakiness. For example, hair spacing has virtually no effect on leakiness at hair $\Re$'s of order 10^{-3} and lower, but as size increases to $\Re$'s of 10^{-2} to 1, spacing has an enormous effect on flow through the array of hairs. At even larger sizes ($\Re > 1$), hair spacing once again has no effect on leakiness (unless hairs are very close together) (Figure 1). Adding more hairs to a row of hairs reduces the leakiness of the array if $\Re < 1$, but has the opposite effect at larger size ($\Re > 1$). Similarly, as size changes, there are transitions in which behaviors can affect the leakiness of hairy appendages. For example, moving the appendage near a wall (such as the body surface) increases leakiness at $\Re$'s of 10^{-2} and lower, but not at larger size. In contrast, speeding up appendage movement only affects leakiness at $\Re$'s between 10^{-2} and one, but not at smaller or larger sizes. Thus, for hair-bearing appendages, a continuous change in size can lead to a discontinuous change in how particular morphological or behavioral traits affect performance.

2.1.2 Streamlining.

The effect of body shape on drag depends on the size of the organism. Drag, a force which tends to push an organism in the direction that water or air flows past it, is due to skin friction (the resistance of the fluid in the boundary layer around the body to being sheared) at low $\Re$'s, but is due to skin friction plus form drag (the net pressure on a body behind which a wake has formed) at high $\Re$'s [57]. Streamlined body shapes (those with a long, tapered downstream end) reduce form drag compared with bluff body shapes of the same width because smaller wakes form behind the streamlined bodies, although the larger area of the long, tapered tail raises skin friction. Drag coefficient (C_D) is a dimensionless index of how drag-inducing a body shape is. For large organisms operating at high $\Re$'s, at which form drag is much greater than skin friction, streamlining reduces C_D, but for small organisms at low $\Re$ that only experience skin friction, streamlining increases C_D. For example, C_D's of globose ammonoid shells are higher than C_D's of flat, streamlined shells at $\Re > 100$, but the reverse is true for smaller shells at lower $\Re$ [23]. Similarly, the C_D of small ($\Re = 1$ to 10) benthic stream invertebrates is lower if their shape is more hemispherical, whereas the C_D of larger animals is lower ($\Re = 1000$) if they are more flattened [55].

2.1.3 Bumpy Skin.

Body size determines whether or not bumps on the skin of an organism affect drag [6, 28, 56, 57]. When organisms are small, surface texture is buried in the boundary layer and has no effect on drag, whereas at very high $\Re$'s surface bumps can protrude through the boundary layer and increase skin friction drag. The net drag on streamlined bodies is simply increased by skin bumps once the critical $\Re$ (i.e., size) is reached. Changes in body size have more complex consequences for organisms that are not streamlined: at small sizes ($\Re$'s) surface texture has no effect on drag, whereas at large sizes surface bumps increase the drag; however, at intermediate sizes surface bumps have the opposite effect and *decrease* the drag (mechanisms explained in Vogel [57] and Koehl [31]).

2.1.4 Gliding.

The wing shape that improves the distance traveled by gliding animals or plant seeds depends on body size. Short, wide wings enhance gliding at small size, whereas long, narrow wings improve performance at large size [14].

2.2   ECOLOGICAL AND EVOLUTIONARY CONSEQUENCES OF SIZE-DEPENDENT CONSEQUENCES OF MORPHOLOGY AND BEHAVIOR

The effects of structure or kinematics on performance can shift as size changes. Therefore, if size changes during the evolution of a lineage, then selection for different morphologies or behaviors can occur for the large species than can occur for the small ones, even if their function does not change. Ecomorphologists and paleontologists use morphological characters of organisms to infer their ecological roles, their function, or their performance of particular tasks relative to other organisms (reviewed in Koehl [31]). In doing so , they should be aware that a particular morphological trait may have very different effects on the performance of small organisms than it does on the functioning of larger ones. Which morphologies correlate with particular ecological roles can depend on size.

3   THE CONSEQUENCES OF SIZE CHANGES DEPEND ON SIZE

3.1   EXAMPLES OF SIZE-DEPENDENT EFFECTS OF SIZE CHANGE

In the introductory chapter of this book, Brown discusses examples of plotting power functions on both linear and logarithmic axes. If we look at linear plots of the performance of some process as a function of body size, it is easy to see that various aspects of performance vary with body size in nonlinear ways. The size ranges for which such nonlinear plots have steep slopes are the ranges in which a *change* in size can have important consequences. In contrast, within the size ranges for which the slope is very shallow, modifications of size make little difference to performance. If the sign of the slope of such a plot changes, then an increase in size can have the opposite effect for small organisms than for large ones.

3.1.1   Exponential Relationships Between Size and Performance.

An example of an exponential curve is shown in Figure 2, a plot of predator handling time as a function of prey size [13]. Differences in the size of small prey have little effect on predator handling time and hence on the prey's likelihood of being eaten, whereas differences in body size between larger prey can have a big effect on the danger of becoming a meal.

Many aspects of mechanical performance also have exponential relationships to size. For example, the deflection of a bending beam bearing a distributed load is proportional to its L^4, so the effect of a 10-cm increase in the height of a sessile, cantileverlike organism of a given width being bent by ambient water flow is small if the organism is short, but is large if an organism of the same width is tall [31]. There are many other examples of exponential relationships between function and size [1, 2, 40, 41, 57, 58], including the

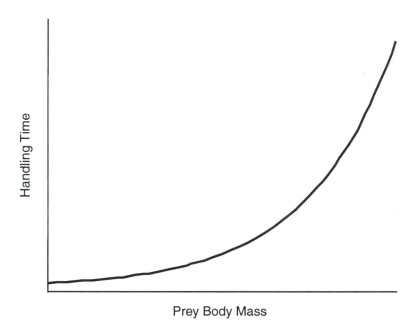

FIGURE 2   Handling time for a predator, plotted as a function of the body size of the prey (redrawn from a curve calculated by Emerson et al. [13]).

volume of fluid per time that can be pushed through a pipe (such as a blood vessel) by a given pressure difference, which is proportional to pipe diameter[4] [4], or the weight that must be borne by a skeleton, which is proportional to body volume (proportional to L^3).

3.1.2   Optimal Sizes.   If the plot of some aspect of an organism's performance as a function of body size goes through a maximum or a minimum, then the effect of increasing size reverses once it passes a critical value. One of many examples of how the consequence of a change in size can reverse as size increases is provided by Sebens [53], who analyzed the energetics of suspension-feeding animals. The energy available for growth and reproduction increases as such animals get larger up to a point, beyond which further increments in size have the opposite effect and reduce the excess energy for growth and gonad production (Figure 3). We usually consider such curves in the context of optimization analyses [3, 47], and we try to relate such maxima and minima to the peaks and troughs in adaptive landscapes [16, 25]. In addition, we might also consider that passing through such an inflection point represents the acquisition of a novel consequence for a size change.

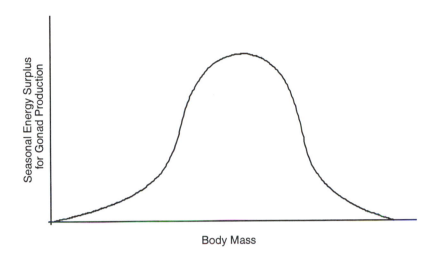

FIGURE 3   Energy surplus available for gonad production for an entire season, plotted as a function of the body mass of a passive suspension feeder, when prey sizes are normally distributed (redrawn from a curve calculated by Sebens [53]).

4   THE CONSEQUENCES OF SIZE CHANGE DEPEND ON MORPHOLOGY

4.1   EXAMPLES OF MORPHOLOGY-DEPENDENT EFFECTS OF SIZE CHANGE

4.1.1   Hairy Legs.   Our recent work using dynamically scaled physical models to study the water flow through the setulose feeding appendages of copepods has shown that morphology affects the size at which the transition occurs between functioning like a nonleaky paddle and working like a leaky sieve. Coarsely meshed appendages become filters at smaller size than do appendages bearing closely spaced hairs [34].

4.1.2   Hydrodynamic Forces on Sessile Marine Organisms in Waves.   The structure of the skeletal support tissues of sessile marine organisms can affect the hydrodynamic consequences of increasing size [31]. The microarchitecture of the support tissues determines their resistance to deformation, which in turn affects the flexural stiffness of the organism. Many marine organisms attached to the substratum, such as stony corals, are supported by stiff skeletal materials. In contrast, others like flexible gorgonians and seaweeds are made of tissues that are less stiff and can bend when subjected to hydrodynamic forces. All these attached organisms risk being dislodged or broken by ocean waves. We have been studying how the stiffness of their tissues affects the magnitude of the hydrodynamic forces they experience in the back-and-forth accelera-

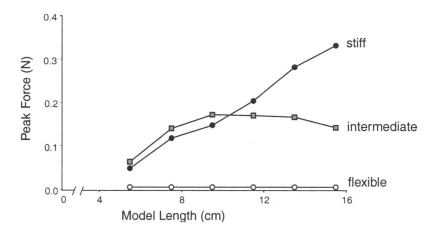

FIGURE 4   Peak hydrodynamic force in oscillatory flow measured on models of upright, planar benthic organisms, plotted as a function of model length. Each line represents a different flexural stiffness. (Redrawn from Koehl [31]).

tional flow they experience in waves. One approach that we have used is to build models of generic organisms, holding shape constant but using different materials to provide a range of flexural stiffness. We measured the hydrodynamic forces on such models of different lengths in a wave tank (Figure 4). As the stiff models "grew," the force increased. In contrast, the very flexible models flopped over into more streamlined shapes and went with the flow such that the force they bore did not measurably increase with length. A third type of size-dependent behavior was shown by the models of intermediate stiffness: lengthening increased hydrodynamic forces on short models, had no effect on models of intermediate length, and decreased forces on long models. Since deflection of a cantilever depends on L^4, the longer the models of intermediate stiffness become, the more they bend over and go with the flow.

4.2   ECOLOGICAL AND EVOLUTIONARY CONSEQUENCES OF MORPHOLOGY-DEPENDENT CONSEQUENCES OF SIZE CHANGE

If the effects of changing size depend on structure, then the consequences of growth or of evolutionary shifts in size can be different for organisms of various morphologies. Ecological modelers should be mindful that a single size-dependent expression may not describe the function of diverse members of a community. Furthermore, when organisms alter their morphology during ontogeny, the quantitative relationship between certain aspects of performance and body size can change. For example, when copepods shift from the body form of the nauplius larva to the morphology of the copepodid stages, the slope of the log-log plot of mass-specific metabolic rate as a function of body mass switches from ~ 1 (indicating little change as size increases) to $\ll 1$ (indicating a decline as size increases) [15]. Thus, a single size-dependent expression may not even describe a single species in a community.

5 THE CONSEQUENCES OF SIZE CHANGE DEPEND ON HABITAT

5.1 EXAMPLES OF HABITAT-DEPENDENT EFFECTS OF SIZE CHANGE

5.1.1 Hydrodynamic Forces on Sessile Organisms.
The effect of size on the hydrodynamic forces on stiff sessile organisms depends on whether they live in habitats exposed to waves or to unidirectional water currents [12]. When exposed to waves, organisms experience acceleration reaction force as well as drag. The acceleration reaction force is proportional to the volume of the organism, and hence increases with L^3, while the attachment area of the organism is proportional to L^2. Therefore, in wave-swept habitats an increase in body size leads to an increase in the probability of being swept off the shore by waves. In contrast, sessile organisms in habitats subjected to unidirectional water currents only experience drag force, which depends on their projected area. Since both drag and attachment strength are proportional to L^2, growth does not impose an increased risk of being ripped off the substratum like it does in wave-swept habitats.

Water-flow habitat also affects the consequences of growth for flexible organisms, but differently from stiff organisms [31]. In unidirectional currents, both the drag and attachment strength of flexible creatures are proportional to L^2, as for stiff organisms. However, flexible organisms in the back-and-forth flow of waves can experience a reduction in hydrodynamic forces as they increase in length beyond the distance the water travels before it stops and flows back the other way, as measured on real kelp in the field as well as on models in a wave tank [33].

5.1.2 Spawning by Sessile Organisms.
Many attached marine organisms spawn gametes into the surrounding water. An increase in body height improves gamete transport and mixing in gentle currents, but has no effect in turbulent waves [31].

5.1.3 Suspension Feeding by Colonial Animals.
The effect of an increase in the size of a colony of suspension-feeding bryozoans can depend on the hydrodynamic environment in which the colony lives [43, 44]. An increase in colony size can lead to a decrease in particle-capture rate per zooid in habitats characterized by slow currents, as upstream zooids deplete the water of food. However, in habitats exposed to rapidly flowing water, colony growth has the opposite effect on feeding rates per zooid: larger colonies are more effective than small

ones at slowing the water flowing through them enough that zooids can catch and hold on to food particles.

5.2 ECOLOGICAL AND EVOLUTIONARY CONSEQUENCES OF HABITAT-DEPENDENT CONSEQUENCES OF SIZE CHANGE

Since the performance consequences of changing size can depend on habitat, analyses of size and scaling should be done in the context of the environment in which the organisms live. Furthermore, the habitat dependence of the effects of size suggests that, when organisms disperse to new habitats or when the environment changes, selection on body size can change. Striking examples of this are provided by the evolutionary size changes exhibited by isolated populations of mammals on islands where resource availability and predation pressure are different from those on the mainland (e.g., Lomolino [37]).

6 EVOLUTIONARY CONSEQUENCES OF SIZE CHANGES

6.1 ANOTHER POSSIBLE MECHANISM FOR THE ORIGIN OF NOVELTY

A variety of mechanisms have been proposed for the origin of novelty during the process of evolution (reviewed by Koehl [31]). Size change should be added to the list. There is ample evidence for selection on body size, and there are many examples in the fossil record of size changes within lineages over evolutionary time (reviewed by Koehl [31]). Many studies of organism size have explored how body allometry permits function to be maintained at different sizes. However, another way to think about allometry is to consider that if organisms do *not* change their form as they change in size, their function *is* altered, and such functional shifts might be a source of evolutionary innovation.

One obvious mechanism by which a change in body size might lead to evolutionary novelty is that a structure acquires a new function once size crosses some threshold. That structure then becomes subject to a different suite of selective pressures than it was when it performed the old function at the former size. A similar mechanism by which size change might contribute to evolutionary novelty is that morphological and kinematic diversity might accumulate at small size without consequences to the performance or fitness of the organisms, but might gain functional significance and thus become subject to natural selection at larger size (reviewed by Koehl [31]).

A third mechanism by which size change might lead to evolutionary innovation is that size differences in developing embryos can affect pattern formation, thereby producing novelties in adult morphology (reviewed by Koehl [31]).

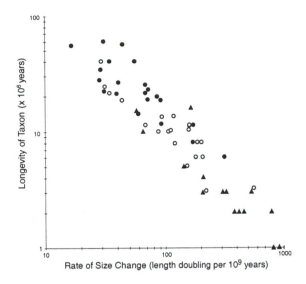

FIGURE 5 Longevity of each taxon in the fossil record, plotted as a function of the rate of change of body size, for Jurassic ammonites (each triangle represents a genus) and bivalved mollusks (each black circle represents a genus; each open circle represents a species). (Redrawn from Koehl [31]; data from Hallam [20].)

6.2 POSSIBLE EFFECTS OF SIZE CHANGE ON RATES OF EVOLUTION

Since shifts in size can be accompanied by alterations in function and changes in the consequences of particular morphological features, a reasonable speculation might be that rapid evolutionary change should tend to occur when size changes within a lineage. If we assume that short taxon longevity is a rough indication of rapid evolutionary change (i.e., high rates of modification or extinction), then Jurassic ammonites and bivalves (Figure 5) provide fossil evidence consistent with this speculation, but obviously this issue requires further study. For example, extinction might be due to acquisition of "poor" function accompanying a size change in the face of a shifting abiotic or biotic environment, or it might be the consequence of random events. Nonetheless, the pattern revealed by Hallam's data is intriguing.

7 CONCLUSIONS

Changes in body size can occur during the ontogeny of an individual or during the evolution of a lineage. The examples of the consequences of such size changes cited in this chapter illustrate that the effects of becoming larger or smaller are complicated, can involve dramatic transitions in function, and can depend on the morphology and habitat of the organisms. My point in raising these examples is *not* that the consequences of changing size are too complex to be understood using simple rules. Quite to the contrary, my point is that basic physical principles permit us to understand and to predict such transitions in the functional consequences of size changes.

ACKNOWLEDGMENTS

I am grateful to the following people for leads into the literature and/or discussions that helped shape my thinking about the issues addressed above: J. Brown, H. Greene, D. Jablonsky, J. Kingsolver, R. Paine, J. Valentine, S. Vogel, S. Wainwright, and the participants in the biomechanics discussion group at the University of California, Berkeley. I thank K. Bishop for preparing the figures. My data presented here were gathered with support from National Science Foundation Grant #OCE-9217338, Office of Naval Research Grant #N00014-96-1-0594, and a MacArthur Fellowship.

REFERENCES

[1] Alexander, R. M. *Size and Shape*. London: Edward Arnold, 1971.

[2] Alexander, R. M. *Locomotion of Animals*. New York & Glasgow: Blackie, 1982.

[3] Alexander, R. M. *Optima for Animals*. London: Edward Arnold, 1982.

[4] Batty, R. S. "Development of Swimming Movements and Musculature of Larval Herring (*Clupea harengus*)." *J. Exp. Biol.* **37** (1984): 129–153.

[5] Berg, H. C., and E. M. Purcell. "Physics of Chemoreception." *Biophys. J.* **20** (1977): 193–219.

[6] Bushnell, D. M., and K. J. Moore. "Drag Reduction in Nature." *Ann. Rev. Fluid Mech.* **23** (1991): 65–79.

[7] Calder, W. A., III. *Size, Function, and Life History*. Cambridge, MA: Harvard University Press, 1984.

[8] Cheer, A. Y. L., and M. A. R. Koehl. "Fluid Flow Through Filtering Appendages of Insects." *I.M.A.J. Math. Appl. Med. Biol.* **4** (1987): 185–199.

[9] Cheer, A. Y. L., and M. A. R. Koehl. "Paddles and Rakes: Fluid Flow through Bristled Appendages of Small Organisms." *J. Theor. Biol.* **129** (1987): 17–39.

[10] Craig, D. A. "Behavioral Hydrodynamics of *Cloeon dipterum* Larvae (Ephemeropter: Baetidea)." *J. N. Am. Benthol. Soc.* **9** (1990): 346–357.

[11] Dadswell, M. J., and D. Weihs. "Size-Related Hydrodynamic Characteristics of the Giant Scallop, *Placopecten magellanicus* (Bivalvia: Pectinidae)." *Can. J. Zool.* **68** (1990): 778–785.

[12] Denny, M. W., T. Daniel, and M. A. R. Koehl. "Mechanical Limits to the Size of Wave-Swept Organisms." *Ecol. Monogr.* **55** (1985): 69–102.

[13] Emerson, S. B., H. W. Greene, and E. L. Charnov. "Allometric Aspects of Predator-Prey Interactions." In *Ecological Morphology: Integrative Organismal Biology*, edited by P. C. Wainwright and S. M. Reilly, 123–139. Chicago, IL: University of Chicago Press, 1994.

[14] Ennos, A. R. "The Effect of Size on the Optimal Shapes of Gliding Insects and Seeds." *J. Zool., Lond.* **219** (1989): 61–69.

[15] Epp, R. W., and W. M. Lewis, Jr. "The Nature and Ecological Significance of Metabolic Changes During the Life History of Copepods." *Ecology* **611** (980):259–264.

[16] Futuyma, D. J. *Evolutionary Biology.* Sunderland, MA: Sinaur Associates, 1986.

[17] Glasheen , J. W. "A Hydrodynamic Model of Locomotion in the Basilisk Lizard." *Nature* **380** (1996): 340–342.

[18] Glasheen, J. W., and T. A. McMahon. "Weight Support on the Water Surface in Basilisk Lizards." *Nature* **380** (1996): 340–342.

[19] Godley, J. S. "Foraging Ecology of the Striped Swamp Snake, *Regina alleni* in Southern Florida." *Ecol. Monogr.* **50** (1980): 411–436.

[20] Hallam, A. "Evolutionary Size Increase and Longevity in Jurassic Bivalves and Ammonites." *Nature* **258** (1975): 493–496.

[21] Hansen, B., and P. Tiselius. "Flow through the Feeding Structures of Suspension Feeding Zooplankton: A Physical Model Approach." *J. Plankton Res.* **14** (1992): 821–834.

[22] Hentschel, B. T. "Spectrofluorometric Quantification of Neutral and Polar Lipids Suggests a Food-Related Recruitment Bottleneck for Juveniles of a Deposit-Feeding Polychaete Population." *Limnol. Oceanogr.* **43** (1998): 543–549.

[23] Jacobs, D. K. "Shape, Drag, and Power in Ammonoid Swimming." *Paleobiology* **18**(1992): 203–220.

[24] Jumars, P. A., L. M. Deming, J. A. Baross, and R. A. Wheatcroft. "Deep-Sea Deposit-Feeding Strategies Suggested by Environmental and Feeding Constraints." *Phil. Trans. Roy. Soc. Lond. A* **331** (1990): 85–101.

[25] Kingsolver, J. G. "Thermoregulation, Flight, and the Evolution of Wing Pattern in Pierid Butterflies: The Topography of Adaptive Landscapes." *Am. Zool.* **28** (1988): 899–912.

[26] Kingsolver, J. G., and M. A. R. Koehl. "Aerodynamics, Thermoregulation, and the Evolution of Insect Wings: Differential Scaling and Evolutionary Change." *Evolution* **39** (1985): 488–504.

[27] Kingsolver, J. G., and M. A. R. Koehl. "Selective Factors in the Evolution of Insect Wings." *Ann. Rev. Entomol.* **39**(1994): 425–451.

[28] Koehl, M. A. R. "Effects of Sea Anemones on the Flow Forces They Encounter." *J. Exp. Biol.* **69** (1977): 87–105.

[29] Koehl, M. A. R. "Mechanisms of Particle Capture by Copepods at Low Reynolds Number: Possible Modes of Selective Feeding." In *A.A.A.S.*

Selected Symposium #85: Trophic Interactions Within Aquatic Ecosystems, edited by D. L. Meyers and J. R. Strickler, 135–160. Boulder, CO: Westview Press, 1984.

[30] Koehl, M. A. R. "Fluid Flow Through Hair-Bearing Appendages: Feeding, Smelling, and Swimming at Low and Intermediate Reynolds Number." In *Biological Fluid Dynamics, Soc. Exp. Biol. Symp.*, edited by C. P. Ellington and T. J. Pedley, 157–182, vol. 49. London: Company of Biologists, 1995.

[31] Koehl, M. A. R. "When Does Morphology Matter." *Ann. Rev. Ecol. System.* **27** (1996): 501–542.

[32] Koehl, M. A. R. "Small-Scale Fluid Dynamics of Olfactory Antennae." *Mar. Fresh. Behav. Physiol.* **27** (1996): 127–141.

[33] Koehl, M. A. R. Unpublished data.

[34] Koehl, M. A. R., and J. Jed. Unpublished data.

[35] Koehl, M. A. R., and J. R. Strickler. "Copepod Feeding Cuments: Food Capture at Low Reynolds Number." *Limnol. Oceanogr.* **26** (1981): 1062–1073.

[36] Koehl, M. A. R., and S. A. Wainwright. "Mechanical Adaptations of a Giant Kelp." *Limnol. Oceanogr.* **22** (1977): 1067–1071.

[37] Lomolino, M. V. "Body Size of Mammals on Islands: The Island Rule Reexamined." *Amer. Natur.* **125** (1985): 310–316.

[38] Loudon, C., B. A. Best, and M. A. R. Koehl. "When Does Motion Relative to Neighboring Surfaces Alter the Flow Through an Array of Hairs?" *J. Exp. Biol.* **193** (1994): 233–254.

[39] Manuel, J. L., and M. J. Dadswell. "Swimming of Juvenile Sea Scallops, *Placopecten magellanicus* (Gmelin): A Minimum Size for Effective Swimming." *J. Exp. Mar. Biol. Ecol.* **174** (1993): 137–175.

[40] McMahon, T. A., and J. T. Bonner. *On Size and Life.* New York: W. H. Freeman, 1983.

[41] Niklas, K. J. *Plant Biomechanics: An Engineering Approach to Plant Form and Function.* Chicago, IL: University of Chicago Press, 1992.

[42] Niklas, K. J. *Plant Allometry: The Scaling of Form and Process.* Chicago, IL: University of Chicago Press, 1994.

[43] Okamura, B. "The Effects of Ambient Flow Velocity, Colony Size, and Upstream Colonies on the Feeding Success of Bryozoa. I. *Bugula stolonifera* (Ryland), an Arborescent Species." *J. Exp. Mar. Biol. Ecol.* **83** (1984): 179–193.

[44] Okamura, B. "The Effects of Ambient Flow Velocity, Colony Size, and Upstream Colonies on the Feeding Success of Bryozoa-II. *Conopeum reticalum* (Linnaeus), an Encrusting Species." *J. Exp. Mar. Biol. Ecol.* **89** (1985): 69–80.

[45] Osenberg, C. W., E. E. Werner, G. C. Mittelbach, and D. J. Hall. "Growth Patterns in Bluegill (*Lepomis macrochirus*) and Pumpkinseed (*Lepomis gibbosus*) Sunfish: Environmental Variation and the Importance

of Ontogenetic Niche Shifts." *Can. J. Fish. Res. Aquatic Sci.* **45** (1988): 17–26.

[46] Osse, J. W. M., and M. R. Drost. "Hydrodynamics and Mechanics of Fish Larvae." *Pol. Arch. Hydrobiol.* **36** (1983): 455–466.

[47] Oster, G. F., and E. O. Wilson. *Caste and Ecology in the Social Insects.* Princeton, NJ: Princeton University Press, 1978.

[48] Paine, R. T. "Size-Limited Predation: An Observational and Experimental Approach with the Mytilus-Pisaster Interaction." *Ecology* **57** (1976): 858–873.

[49] Paine, R. T. *Marine Rocky Shores and Community Ecology: An Experimentalist's Perspective*, edited by O. Kinne. Oldendorf/Luhe: Ecology Institute, 1994.

[50] Penry, D. L., and P. A. Jumars. "Gut Architecture, Digestive Constraints and Feeding Ecology of Deposit-Feeding and Carnivorous Polychaetes." *Oecologia.* **82** (1990): 1–11.

[51] Rubenstein, D. I., and M. A. R. Koehl. "The Mechanisms of Filter Feeding: Some Theoretical Considerations." *Amer. Natur.* **111** (1977): 981–994.

[52] Schmidt-Nielsen, K. *Scaling: Why Is Animal Size So Important?* Cambridge, MA: Cambridge University Press, 1984.

[53] Sebens, K. P. "The Limits to Indeterminate Growth: An Optimal Size Model Applied to Passive Suspension Feeders." *Ecology* **63** (1982): 209–222.

[54] Statzner, B. "Growth and Reynolds Number of Lotic Macroinvertebrates: A Problem for Adaptation of Shape to Drag." *Oikos* **51** (1988): 84–87.

[55] Statzner, B., and T. F. Holm. "Morphological Adaptation of Shape to Flow: Microcurrents Around Lotic Macroinvertebrates with Known Reynolds Numbers at Quasi-Natural Flow Conditions." *Oecologia* **78** (1989): 145–157.

[56] Videler, J. J. "Body Surface Adaptations to Boundary-Layer Dynamics." In *Biological Fluid Dynamics, Soc. Exp. Biol. Symp. 49*, edited by C. P. Ellington and T. J. Pedley, 1–20. London: Company of Biologists, 1995.

[57] Vogel, S. *Life in Moving Fluids.* Princeton, NJ: Princeton University Press, 1994.

[58] Wainwright, S. A., W. D. Biggs, J. D. Currey, and J. W. Gosline. *Mechanical Design in Organisms.* Princeton, NJ: Princeton University Press, 1976.

[59] Weihs, D. "Energetic Significance of Changes in Swimming Modes During Growth of Larval Anchovy *Engraulis mordax.*" *Fishery Bull. Fish Wildl. Serv. US* **77** (1980): 597–604.

[60] Williams, T. A. "Locomotion in Developing Artemia Larvae: Mechanical Analysis of Antennal Propulsors Based on Large-Scale Physical Models." *Biol. Bull.* **187** (1994): 156–163.

[61] Williams, T. A. "A Model of Rowing Propulsion and the Ontogeny of Locomotion in Artemia Larvae." *Biol. Bull.* **187** (1994): 164–173.

The Origin of Universal Scaling Laws in Biology

Geoffrey B. West
James H. Brown
Brian J. Enquist

1 INTRODUCTION

Life is the most complex physical system in the universe. One of its most salient features is its amazing diversity spanning more than 21 orders of magnitude in size. In spite of this variation, organisms from microbes to whales obey a host of remarkably simple and systematic empirical scaling laws that dictate how biological features change with size [2, 22, 27, 29]. These scaling laws apply to such fundamental quantities as metabolic rate (the rate at which energy must be supplied in order to sustain an organism), time scales (such as life span and heart rate), and sizes and shapes of component parts (such as length of the aorta or height of a tree). It is a remarkable fact that all of these can be expressed as power-law relationships with body-mass exponents that are simple multiples of 1/4 (e.g., 1/4, 3/4, 3/8). They appear to be valid for almost all forms of life, including mammals, birds, reptiles, fish, various groups of terrestrial, freshwater, and marine invertebrates, vascular plants, and unicellular algae and protists. Clearly the universal character of these "laws" is telling us something important about the way life is organized and the constraints under which it has evolved. The origin of these so-called allometric scaling relationships (a term coined by Julian Huxley) and, in par-

Scaling in Biology, edited by J. H. Brown and G. B. West.
Oxford University Press, 2000.

ticular, why their exponents are almost always simple multiples of $1/4$, have been longstanding fundamental problems in biology.

Allometric relationships are usually expressed in the form:

$$Y = Y_0 M^b, \tag{1}$$

where Y is some biological observable, Y_0 a normalization constant, M the mass of the organism, and b the scaling exponent. Some specific examples are heart rate ($b = 1/4$), life span ($b = -1/4$), and the radius ($b = 3/8$) and length ($b = 1/4$) of *both* mammalian aortae and tree trunks. One of the best-known and most fundamental allometric scaling "laws," first detailed by Max Kleiber in 1932, relates how basal metabolic rate changes with mass. Figure 1 shows such a plot for endothermic vertebrates (mammals and birds) spanning six orders of magnitude in mass; it yields $b = 0.75 \pm 0.01$. In nearly all taxonomic groups spanning a wide range of body masses from unicellular organisms to mammals and vascular plants, the same three-quarter power dependence is manifested [12], as illustrated in Figure 2. The same relationship appears to hold for the components of the metabolic system *within* organisms, down to the organelle and even molecular levels [44]. Figure 3 shows data on mammalian metabolic rates plotted versus mass from shrews to elephants spanning almost six orders of magnitude; a best-fit regression gives an exponent of 0.751 ± 0.005. When extrapolated back a further 20 orders of magnitude, this fit coincides with the respiratory rate not only of mammalian

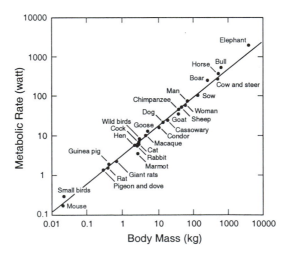

FIGURE 1   Metabolic rate (in watts) for a series of mammals and birds as a function of mass (in kg); the scale is logarithmic and exemplifies the 3/4-power scaling law discovered by Kleiber [2, 22, 27, 29].

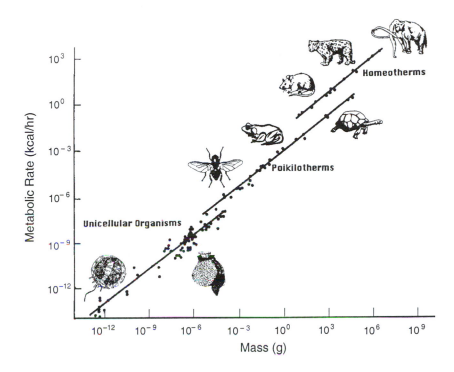

FIGURE 2 Metabolic rate (in kcal/hr) for a series of organisms ranging from the smallest microbes to the largest mammals as a function of mass (in g), exemplifying the persistence of the 3/4-power scaling law (the solid lines) over 20 orders of magnitude (Hemmingsen [12]).

mitochondria but even with that of the molecular respiratory complex and terminal oxidase molecular units within mitochondrial membranes! Thus, for the metabolic rate of mammals, an $M^{3/4}$-power law extends over almost 27 orders of magnitude from the largest mammal down to the individual molecules catalyzing metabolism.

Notice that Kleiber's law implies that the specific metabolic rate, i.e., the power required to sustain a unit mass of an organism, *decreases* with increasing body size like $M^{-1/4}$. Thus, to support one gram of mouse for a given period of time requires three times the rate of energy consumption needed for a dog and nine times that for an elephant! In this sense it is clearly more efficient to be larger. At first glance this is surprising since all mammals are built from essentially the same "fundamental" cellular materials with the number of cells increasing linearly with body mass. One might therefore have expected that metabolic rate (B) also increases linearly with mass so that $b = 1$. On

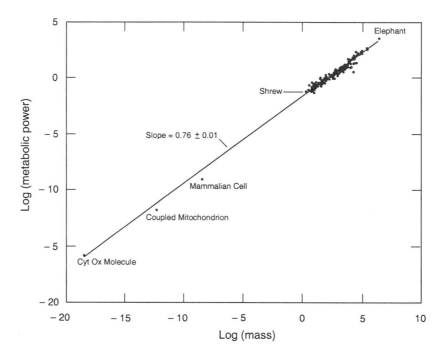

FIGURE 3   The straight line is the best fit to the metabolic rate of 389 mammals shown at the top versus their mass; it has a slope of 0.751 ± 0.005. When extrapolated 20 orders of magnitude, it agrees well with similar *in vitro* data on the mammalian cell and mitochondrion, and, finally, the cytochrome oxidase molecule of the respiratory complex inside mitochondria (Woodruff et al. [44]).

the other hand, since the metabolic heat generated in the body must be dissipated through its surface, we might have expected that $B \propto A$, the body surface area. If mammals scale purely geometrically, then $A \propto M^{2/3}$, suggesting $b = 2/3$. It is instructive to compare this with the analogous scaling for mechanical engines, which do indeed scale isometrically [22]. For example, over nearly six orders of magnitude the power output ("metabolic rate") of internal combustion engines scales linearly with mass ($b = 1$) while the revolution rate ("heart rate") scales as $M^{1/3}$. The reason why $b = 3/4$ for biological energy transforming systems rather than a simple linear relationship, $b = 1$, or a naïve surface-to-volume relationship, $b = 2/3$, has been sought for decades.

One of the most intriguing consequences of these scaling laws is the emergence of invariant quantities governing longevity. In mammals, for example, life span increases like $M^{1/4}$ whereas heart rate decreases like $M^{-1/4}$, so the number of heartbeats in a lifetime is the same (about 1.5×10^9) for all mammals regardless of size. Similarly, since specific metabolic rate also decreases like $M^{-1/4}$, the total energy needed to support a given mass of an organism during its lifetime is also the same for all mammals (about 70 kJ/gm). A similar relationship holds at the molecular level; these can be put slightly differently: the number of turnovers of the molecular respiratory complex per cell during a typical lifetime is also an invariant ($\approx 10^{16}$) [44].

To summarize: Allometric scaling laws are special because they express a systematic and universal simplicity in the most complex of all complex systems. They provide one of the rare examples of universal quantitative laws in biology. Their origin presents a major challenge because their very existence, not to mention their common quarter-powers, suggests the operation of a set of principles that are fundamental to all life.

Recently, we have proposed a general model for the origin of these universal quarter-power scaling laws [41]. It is based on the observation that biological structures and functions are ultimately limited by the rate at which the essential resources that sustain them can be supplied. The model provides a theoretical framework for why scaling exponents for almost all biological phenomena are simple multiples of $1/4$ and, in particular, accounts for the $3/4$-exponent for metabolic rate. Its central hypothesis is that, in order to supply the huge number of metabolizing fundamental units of an organism (cells in the case of whole organisms, the mitochondrial respiratory molecules in the case of cells), a linear, branching hierarchical transport network is required. Examples of such networks include the circulatory and respiratory systems of mammals or the vascular systems of plants. We anticipate that the model should also apply to less well understood systems, such as insect tracheae and intracellular transport. Indeed, the fact that scaling persists down to the molecular level strongly suggests that the same principles that govern organismal scaling are at work inside the cell. It is important, therefore, that any model incorporates principles that are sufficiently general that they are not sensitive to details of specific kinds of organisms or levels of organization. We propose the following three basic general principles for the design of biological network transport systems:

a. in order for the network to supply the entire volume of the organism, a space-filling hierarchical branching pattern is required;
b. the final branch of the network, where nutrients are exchanged (e.g., the capillary of the circulatory system or the petiole of a plant), is a size-invariant unit; and
c. organisms have evolved so that the energy required to transport materials through the network is minimized.

Scaling laws arise from the interplay between the physical and geometric constraints implicit in these three principles. Below we show that they predict that these networks must be fractal-like structures with self-similar properties dominated by area-preserving branching. Quarter-power scaling necessarily follows even though the various transport systems, and the pumps that drive them, have quite different characteristics. Thus, for example, the architecture of both mammalian arteries and plant xylem are both fractal-like branching "trees" even though the former is driven by a pulsatile pump (the heart), which propels blood through vessels of decreasing size, while the latter consists of an osmotic pump, which draws fluid through many parallel microcapillary tubes.

A particularly salient feature of these networks is that their hydrodynamic resistance decreases with size, typically as $M^{-3/4}$. This explains why less power is required to sustain a unit mass of a larger animal. Our model enables us to predict many scaling laws, not only *between* organisms but also *within* a given organism (e.g., it correctly predicts the length and speed of blood flow in the aorta relative to that in a capillary). In addition, the model shows how biological imperatives of resource supply put severe physical constraints on the structural design of organisms. For plants, and more generally for large terrestrial organisms, gravity may play an important role and biomechanical considerations must be incorporated [9, 20, 21, 26]. The combination of hydrodynamic and biomechanical constraints has clearly played an important role in the evolution of large mammals and trees. Indeed, we show how together these two constraints set an upper limit on the maximum height of trees.

It is important to appreciate that the model treats the system as a single, completely integrated entity. Indeed, almost none of its successes would hold if individual parts of the system were treated in isolation. As already emphasized, the model is quantitative and provides precise predictions for the values of many variables. Where data exist, excellent agreement with predictions are generally found. Where data are not available, the model provides a priori testable predictions. Although the model was constructed with macroscopic organisms in mind, it is much more general because it relies mostly on general principles of energy transport in complex systems. The actual nature of the transport process can be very broad; what is required is the idea of hierarchically branching pathways.

2 THE MODEL

2.1 GENERAL DESCRIPTION AND TERMINOLOGY

As presented here, the model should be viewed as an idealized representation of typical biological distribution networks. We assume, for example, that all vessels have cylindrical symmetry and that turbulence and nonlinear effects at junctions do not play a crucial role in the fluid flow. Otherwise, the model is quite realistic, incorporating the most salient features of these systems. It can be used as a point of departure for more detailed analyses and refined versions, which incorporate additional features of specific systems.

All systems can be described by a branching network in which the sizes of tubes regularly decrease (Figure 4). A familiar example is exhibited by the vertebrate circulatory and respiratory systems (Figure 4(a)); another is the "vessel bundle" structure of multiple parallel tubes which is characteristic of

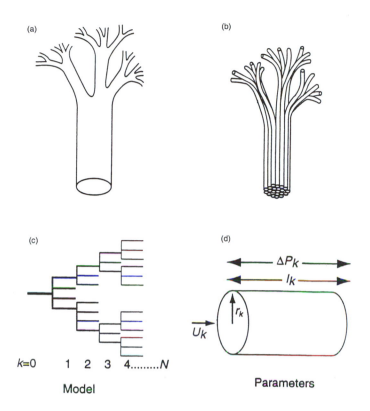

FIGURE 4 Diagrammatic examples of segments of biological distribution networks: (a) mammalian circulatory and respiratory systems comprised of branching tubes; (b) plant vessel-bundle vascular system comprised of diverging vessel elements (the "pipe model"); (c) topological representation of such networks, where k specifies the order of the level, beginning with the aorta ($k = 0$) and ending with the capillary ($k = N$); and (d) parameters of a typical tube at the kth level.

plant vascular systems (Figure 4(b)) [31, 32, 39, 46]. Although these systems can be represented by a similar hierarchical network illustrated in Figure 4(c), they are physically quite different. They also vary in the nature of the pump that drives the system: the cardiovascular system is driven by a pulsatile compression pump, the respiratory system by a pulsatile bellows pump, and the plant vascular system by an almost constant osmotic and vapor pressure gradient. The fluids transported also differ widely in density and viscosity, and from liquids to gases. We will show that, in spite of these differences, all of these systems exhibit essentially the same set of scaling laws. We first develop a model based on the mammalian cardiovascular system, although reference to other systems will be made at appropriate junctures. Later we shall briefly discuss how the model can be modified to apply to other systems, especially the mammalian respiratory system and the xylem of vascular plants.

We can characterize the network as being composed of N branchings, or generations, arranged hierarchically beginning with the aorta (level 0) and ending with the capillaries (level N). Figure 4(d) shows a typical branch at some intermediate level, k: its length is denoted by l_k, its radius by r_k, and the pressure drop across it by Δp_k. The volume rate of flow is $\dot{Q}_k = \pi r_k^2 \overline{u}_k$, where $\overline{u}_k$ is the velocity averaged over the cross section and, if necessary, over time. If each such tube branches into n_k smaller ones,[1] then the total number of tubes at level k is just $N_k = n_0 n_1 \ldots n_k$. Since fluid is conserved as it flows through the system

$$\dot{Q}_0 = N_k \dot{Q}_k = N_k \pi r_k^2 \overline{u}_k = N_N \pi r_N^2 \overline{u}_N , \qquad (2)$$

which holds for any k.

We next introduce the important principle, b. above, that no matter how many generations (N) of branchings there are in the network, the terminal units (capillaries) are invariant, so that r_N, l_N, $\overline{u}_N$ and, consequently Δp_N, are all independent of body size. Since the oxygen and nutrients transported by the fluid supply the metabolism, $\dot{Q}_0 \propto B$; indeed, metabolic rate is typically measured by determining oxygen consumption, since it is directly proportional to the rate of fluid flow. Thus, if $B \propto M^a$ (where a will later be determined to be 3/4), then $\dot{Q}_0 \propto M^a$. Eq. (2) immediately leads to the important prediction that the number of capillaries must scale as B, i.e., $N_N \propto M^a$. Thus, when $a = 3/4$, $N_N \propto M^{3/4}$. This implies that the number of cells and volume of tissue serviced by a single capillary increases with M.

In order to describe the network, we need to determine how the number of branches per node (the "branching ratio," n_k) and the radii, r_k, and lengths, l_k, of the tubes change throughout the network. To characterize the branching we introduce scale factors via the ratios $\beta_k \equiv r_{k+1}/r_k$ and $\gamma_k \equiv l_{k+1}/l_k$. In the following subsection we shall show that volume filling (principle a., above) requires $\gamma_k = \gamma$, a constant independent of the branching level, k. Later we shall prove from principle c. above (the minimization of energy expended) that, subject to an important exception, $\beta_k = \beta$ and $n_k = n$, both independent of k. The exception is a crucial modification in β_k for pulsatile systems, which will be discussed in Section 3 below. If β_k, γ_k, and n_k are all independent of k, then the network is a conventional self-similar fractal.[2] Before proving this it is worth noting with $n_k = n$, the number of branches increases geometrically from level 0 to level N, ($N_k = n^k$), as their size similarly decreases. Furthermore, since $N_N = n^N$, the number of generations of branches

---

[1]Note that we are consciously *not* using the Strahler method [34] commonly employed to describe the morphology of networks [3, 7]. In that system, branches at the same generational level can have significantly different physical characteristics, which make the method unsuitable for a dynamic analysis such as we develope.

[2]For an overview of fractals and self-similarity see Mandelbrot [19].

scales only logarithmically with mass:

$$N = \frac{a \ln (M/M_0)}{\ln n},\tag{3}$$

where M_0 is a normalization scale for M. This means, for example, that in going from a mouse to a whale, an increase in mass by a factor of 10^7, the number of branchings from aorta to capillary increases by only about 70%.

2.2 VOLUME-FILLING AND VOLUME-SERVICING BRANCHING

Our first principle expresses the notion that a volume-filling network is a natural structure for ensuring that all tissues are supplied by capillaries. The organism is composed of many groups of cells, referred to here as "service volumes," v_N, which are supplied by a single capillary. The network must branch so as to reach all such service volumes. The total volume to be filled, or serviced, is given by $V = N_N v_N$. For a network with a large number of branchings, N, complete volume filling implies that this same volume is filled by an analogous volume throughout the network, v_k, defined by branches at each level k. Since $r_k \ll l_k$, $v_k \propto l_k^3$, so volume filling constrains only branch lengths, l_k. Thus, $V \approx N_k v_k \propto N_k l_k^3$, independent of k. In other words, complete volume filling implies that the volume filled, or serviced, does not depend on the level used to estimate it. We shall refer to this property as "volume-servicing branching." Following our first principle, we assume its validity throughout the network, although it becomes less realistic for small values of k (or N). This then leads to $\gamma_k^3 \equiv (l_{k+1}/l_k)^3 \approx N_k/N_{k+1} = 1/n$, so that $\gamma_k \approx n^{-1/3} \equiv \gamma$, independent of k. Note that this can be straightforwardly generalized to d dimensions to give $\gamma \approx n^{-1/d}$. This result for γ_k, which is the quantitative expression for the volume-filling nature of these networks, will be taken to be a general property of all systems that we consider.

2.3 AREA-PRESERVING BRANCHING AND SELF-SIMILAR FRACTALS

Below we shall show how an analogous result for β_k can be derived from dynamical considerations based on the energy minimization principle. For many systems this leads to area-preserving branching, meaning that the sum of the cross-sectional areas of the daughter branches is equal to that of the parent: in other words, $\pi r_k^2 = n \pi r_{k+1}^2$. Thus, $\beta_k \equiv r_{k+1}/r_k = n^{-1/2} \equiv \beta$, independent of k. Proving this is quite technical, especially for the circulatory system. Before doing so we first explore its consequences and show how it is a key ingredient in deriving quarter-power scaling. It is worth noting, however, that in the simple, classic, rigid pipe model for plants, illustrated in Figure 4(b), area-preserving branching is automatic. In that model a plant is simulated by a sheath containing tightly bound, equally sized tubes of fixed radius, analogous to wires in an electrical cable, running continuously from a rootlet to a petiole. However, these microscopic tubes diverge at the macroscopic branch nodes, thereby supplying different parts of the plant. The unraveling of the sheath of "wires," as illustrated in Figure 4(b), defines the above- and below-ground branching architecture of the plant. When actually modeling the vascular system, we shall abandon this model and allow for the existence of nonconducting heartwood as well as for the conducting tubes to be tapered and loosely packed. In this case, area-preserving macroscopic branching is still maintained because of biomechanical constraints (see section 4.2 and Enquist et al. this volume).

As implied earlier, once we show that β_k, γ_k, and n_k are all independent of k, then we have proven that the network is a conventional self-similar fractal. This has been tested empirically for plants by measuring the fractal dimension of roots and shoots using variants of the "box counting" method [19]. This involves measuring the length of the boundary of their images at different resolutions, as illustrated in Figure 5 [6, 24, 35]. For a self-similar fractal these are related by a power relationship whose exponent defines the fractal dimension as in Figure 6. Although there is some variation in the empirical measurements depending on details of age and growth conditions [1],

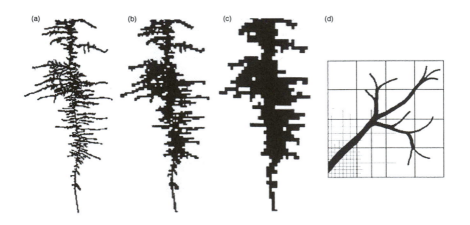

FIGURE 5 Illustration of the "box counting" method showing images of two-dimensional projections of a complete root system taken at different resolutions (a)–(c), and of a terminal shoot system (d). In (a)–(c) the resolution is defined by the varying size, ϵ, of the pixels used to create the image, and in (d) by the analogous varying size of the grid squares. The fractal dimension, d, is determined from the total number of pixels, or boxes, $\mathcal{N}$, intersecting the boundary of the image: $d = -\log \mathcal{N} / \log \epsilon$.

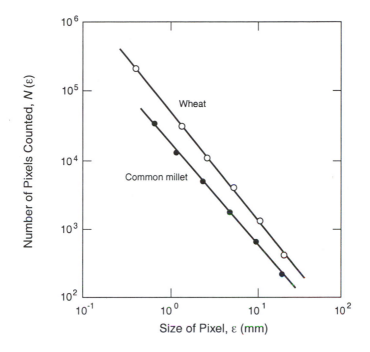

FIGURE 6 A plot of $\log \mathcal{N}$ vs. $\log \epsilon$ for a root system; the straight line indicates a self-similar fractal structure with dimension $d \approx 1.5$ [6, 24, 35].

the observations do indeed support a self-similar fractal structure. It is probably not practical to carry out an analogous empirical study of mammalian network systems; it might, however, be possible to use casts of the lung to make analogous measurements for the respiratory system. Furthermore, measurements of tube radii and lengths of the cardiovascular system do support the geometric progression of vessel sizes and branching ratios [3, 7] (see also Zamir this volume).

2.4 DERIVATION OF THE 3/4-POWER EXPONENT

To derive allometric relations we need to relate the scaling of vessel size within an organism to its body mass. A natural vehicle for this is the total volume of fluid in the network, V_b. Below we show that the network that minimizes energy (our third principle) not only has area-preserving branching but also requires $V_b \propto M$. Now, V_b is straightforwardly given by

$$V_b = \sum_{k=0}^{N} N_k V_k = \sum_{k=0}^{N} \pi r_k^2 l_k n^k = \frac{(n\gamma\beta^2)^{-(N+1)} - 1}{(n\gamma\beta^2)^{-1} - 1} n^N V_N. \tag{4}$$

Since γ_k and β_k are both independent of k, the above sum reduces to a simple geometric series; the result therefore reflects the self-similar fractal nature of the system. Since $n\gamma\beta^2 < 1$ and $N \gg 1$, an excellent approximation to Eq. (4) is $V_b = V_0/(1 - n\gamma\beta^2) = V_N(\gamma\beta^2)^{-N}/(1 - n\gamma\beta^2)$. From our second principle, that capillaries are invariant units, V_N is independent of M, from which it now follows that $(\gamma\beta^2)^{-N} \propto M$. Recall that, if $B \propto M^a$, then the number of capillaries $N_N \propto M^a$, so Eq. (3) then gives

$$a = -\frac{\ln n}{\ln(\gamma\beta^2)}. \tag{5}$$

When the area-preserving branching relation $\beta = n^{-1/2}$ is combined with the volume-filling volume-servicing branching result $\gamma = n^{-1/3}$, one immediately obtains from Eq. (5) that $a = 3/4$ (*independent* of the value of the branching ratio, n). Consequently, $B \propto M^{3/4}$.

Many other scaling laws follow from this. For example, for the aorta, $r_0 = \beta^{-N}r_N = N_N^{1/2}r_N$ and $l_0 = \gamma^{-N}r_N = N_N^{1/3}l_N$, yielding $r_0 \propto M^{3/8}$ and $l_0 \propto M^{1/4}$. Notice that γ and β play a dual scaling role: they not only determine how physical quantities scale from the aorta to the capillary within a single organism, but they also determine how that quantity scales when organisms of different masses are compared. Although these allometric scaling results are in good agreement with data, there are some problems with scaling results within an individual organism that need to be resolved. The first is that, for humans where $N_N \approx 10^{10}$ (estimated [3, 7]), the above gives $r_0/r_N \approx 10^5$, in disagreement with the observed value of 10^4. The second is more serious: area-preserving branching implies that the fluid velocity remains constant throughout the network, i.e., $u_0 = u_k = u_N$. This may not be serious for plants but would be disastrous (and obviously wrong!) for mammals. Indeed, for the efficient transfer of oxygen and nutrients to tissue across the walls of capillaries, the blood, which leaves the heart at over $100\,\mathrm{cm/sec}$, must slow down to less than $1\,\mathrm{cm/sec}$ by the time it reaches the capillaries. These, and other problems, can now be solved by applying the energy minimization principle to the dynamics of these systems. In so doing we derive an expression for β_k and show in what sense area-preserving branching is valid.

3   MINIMIZATION: ENERGY LOSS AND IMPEDANCE

In this section we consider the dynamics of the circulatory (and, by extension, the respiratory) system and examine the consequences of the assumption that biological networks are designed to minimize energy dissipation. We shall show how the pulsatile nature of blood flow, which dominates the larger vessels (aorta and major arteries), must have the crucial area-preserving branching, so that $\beta = n^{-1/2}$ which leads to quarter-power scaling. The smaller vessels, on the other hand, have the classic "cubic law" branching,[3] where $\beta = n^{-1/3}$, and play a relatively minor role in allometric scaling. Thus, the origin of most allometric scaling laws for these systems lies hidden in the dynamics.

3.1 NONPULSATILE FLOW

In order to illustrate our methodology, we first consider the simpler problem of nonpulsatile flow. For steady laminar flow of a Newtonian fluid, the viscous resistance of a single tube is given by the well-known Poiseuille formula: $R_k = 8\mu l_k / \pi r_k^4$, where μ is the viscosity of the fluid. Ignoring effects such as turbulence and nonlinearities at junctions, which are expected to be small, gives, for the total resistance of the network,

$$Z = \sum_{k=0}^{N} \frac{R_k}{N_k} = \sum_{k=0}^{N} \frac{8\mu l_k}{\pi r_k^4 n^k} = \frac{[1 - (n\beta^4/\gamma)^{N+1}]R_N}{(1 - n\beta^4/\gamma)n^N}. \tag{6}$$

Now, $n\beta^4/\gamma < 1$ and $N \gg 1$, so an excellent approximation to this is $Z = R_N/(1 - n\beta^4/\gamma)N_N$. Since R_N is an invariant, $Z \propto N_N^{-1} \propto M^{-a}$, so the total resistance *decreases* with body size. It is in this sense that a larger organism is more efficient. This result leads to two important scaling laws:

1. Blood pressure, $\Delta p = \dot{Q}_0 Z$, must be independent of body size.
2. The power dissipated in the system (i.e., cardiac output), $W = \dot{Q}_0 \Delta p \propto M^a$, so that the power expended by the heart in overcoming viscous forces is a size-independent fraction of the metabolic rate.

Notice that neither of these results, both of which are in agreement with data [3, 7], depends on detailed knowledge of n, β, or γ or, therefore, a. In addition Eq. (2), $\dot{Q}_0 = \pi r_0^2 \bar{u}_0$, implies that the velocity of blood in the aorta $\bar{u}_0 \propto M^0$, which is also in agreement with data [2, 3, 7, 22, 27, 29]. The two results, that $\bar{u}_0$ and Δp are independent of M, are both quite surprising and counter-intuitive. After all, the aorta of a whale has a radius of 30 cm whereas

---

[3]This is often referred to as Murray's law (see Murray [25]), though it is mentioned earlier by Thompson [38]. Also, Murray introduced the "cost function" as the total energy required to sustain the system, which he took to be the sum of the cardiac output and the energy required to maintain the blood. The latter was assumed to depend linearly on only the volume of blood, V_b, so his cost function effectively reduces to the first two terms of our Eq. (7). This was applied to nonpulsatile Poiseuille flow through just a single vessel and later extended to a single branching. Minimization with respect to the vessel radius leads to the cubic branching law. However, if the cost function is also minimized with respect to its length, no consistent result can be derived from Murray's formulation. Our method is somewhat different: we minimize the work expended subject to specific physical and geometrical constraints using the method of Lagrange multipliers and apply it to the network as a whole, rather than to just a single branching. The inconsistency in Murray's formulation can be traced to the absence in the cost function of any analog to the space-filling constraint term in our Eq. (7). In addition, we generalize to include the more realistic case of pulsatile flow, as discussed in subsection 3.2.

that of a shrew is a microscopic $0.01\,\mathrm{cm}$, yet both sustain the same blood pressure and velocity!

In spite of these successful predictions, we still have the problem, noted above, that area preserving implies that blood does not slow down in going from the aorta to the capillary. We will show how the more complicated pulsatile nature of the system solves this problem while managing to maintain the same set of scaling relations. First, however, we illustrate the importance of the energy minimization constraint within the context of the more simple nonpulsatile situation.

3.2 ENERGY MINIMIZATION

The basic idea here is that the trial-and-error feedback implicit in evolutionary adaptation (see Bonner and Horn this volume) has led to network transport systems that, on average, minimize the energy required to run them. Thus, to sustain a given metabolic rate in an organism of fixed mass, M, with a given volume of blood, V_b, the cardiac output must be minimized subject to a volume-filling geometry. To enforce such a constraint it is natural to use the classic method of Lagrange multipliers. Consider the cardiac output, $W(r_k, l_k, n_k, M)$, as a function of all relevant variables characterizing the network. We need to minimize the auxiliary function

$$F(r_k, l_k, n_k) = W(r_k, l_k, n_k, M) + \lambda V_b(r_k, l_k, n_k, M) + \sum_{k=0}^{N} \lambda_k N_k l_k^3 + \lambda_M M.$$

$$(7)$$

The constants, λ, λ_k, and λ_M (the Lagrange multipliers), multiply quantities that are to be kept fixed when making variations on the system. Since $B \propto Q_0$ and $W = \dot{Q}_0^2 Z$, this is tantamount to minimizing the total resistance, Z, which can therefore be used in Eq. (7) in place of W. For illustrative purposes, consider the case where $n_k = n$ so that we can use Eqs. (4) and (6) for V_b and Z, respectively. Then, by demanding $\partial F/\partial l_k = \partial F/\partial r_k = \partial F/\partial n = 0$, and solving the resulting equations, one obtains $\beta_k = n^{-1/3}$. More generally, by allowing variations with respect to n_k, one can show that $n_k = n$, independent of k. The result, $\beta_k = n^{-1/3}$, is a generalization to the complete network of Murray's cubic branching [25, 38]. He derived this for a single branching, based on minimizing a somewhat arbitrarily defined "cost function" for a single tube [25]. Our method can be viewed as giving a unique, well-defined meaning to Murray's concept of a "cost function" for the entire network in terms of biophysical constraints on the whole system. Minimizing F with respect to M (i.e., $\partial F/\partial M = 0$) now gives $V_b \propto M$ which is just the relationship needed to derive Eq. (5).

Although the result $\beta_k = n^{-1/3}$ is independent of k, it is obviously *not* area preserving and, therefore, does not give $a = 3/4$ when used in Eq. (5); instead it gives $a = 1$. It does, however, have the virtue of solving the problem of slowing down blood in the capillaries: Eq. (2) gives $\bar{u}_N/\bar{u}_0 = (n\beta^2)^{-N} =$

$N_N^{-1/3}$. For humans, $N_N \approx 10^{10}$ so $\bar{u}_N/\bar{u}_0 \approx 10^{-3}$, which is in reasonable agreement with data [3, 7]. On the other hand, it leads to an incorrect scaling law for this ratio: $\bar{u}_N/\bar{u}_0 \propto M^{-1/4}$. We now show that incorporating pulsatile flow not only solves all of these problems, giving the correct scaling relations (e.g., $a = 3/4$ and $\bar{u}_c/\bar{u}_0 \propto M^0$), but also gives the correct magnitude for $\bar{u}_N/\bar{u}_0$.

3.3   PULSATILE FLOW

A detailed treatment of pulsatile flow is complicated. Here, we present a condensed version that contains the essential features needed for the scaling problem. Most importantly, blood vessels are no longer taken to be rigid but allowed to expand and contract as the pulse wave generated by the contraction of the heart propagates along them; see Figure 7. The classic Poiseuille resistance of the rigid tube, relating the fluid volume flow rate to the driving pressure gradient, is thereby generalized to a complex impedance signifying attenuated wave propagation [3, 7, 43, 42].

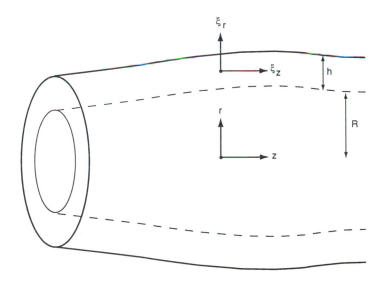

FIGURE 7   A typical tube in the circulatory system, showing distortion as the blood pressure wave propagates through it. Fluid coordinates are denoted by r (radial distance) and z (longitudinal distance) and corresponding tube wall coordinates by η_r and η_z. The equilibrium inner radius of the tube is R and the wall thickness h.

The equation of motion governing fluid flow is the Navier-Stokes equation. Neglecting nonlinearities responsible for turbulence, this reads:

$$\rho\frac{\partial \mathbf{v}}{\partial t} = \mu\nabla^2\mathbf{v} - \nabla p. \tag{8}$$

Here, the vector $\mathbf{v}$ is the local fluid velocity at some time t, p the local pressure, and ρ the fluid density. If the fluid is incompressible, then the local conservation of fluid requires $\nabla\cdot\mathbf{v} = 0$. When combined with Eq. (8) this gives the subsidiary condition

$$\nabla^2 p = 0. \tag{9}$$

The analogous equation governing the elastic motion of the tube is the Navier equation. Again neglecting nonlinearities, this reads:

$$\rho_W\frac{\partial^2 \xi}{\partial^2 t} = B\nabla^2\xi - \nabla p, \tag{10}$$

where the vector ξ is the local displacement of the tube wall, ρ_W its density, and B its modulus of elasticity. These three coupled equations—Eqs. (8), (9), and (10)—are to be solved subject to boundary conditions requiring the velocity and force to be continuous at the tube wall interfaces.

It is natural to solve these equations in cylindrical coordinates using a Fourier-Bessel series. For example, the pressure can be expressed as a decomposition into waves of a given angular frequency, ω ($= 2\pi\nu$, where ν is the corresponding frequency) and wave number K ($= 2\pi/\lambda$, where λ is the corresponding wavelength):

$$p(R, z, t) = \int_{-\infty}^{\infty} d\omega \int_{-\infty}^{\infty} dK e^{i(\omega t - Kz)} P(\omega, K) J_0(iKR). \tag{11}$$

Here, z is the distance along the tube, and R is the radial coordinate, as shown in Figure 7. Analogous representations can be made for the longitudinal and transverse components of both $\mathbf{v}$ and ξ. Because of the viscous nature of blood, this is a dissipative system so K will, in general, be a complex number. Each Fourier component is thereby damped and attenuated as it travels down the tube. Notice that the wave velocity, $c = \omega/K = 2\pi\omega\lambda$, is also complex.

Both Z and c are determined using these Fourier-Bessel representations in Eqs. (8), (9), and (10). This was first carried out by Womersley [43, 42]. In the approximation where the tube wall thickness $h \ll r$, r being the static equilibrium value of its radius, the problem can be solved analytically to give

$$\left(\frac{c}{c_0}\right)^2 \approx -\frac{J_2(i^{3/2}\alpha)}{J_0(i^{3/2}\alpha)} \quad \text{and} \quad Z \approx \frac{c_0^2\rho}{\pi r^2 c}. \tag{12}$$

Here $\alpha \equiv (\omega\rho/\mu)^{1/2}r$ is a dimensionless parameter, the Womersley number, and $c_0 \equiv (Eh/2\rho r)^{1/2}$, the classic Korteweg-Moens velocity [3, 7]. As already noted, c and, therefore, Z are both complex functions of ω, so the wave is attenuated and dispersed as it propagates. Let us examine the consequences of these formulae as the blood flows through progressively smaller tubes, forcing α to decrease. The crucial point is that the character of the wave depends critically on whether $|\alpha|$ is less than or greater than 1. This can be seen explicitly in Eq. (12), where the behavior of the Bessel functions changes from a simple power expansion for small $|\alpha|$ to an oscillatory behavior when $|\alpha|$ is large. In humans, typical values of α range from around 15 in the aorta, to 5 in the arteries, 0.04 in the arterioles, and to about 0.005 in capillaries. Furthermore, since the volume of blood $V_b \propto M$, we expect the tidal volume of the heart, V_H, to scale likewise. Now, the overall volume flow rate $Q_0 = V_H \nu$, so heart rate, ν, scales as $M^{-1/4}$, in good agreement with data. Consequently, $\alpha \sim M^{1/4}$. So, in the smallest mammals $|\alpha|$ is barely larger than 1 even in their aorta.

1. For large tubes, where α is large (> 1), Eq. (12) gives $c = c_0$, the classic Korteweg-Moens velocity. Numerically this gives $c \approx 580\,\mathrm{cm/sec}$ in good agreement with measurements [3, 7]. Since this is a purely real quantity, the wave suffers neither attenuation nor dispersion, reflecting the fact that, in these large vessels, viscosity plays almost no role. The corresponding impedance is, from Eq. (12), given by $Z = \rho c_0/\pi r^2$. Notice that its r dependence has dramatically changed from the r^{-4} Poiseuille behavior to r^{-2}. Using this minimize energy loss now leads to h_k/r_k (and, therefore, c_k) being independent of k and, most importantly, to an area-preserving law at the junctions, so $\beta_k = n^{-1/2}$. Physically this ensures that when pulse waves traveling in a vessel come to a branch point no energy is reflected back; this is the exact analog of impedance matching at the junctions of electrical transmission lines [3, 7].

2. For small tubes where $|\alpha| < 1$, the role of viscosity becomes increasingly important until it eventually dominates the flow. Eq. (12) gives $c \approx 1/4i^{1/2}\alpha c_0 \to 0$, in quantitative agreement with observation [3, 7]. Because c now has a significant imaginary part, the traveling wave is heavily damped, leaving an almost steady oscillatory flow whose impedance is, from Eq. (12), given by the original Poiseuille formula; thus, the r^{-4} behavior is restored! Following our previous argument we now find that for large k, corresponding to small vessels, $\beta_k = n^{-1/3}$, so area-preserving branching is lost. It is interesting to note that, in this regime, the traveling wave is 99.8% attenuated per wavelength, its velocity being reduced to the invariant value $c \approx (\nu Bh/16\mu r)^{1/2} \approx 10\,\mathrm{cm/sec}$, in good agreement with observations.

We have thus shown that for the realistic case of pulsatile flow, β_k is *not* independent of k but, rather, has a steplike behavior as illustrated in Figure 8(a). This feature is well supported by empirical data. For example, Figure 8(b) shows the total cross-sectional area of the bed versus vessel diam-

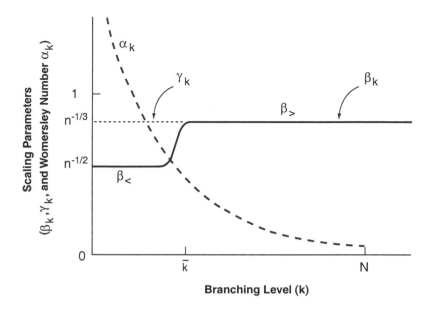

FIGURE 8   (a) Schematic variation of the Womersley number, α_k, and the scaling parameters, β_k and γ_k, with level number (k) for pulsatile systems. Note the step-like change in β_k at $k = \overline{k}$ from area-preserving pulse-wave flow in major vessels to area-increasing Poiseille-type flow in small vessels. (b) Total cross section of the vascular bed from the aorta to the capillaries versus size of the corresponding vessels. Note the change from area preserving for the large tubes (no variation in the total cross-section) to area increasing as the vessels decrease in size.

eter; the crossover from area preserving to area increasing is quite apparent [3, 7].[4] As discussed below this solves the problem of slowing blood down at the capillary level while maintaining the various scaling laws. The crossover from one behavior to the other occurs over the region where the wave and Poiseuille impedances are comparable in size. The approximate value of k where this oc-

---

[4]See, for example, articles by Iberall [14] and Sherman [30], which contain summaries of earlier data; see also Zamir et al. [45] and Li [17]. Care must be taken in comparing measurements with prediction, particularly if averages over many successive levels are used. For example, if $A_k \equiv \sum_k \pi r_k^2$ is the total cross-sectional area at level k, then, for the aorta and major arteries where $k < \overline{k}$ and the branching is area preserving, we predict $A_0 = A_k$. Suppose, however, the first K levels are grouped together. Then, if the resulting measurement gives $\overline{A}_K$, area preserving predicts $\overline{A}_K = KA_0$ (but not $\overline{A}_K = A_0$). It also predicts $r_0^3 \approx n^{1/2} \sum N_k r_k^3$. Using results from LaBarbera [16], who used data averaged over the first 160 vessels (approximately the first 4 levels), gives, for human beings, $A_0 \approx 4.90\,\mathrm{cm}^2$, $\overline{A}_K \approx 19.98\,\mathrm{cm}^2$, $r_0^3 \approx 1.95\,\mathrm{cm}^3$, and $\sum N_k r_k^3 \approx 1.27\,\mathrm{cm}^3$ in agreement with area preserving. LaBarbera, unfortunately, took the fact that $\overline{A}_K \neq A_0$ and $r_0^3 \approx N_k r_k^3$ as evidence supportive of cubic branching and against area preserving. For small vessels, where $k > \overline{k}$, convincing evidence for the cubic law can be found in the analysis of the arteriolar system by Ellsworth et al. [5].

curs ($\overline{k}$, say) is therefore given by $\rho c_0 / \pi r_{\overline{k}}^2 \approx 8 \mu l_k / \pi r_k^4$; i.e., $r_{\overline{k}}^2 / l_{\overline{k}} \approx 8\mu / \rho c_0$, leading to $N - \overline{k} \equiv \overline{N} \approx \ln\left(8\mu l_c / \rho c_0 r_c^2\right) / \ln n$, independent of M. Thus, the number of generations where Poiseuille flow dominates should be independent of body size. On the other hand, the crossover point itself grows logarithmically: $\overline{k} \propto N \propto \ln M$. For humans, with $n = 3$ (the approximate empirical value [3, 7]), $\overline{N} \approx 15$ and $N \approx 22$ (assuming $N_c \approx 2 \times 10^{10}$), whereas with $n = 2$, $\overline{N} \approx 24$ and $N \approx 34$. This means that in humans Poiseuille-type flow becomes appreciable after about seven branchings, whereas in mice after only two to three. Our formula predicts $\overline{k} > 0$ for all mammals, thereby ensuring wave propagation at least through the major arteries. Interestingly, a mammal not much smaller than a shrew ($M \approx 3\,\text{g}$) could not support a pulse wave much beyond the aorta! This may well be the reason why this is the lower limit on the size of a mammal for, if impedances cannot be matched, the energy dissipated is no longer minimal.

3.4 SOME CONSEQUENCES

Though considerably more involved, scaling laws based on β_k can be derived using the "exact" expression for Z, Eq. (12), in the minimization constraint, Eq. (7). This leads to essentially the same results as before, but without the attendant problems. To see how this works, let us assume, for simplicity, that the crossover region is sharp; using a smooth transition as in Figure 8(a), which is obtained by solving the equations exactly, does not change the resulting scaling laws. So, for $k > \overline{k}$, take $\beta_k \equiv \beta_> = n^{-1/3}$ and, for $k < \overline{k}$, $\beta_k \equiv \beta_< = n^{-1/2}$ with $\overline{k}$ given as above. Notice that this implies that area-preserving branching only persists in the pulsatile region from the aorta through the large arteries until $k \approx \overline{k}$. Let us now examine some consequences of this new form for β_k.

1. As a first example consider r_0, the radius of the aorta: its scaling behavior is now given by $r_0 = r_N \beta_>^{\overline{k}-N} \beta_<^{\overline{k}} = r_N n^{1/3N + 1/6\overline{k}} = r_N n^{1/2N - 1/6\overline{N}}$. This gives $r_0 \propto M^{3/8}$, as before and, for humans, $r_0 / r_N \approx 10^4$, in good agreement with data [2, 22, 27, 29]. Recall that the nonpulsatile analysis, with area preserving throughout the network, gave 10^5 for this ratio. In addition, when combined with Eq. (3), this gives, for the ratio of fluid velocity in the aorta to that in the capillary, $\overline{u}_0 / \overline{u}_N = N_N (r_N / r_0)^2 = n^{\overline{N}/3}$, which implies that $\overline{u}_0 / \overline{u}_N \approx 250$, independent of M, again in excellent agreement with data. This accounts quantitatively for the problem of slowing blood down in the capillaries. It is worth noting that since γ reflects the volume-filling geometry, it remains unchanged so we still have $l_0 \propto M^{1/4}$.

2. Blood volume, V_b, however, is more complicated:

$$V_b = \frac{V_c}{(\beta_>^2)^N} \left[\left(\frac{\beta_>}{\beta_<} \right)^{2\overline{k}} \frac{1 - (n\beta_<^2 \gamma)^{\overline{k}}}{1 - (n\beta_<^2 \gamma)} + \left\{ \frac{1 - (n\beta_>^2 \gamma)^N}{1 - (n\beta_>^2 \gamma)} - \frac{1 - (n\beta_>^2 \gamma)^{\overline{k}}}{1 - (n\beta_>^2 \gamma)} \right\} \right]$$

(13)

This is a generalization of Eq. (4). The ratio of the first set of terms within square brackets to the second is approximately $n^{\overline{k}/3}$, so V_b is dominated by the first term, which represents the contribution of the large tubes (aorta and major arteries). Thus, $V_b \propto n^{N+(1/3)\overline{k}} \propto n^{(4/3)N}$. From the minimization constraint this must scale as M, which leads, as before, to $a = 3/4$. An analogous expression to Eq. (13) can be derived for the total impedance of the network, Z. Not surprisingly it is dominated by the Poiseuille resistance of the small tubes (arterioles and capillaries) and therefore leads to our previous results such as blood pressure, Δp, and blood velocity in the aorta, $\overline{u}_0$, being independent of M.

3. Time scales in the network, such as blood circulation time, typically vary like $M^{1/4}$. We have already seen that heart rate scales like $M^{-1/4}$.

4. Since the number of capillaries scales as $M^{3/4}$, whereas the total body volume scales as M, the volume serviced by each capillary increases as $M^{1/4}$.[5]

5. The model can be generalized to d-dimensional organisms. The only significant change is that, since the network must now fill a d-dimensional volume, γ generalizes to $\gamma = n^{-1/d}$. Repeating the previous derivation leads, as before, to Eq. (5) from which we obtain $a = 1/(1 + 1/d) = d/(d + 1)$. This shows that the "3" in 3/4 represents the dimensionality of space that most organisms must supply. Nearly two-dimensional organisms, such as flatworms, might therefore be expected to have $a \approx 2/3$. Note the "1" in $(d + 1)$ arises from $2 \times 1/2$, the "1/2" coming from the area-preserving exponent in β and the "2" from converting this from radii to cross-sectional areas.

4   EXTENSIONS

Other analogous network systems can be analyzed in a similar fashion. Here we shall briefly describe the extension to the mammalian respiratory and plant vascular systems. More details of the latter can be found in the chapter by Enquist et al. (this volume).

4.1   RESPIRATORY SYSTEM

A minor variant of the model provides an integrated description of the mammalian respiratory system giving similar scaling relationships. Although pulse

---

[5]If the volume served is roughly spherical in shape, then its radius scales as $M^{1/12}$ whereas, if it is identified with the Krogh cylinder, then its radius scales as $M^{1/8}$ [15].

waves are irrelevant since the tubes are not elastic, the formula for Z is quite similar to Eq. (12). The fractal bronchial tree terminates in $N_A \propto M^{3/4}$ alveoli. The network is space filling and the alveoli (or, more properly, the acini) play the role of the service volume. They account for most of the total volume of the lung, which scales as M. Thus the volume of an alveolus $V_A \propto M^{1/4}$, its radius $r_A \propto M^{1/12}$, and its surface area $A_A \propto r^2 \propto M^{1/6}$. Hence, the total surface area of the lung $A_L = N_A A_A \propto M^{11/12}$. This explains the paradox noted by Weibel [8] that A_L scales with a higher exponent ($11/12 \approx 0.92$) than the $3/4$ seemingly needed to supply oxygen. The rate of oxygen diffusion across an alveolus, which must be independent of M, is $\propto \Delta p_{O_2} A_A / r_A$. Thus $\Delta p_{O_2} \propto M^{-1/12}$ which must be compensated for by a similar scaling of the oxygen affinity of hemoglobin. To the extent that data are available, they support these predictions (Table 1) [2, 22, 27, 29].

4.2 PLANT VASCULAR SYSTEMS

A more detailed discussion of the plant vascular network can be found in the chapter by Enquist et al. (this volume). Here we present only a very brief overview with some salient results that relate to the general model.

Plant vascular systems are essentially a bundle of parallel vascular tubes, analogous to an electrical cable. In trees, some old tubes no longer conduct fluid; they form the heartwood which gives biomechanical stability to the

TABLE 1 Values of allometric exponents for variables of the mammalian cardiovascular and respiratory systems predicted by the model compared with empirical observations. Observed values of exponents are taken from [2, 22, 27, 29]; ND denotes that no data are available.

VARIABLE Cardiovascular	SCALING EXPONENT Predicted	Observed	REFERENCE
Radius of Aorta, r_0	$3/8 =$ 0.375	0.36	Holt et al. [13]
Pressure in Aorta, Δp_0	$0 =$ 0.00	0.032	Gunther & de la Barra [11]
Velocity of Blood in Aorta, u_0	$0 =$ 0.00	0.07	Milnor [23]
Blood Volume, V_b	$1 =$ 1.00	1.00	Prothero [28]
Circulation Time	$1/4 =$ 0.25	0.25	Schmidt-Nielsen [29]
Circulation Distance, l	$1/4 =$ 0.25	ND	–
Cardiac Stroke Volume	$1 =$ 1.00	1.03	Gunther [10]
Cardiac Frequency, ω	$-1/4 = -0.25$	−0.25	Stahl [33]
Cardiac Output, $\dot{E}$	$3/4 =$ 0.75	0.74	Gunther [10]
Number of Capillaries, N_c	$3/4 =$ 0.75	ND	–
Supply Radius of Cells	$1/12 =$ 0.083	ND	–
Radius of Krogh Cylinder	$1/8 =$ 0.125	ND	–
Density of Capillaries	$-1/12 = -0.083$	−0.095	Hoppeler et al.
Oxygen Affinity of Blood, P_{50}	$-1/12 = -0.083$	−0.089	Dhindsa et al. [4]
Total Peripheral Resistance, Z	$-3/4 = -0.75$	−0.76	Gunther [10]
Womersley Number, α	$1/4 =$ 0.25	0.25	–
Metabolic Rate, B	$3/4 =$ 0.75	0.75	Stahl [33]

trunk and branches. This leads to a relationship between length and radius: $l_k \propto r_k^\alpha$. Analyses based on scale-invariant solutions to the bending moment equations for beams (elastic similarity) give $\alpha = 2/3$ [9, 20, 21, 26]. This is most important for the trunk and large branches and agrees well with data for these segments. It leads precisely to area-preserving branching; thus, biomechanical stability implies that branches behave *as if* they were tightly packed vascular bundles of constant-diameter tubes. This constraint is assumed to hold throughout the plant. It leads to a number of interesting geometrical results such as the leaf area distal to the kth branch, $A_k^L = C_L r_k^2$, where $C_L \equiv a_L / r_N^2$ is invariant and a_L is the area of a leaf.

Since hydrodynamic resistance of uniform conducting tubes increases linearly with length, resource supply to apical meristems and forest canopies would be seriously limited. Plants have had to circumvent this problem in order for trees and other erect life forms to evolve. It is solved in the model by allowing vessels to have a small uniform taper. Minimizing the vessel resistance determines the magnitude of this taper, which turns out to be quite small: for example, over 12 orders of magnitude variation in plant mass, the radius of the vessel in the trunk is predicted to change by only about 60%, in agreement with observation. Remarkably, this taper has the consequence that the vessel resistance is *independent of the total length of the tube*, thereby equalizing resource supply to all leaves, especially those on the most distal branches of the tallest trees. The invariance of vessel resistance (which follows from the minimization principle, c.), coupled with area-preserving branching (derived from biomechanical stability), and a volume-filling network leads to many scaling laws including the 3/4-power scaling for metabolic rate. Predictions for conductivity, pressure gradients, fluid velocity, and relative amount of heartwood are all in excellent agreement with data. An added bonus is that we can also show why the maximum height of a tree is of the order of 100 m, rather than 1 m or 1000 m. This follows because the taper cannot continue indefinitely; the trunk, whose size is constrained biomechanically, simply cannot contain enough conducting vessels if it is allowed to grow without bound. This therefore provides an explanation from fundamental principles why the size of trees is limited, and how this size is related to basic parameters which depend on both mechanical and hydrodynamic constraints.

It is also possible to use the model to calculate the fractal dimension of plants as defined by the box-counting method illustrated in Figure 5. As long as the resolution is not too fine, then the model predicts $d = 3/2$, in excellent agreement with data for both shoots and roots [6, 24, 35]. Amusingly, in order to minimize their disturbance, the measurements of d have been repeated on roots grown between glass plates separated by only 3 mm instead of on roots grown conventionally in a three-dimensional medium. These gave $d \approx 1.3$, significantly different from 1.5. However, by growing roots in this way, the network has been restricted to two dimensions rather than three, and so a different value of d is predicted. With the constraint that the network fills only two dimensions, the model predicts $d = 4/3$! This is indirect evidence of the volume-filling principle for these networks.

5 CONCLUSIONS

We have developed a general model that describes the structure and function of biological supply networks based on three fundamental principles: (i) these networks have a hierarchical branching structure, (ii) the terminal units are invariant, and (iii) energy dissipation is minimized. The model provides a complete quantitative description of the structure and dynamics of these networks. As such, it can potentially predict the normalizations, Y_0, as well as the exponents, b, of allometric equations, Eq. (1). The predicted scaling properties do not depend on most details of system design, such as the nature of the fluid in the vessels or the pump that propels it. In addition, the limit of a large number of branchings (N), which determines the allometric scaling properties, depends only weakly, if at all, on the exact branching pattern of the network, provided that it has a hierarchical, fractal-like structure.

The model predicts many scaling relationships. For many of these, data are available and compare favorably with predicted values (see Tables 1 and 2). It should be emphasized that, in order to understand allometric scaling for all structural and functional characteristics, it is necessary to formulate an integrated model for the entire system. In this respect our model represents

TABLE 2   Values of allometric exponents for variables of the mammalian respiratory system predicted by the model compared with empirical observations.

VARIABLE	SCALING EXPONENT		REFERENCE
Respiratory	Predicted	Observed	
Lung Volume	$1 =$ 1.00	1.05	Weibel [40]
Respiratory Frequency	$-1/4 = -0.25$	−0.26	Stahl [33]
Volume Flow to Lung	$3/4 =$ 0.75	0.80	Stahl [33]
Interpleural Pressure	$0 =$ 0.00	0.004	Gunther & de la Barra [11]
Diameter of Trachea	$3/8 =$ 0.375	0.39	Tenney & Bartlett [36] and pers. comm. in Calder [2]
Air Velocity in Trachea	$0 =$ 0.00	0.02	Calder (calculated) [2]
Tidal Volume	$1 =$ 1.00	1.041	Maina & Settle [18]
Power Dissipated	$3/4 =$ 0.75	0.78	Stahl [33]
Number of Alveoli, N_A	$3/4 =$ 0.75	ND	–
Volume of Alveolus, V_A	$1/4 =$ 0.25	ND	–
Radius of Alveolus, r_A	$1/12 =$ 0.083	0.13	Tenney & Remmers [37]
Surface Area of Alveolus, A_A	$1/6 =$ 0.083	ND	–
Surface Area of Lung, A_L	$11/12 =$ 0.92	0.95	Gehr et al. [8]
Oxygen Diffusing Capacity	$1 =$ 1.00	0.99	Gehr et al. [8]
Total Airway Resistance	$-3/4 = -0.75$	−0.70	Stahl [33]
Oxygen Consumption Rate	$3/4 =$ 0.75	0.76	Stahl [33]

a major advance over most previous treatments of mammalian cardiovascular and plant vascular systems.

The paradigm and principles embodied in the general model offer a novel way to explore these types of networks and to understand their scaling properties. On the one hand ours is, by necessity, a "zeroth-order model" incorporating only the essential features of biological systems. It can, however, serve as a point of departure for more detailed analyses. On the other hand, the ability of the model to capture the essence of these systems and to make accurate predictions inevitably suggests applications to other interesting and related problems. Obvious areas deserving serious investigation include intracellular transport, aging, and longevity, and the extension to ecological consequences of body size. Quarter-power scaling has been observed in all of these cases, so it is natural to try to develop rigorous analytical models of them. The success of our general model should be viewed as a beginning rather than an end.

ACKNOWLEDGMENTS

REFERENCES

[1] Bernston, G. M., and P. Stoll. "Correcting for Finite Spatial Scales of Self-Similarity when Calculating the Fractal Dimensions of Real-World Structures." *Proc. Roy. Soc. Lond. B* **264** (1997): 1531–1537.

[2] Calder, W. A., III. *Size, Function and Life History.* Cambridge, MA: Harvard University Press, 1984.

[3] Caro, C. G., et al. *The Mechanics of Circulation.* Oxford: Oxford University Press, 1978.

[4] Dhindsa, D. S., A. S. Hoversland, and J. Metcalfe. "Respiratory Functions of Armadillo Blood." *Resp. Physiol.* **13** (1971): 198–208.

[5] Ellsworth, M. L., et al. *Microvasc. Res.* **34** (1987): 168.

[6] Fitter, A. H., and T. R. Strickland. "Fractal Characterization of Root System Architecture." *Functional Ecology* **6** (1992): 632–635.

[7] Fung, Y. C. *Biodynamics: Circulation.* New York, Berlin, Heidelberg, Tokyo: Springer-Verlag, 1984.

[8] Gehr, P., D. K. Mwangi, A. Ammann, G. M. O. Maloiy, C. R. Taylor, and E. R. Weibel. "Design of the Mammalian Respiratory System. V. Scaling Morphometric Pulmonary Diffusing Capacity to Body Mass: Wild and Domestic Mammals." *Resp. Physiol.* **44** (1981): 61–86.

[9] Greenhill, G. "Determination of the Greatest Height Consistent with Stability that a Vertical Pole or Mast Can Be Made, and the Greatest Height to Which a Tree of Given Proportions Can Grow." *Proc. Cambridge Phil. Soc.* **4** (1881): 65–73.

[10] Gunther, B. "Dimensional Analysis and Theory of Biological Similarity." *Physiol. Rev.* **55** (1975): 659–699.

[11] Gunther, B., and B. L. de la Barra. "Physiometry of the Mammalian Circulatory System." *Acta Physiol. Lat. Am.* **16** (1966): 32–42.

[12] Hemmingsen, A. M. "Energy Metabolism as Related to Body Size and Respiratory Surfaces, and Its Evolution." *Rep. Steno Mem. Hosp. (Copenhagen)* **9** (1960): 1–110.

[13] Holt, J. P., E. A. Rhode, W. W. Holt, and H. Kines. "Geometric Similarity of Aorta, Venae Cavae, and Certain of Their Branches in Mammals." *Amer. J. Physiol.* **241** (1981): R100.

[14] Iberall, A. S. "Anatomy and Steady Flow Characteristics of the Arterial System with an Introduction to Its Pulsatile Characteristics." *Math Biosci.* **1** (1967): 375–395.

[15] Krogh, A. *The Anatomy and Physiology of Capillaries*, 2nd ed. New Haven, CT: NYale University Press, 1929.

[16] LaBarbera, M. "Principles of Design of Fluid Transport Systems in Zoology." *Science* **249** (1990): 992–999.

[17] Li, J. K-J. *Comparative Cardiovascular Dynamics of Mammals*. Boca Raton: CRC Press, 1996.

[18] Maina, J. N., and J. G. Settle. "Allometric Comparisons of Some Morphometric Parameters of Avian and Mammalian Lungs." *J. Physiol. (Lond.)* **330** (1982):28.

[19] Mandelbrot, B. B. *The Fractal Geometry of Nature*. New York: W. H. Freeman, 1977.

[20] McMahon, T. A. "Size and Shape in Biology." *Science* **179** (1973): 1201–1204.

[21] McMahon, T. A., and R. E. Kronauer. "Tree Structures: Deducing the Principle of Mechanical Design." *J. Theor. Biol.* **59** (1976): 443–466.

[22] McMahon, T. A., and J. T. Bonner. *On Size and Life*. New York: Scientific American Library, 1983.

[23] Milnor, W. R. "Aortic Wavelength as a Determinant of the Relation Between Heart Rate and Body Size in Mammals." *Am. J. Physiol.* **237** (1979): R3–R6.

[24] Morse, D. R., J. H. Lawton, J. H. Dodson and M. M. Williamson. "Fractal Dimension of Vegetation and the Distribution of Arthropod Body Lengths." *Nature* **314** (1985): 731–733.

[25] Murray, C. D. "The Physiological Principle of Minimum Work. I. The Vascular System and the Cost of Blood Volume." *Proc. Natl. Acad. Sci. USA* **12** (1926): 207–214.

[26] Niklas, K. J. *Plant Biomechanics: An Engineering Approach to Plant Form and Function*. Chicago, IL: University of Chicago Press, 1992.

[27] Peters, R. H. *The Ecological Implications of Body Size*. Cambridge, MA: Cambridge University Press, 1983.

[28] Prothero, J. "Scaling of Blood Parameters in Mammals." *Comp. Biochem. Physiol.* **67A** (1980): 649–657.

[29] Schmidt-Nielsen, K. *Scaling; Why Is Animal Size so Important?* Cambridge, MA: Cambridge University Press, 1984.

[30] Sherman, T. F. "On Connecting Large Vessels to Small. The Meaning of Murray's Law." *J. Gen. Physiol.* **78** (1981): 431–453.

[31] Shinozaki, K., K., K. Yoda, Hozumi, and T. Kira. "A Quantitative Analysis of Plant Form—the Pipe Model Theory: I. Basic Analysis." *Jap. J. Ecol.* **14** (1964): 97–105.

[32] Shinozaki, K., K. Yoda, K. Hozumi, and T. Kira. "A Quantitative Analysis of Plant Form—The Pipe Model Theory II. Further Evidence of the Theory and Its Application to Forest Ecology." *Jap. J. Ecol.* **14**(4) (1964): 133–139.

[33] Stahl, W. R. "Scaling of Respiratory Variables in Mammals." *J. Appl. Physiol.* **22** (1967): 453–460.

[34] Strahler, A. N. "Revisions of Horton's Quantitative Factors in Erosional Terrain." *Trans. Am. Geophys. Union* **34** (1953): 345–365.

[35] Tatsumi, J. A., A. Yamauchi, and Y. Kono. "Fractal Analysis of Plant Root Systems." *Ann. Botany* **64** (1989): 499–503.

[36] Tenney, S. M., and D. Bartlett, Jr. "Comparative Quantitative Morphology of the Mammalian Lung: Trachea." *Resp. Physiol.* **3** (1967): 130–135.

[37] Tenney, S.M., and D. H. Remmers. "Comparative Quantitative Morphology of the Mammalian Lung: Diffusing Area." *Nature* **197** (1963): 54–56.

[38] Thompson, D'Arcy W. *On Growth and Form.* Cambridge, MA: Cambridge University Press, 1917.

[39] Tyree, M. T., and F. W. Ewers. "The Hydraulic Architecture of Trees and Other Woody Plants." *New Phytologist* **119** (1991): 345–360.

[40] Weibel, E. R. "Morphometric Estimation of Pulmonary Diffusion Capacity. V. Comparative Morphometry of Alveolar Lungs." *Resp. Physiol.* **14** (1973): 26–43.

[41] West, G. B., J. H. Brown, and B. J. Enquist. "A General Model for the Origin of Allometric Scaling Laws in Biology." *Science* **276** (1997): 122–126.

[42] Womersley, J. R. "Method for the Calculation of Velocity, Rate of Flow, and Viscous Drag in Arteries When the Pressure Gradient is Known." *J. Physiol.* **127** (1955): 553–563.

[43] Womersley, J. R. "Oscillatory Motion of a Viscous Liquid in a Thin-Walled Elastic Tube. I. The Linear Approximation for Long Waves." *Phil. Mag.* **46** (Series 7) (1955): 199–221.

[44] Woodruff, W. H., G. B. West, and J. H. Brown. To be published.

[45] Zamir, M., P. Sinclair, and T. H. Wonnacott. "Relation Between Diameter and Flow in Major Branches of the Arch of the Aorta. *J. Biomechanics* **25** (1992): 1303–1310.

[46] Zimmerman, M. H. *Xylem Structure and the Ascent of Sap.* Berlin: Springer-Verlag, 1983.

Scaling and Invariants in Cardiovascular Biology

John K-J. Li

1 INTRODUCTION

The optimal design features of the mammalian cardiovascular system have been marveled at by us *Homo sapiens* for many decades. The complexity of the control aspects of the circulatory function has been unraveled only recently, but to a somewhat limited extent. The complexity arises from the constant biological transformations that reflect structural adaptations to functional demands. Across mammalian species, this complexity can be better appreciated from the beat-to-beat dynamic performance of the heart and its interaction with the vascular system. The complexity can be substantially reduced when appropriate biological scaling laws are imposed and relevant invariant features are identified across species.

The cardiovascular system is selected here to illustrate wonderful features of biological design and scaling. In terms of structure and function, there are some characteristics that must vary with size, and they are consequently scaled with respect to body mass. Some examples are: size of the heart, volume of blood, length of the aorta, and cardiac output. Other characteristics, however, are invariant including: blood pressure, ejection fraction, heart beats per lifetime, and capillary and red blood cell sizes. Many of these can be written as dimensionless numbers, and they reflect properties of the system that are dif-

Scaling in Biology, edited by J. H. Brown and G. B. West.
Oxford University Press, 2000.

ficult to change, for instance, because of constraints on the intrinsic properties of constituent biological materials or the ontogeny of the individual.

Ancient Chinese practitioners often felt palpable pulsations in the radial arteries of their patients as a means to diagnose the states of their hearts. Thus that the contraction of the heart generates the pulsation and that the pulse is subsequently transmitted throughout the body was already known in antiquity. It was not until the sixteenth century, that Harvey, in his now famous *De Mortu Cordis* [13], explained the intermittent pumping function of the heart as a consequence of systolic contraction and diastolic filling. He also made comparisons of the circulatory function from his many "Anatomical Exercises Concerning the Motion of the Heart and Blood in Living Creatures" [14] which he performed on several mammalian species, avians, and amphibians. The quantification of the blood pressure amplitudes and cardiac output in mammalian species, however, was first provided by Hales [12] a century later. Modern assessment of circulatory function continues to consider blood pressure and cardiac output as the most pertinent variables.

Taking a different perspective, D'Arcy Thompson's *On Growth and Form* [43] paved yet another path to modern comparative biological studies by examining diverse collections of biological specimens. In step with biodiversity and looking at growth in relation to form, Huxley [17] introduced the concept of differential growth to explain many of the observed biological transformations. He made clear that different organs and organisms may grow at differential rates in relation to body weight. Much of his interpretation was based on the use of allometric relations.

Allometry is defined as the change of proportions with increase of size both within a single species and between adults of related groups. The allometric formula relates any measured physical quantity Y to body mass M, with a and b as derived or measured empirical constants. Quantitatively, this results in a power law,

$$Y = aM^b. \tag{1}$$

This formula expresses simple allometry. In the special case when the exponent is 0, Y is independent of body mass M. When b is $1/3$, then the variable is said to be dependent on body length dimensions; when b is $2/3$, Y is dependent on body surface area, and when $b = 1$, Y is simply proportional to body mass. This provides what's known as the basis of the "one-third power law" or geometric scaling [21]. This has recently been challenged by the "one-fourth power law" as the basis of biological allometric formulation [44].

The allometric equation has been proven to be powerful for characterization of similarities among species. It is effective in relating a physiological phenomenon, either structural or functional, among mammals of grossly different body mass. A similarity criterion is established when Y, formulated in terms of either product(s) or ratio(s) of physically measurable variables, remains constant despite changes in body mass, and is dimensionless. Thus, the exponent b is necessarily zero. In other words, similarity is present whenever any two dimensionally identical measurements occur in a constant ratio to each other. If such a ratio exists among different species, then a similarity criterion is established as the scaling law. This approach of establishing biological similarity criteria has been very useful [9, 10, 26, 27, 28, 39, 40, 41].

2 BIOLOGICAL DIMENSIONAL ANALYSIS

The importance of dimensional homogeneity in solving equations involving physical variables was the principal emphasis of Fourier, better known for the Fourier series. But the theorem of dimensional analysis was introduced by Buckingham [3]. It states that if a physical system can be properly described by a certain set of dimensional variables, it may also be described by a lesser number of dimensionless parameters which incorporate all the variables. This principle apparently was used to establish similarity rules at about the same time when Lord Raleigh proposed the well-known Raleigh indices. It can be illustrated by applying Laplace's law to the mammalian hearts.

The beat-to-beat pumping ability of the mammalian heart is determined by its force-generating capability and the lengths of its constituent muscle fibers, as was recognized by Starling (hence Starling's law of the heart). The formula for calculating force or tension, however, has been based on the law of Laplace

$$T = pr \,, \tag{2}$$

which states that the pressure difference, p, across a curved membrane in a state of tension is equal to the tension in the membrane, T, divided by its radius of curvature, r [46]. This law has been applied to both blood vessels [4] and the heart [24]. To apply this formula, a certain geometric shape of the heart has to be assumed in order to arrive at the radius or radii of curvature. Therefore the ventricle has been described geometrically as either a thin-walled or thick-walled sphere or ellipsoid. In actuality, the myocardium which encloses the ventricular chamber has finite wall thickness. Also, the long-axis diameter or the base (aortic valve) to apex distance is greater than the short-axis diameter or the ventricular septum to free-wall distance.

When the left ventricle is considered as an ellipsoid, there are two principle radii of curvature, r_1 and r_2. Laplace's law dictates:

$$p = T \left(\frac{1}{r_1} + \frac{1}{r_2} \right) . \tag{3}$$

For the ventricle as a sphere, $r_1 = r_2$ so that

$$p = \frac{2T}{r} \,. \tag{4}$$

In a cylinder such as the blood vessel, one radius is infinite, so that

$$p = \frac{T}{r},$$ (5)

which indicates a lesser tension in the wall is needed to balance the distending pressure. Both arterial pressure and ventricular pressure are invariant in magnitude and in waveforms in many mammalian species.

The larger the size of the mammalian heart, the greater the tension exerted on the myocardium. To sustain this greater amount of tension, the wall must thicken proportionally with the increasing radius of curvature. This results in a larger heart weight. Thus, ventricular wall thickness also needs to be explained. This modifies Laplace's law to the Lamé relation,

$$T = \frac{pr}{h}.$$ (6)

A dimensional matrix can be readily formed by first expressing T, r, p, and h in the mass (M), length (L), and time (T) system:

$$
\begin{array}{c}
 & \begin{array}{cccc} T & r & p & h \end{array} \\
\begin{array}{c} M \\ L \\ T \end{array} & \left(\begin{array}{cccc} 1 & 0 & 1 & 0 \\ 0 & 1 & -1 & 1 \\ -2 & 0 & -2 & 0 \end{array} \right).
\end{array}
$$ (7)

In order to derive dimensionless parameters (π_i), Buckingham's pi-theorem is utilized. To reiterate, the number of pi-numbers (j) is equal to the number of physical quantity considered (n) minus the rank (r) of the matrix [31, 34]. Thus, there will be two pi-numbers, π_1 and π_2.

$$\pi_1 = \frac{T}{ph} \quad \text{and} \quad \pi_2 = \frac{h}{r}.$$ (8)

They provide a description of the geometric and mechanical relations of the mammalian hearts and Laplace's Law is implicit in the ratio of the two,

$$I = \frac{\pi_1}{\pi_2} = \frac{T}{pr}.$$ (9)

3   INVARIANT NUMBERS AND APPLICATIONS

Both π_1 and π_2 and their ratio, I, are not only dimensionless, they are also independent of mammalian body mass. That is, π_2 indicates that ratio of ventricular wall thickness to its radius, h/r, is invariant among mammals. This also establishes a scaling factor. They are thus considered invariant numbers, i.e., $[M]^0[L]^0[T]^0$ = dimensionless constant. This invariance implies that Laplace's Law applies to all mammalian hearts [24, 34].

In the case of pathological cardiac hypertrophy, the h/r ratio is significantly altered as a consequence of increased wall thickness. This latter increase has been suggested as the result of an adaptation process by which the wall tension is normalized (Eq. (6)). In a failing and enlarged heart, the greater tension due to a larger radius of curvature results in excess myocardial oxygen demand. Cardiac size reduction via partial left ventriculectomy, or the "Batista procedure," to normalize the h/r ratio and wall tension has met with success to reverse this detriment. Another example of scaling invariance can be found in blood flow in arteries. A dimensional matrix is first formed by incorporating parameters that are considered pertinent. These are the density (ρ) and viscosity (η) of the fluid, diameter (D) of the blood vessel, and velocities of the flowing blood (v) and of the pulse wave (c).

$$
\begin{array}{c}
\begin{array}{ccccc}
\rho & c & D & \eta & v \\
(g/cm^3) & (cm/s) & (cm) & (poise) & (cm/s)
\end{array} \\
\begin{array}{c} M \\ L \\ T \end{array}
\left(
\begin{array}{ccccc}
1 & 0 & 0 & 1 & 0 \\
-3 & 1 & 1 & -1 & 1 \\
0 & -1 & 0 & -1 & -1
\end{array}
\right) \\
\begin{array}{ccccc}
k_1 & k_2 & k_3 & k_4 & k_5
\end{array}
\end{array}
\tag{10}
$$

where k_n's are Rayleigh indices referring to the exponents of the parameters. The pi-numbers can readily be obtained. Two of these are the well-known Reynold's number, essential for identifying viscous similitude and laminar to turbulent flow transitions [27], $\Re = \rho v D/\eta$, and the Mach number, $\text{Ma} = v/c$, or the ratio of blood velocity to pulse wave velocity. Allometric relation gives

$$
\text{Ma} = 0.04\, M^{0.0}
\tag{11}
$$

which is invariant with respect to mammalian body mass. The Reynold's number is dimensionless, but is not an invariant function of mammalian body mass,

$$
\Re = 260.76\, M^{0.42}.
\tag{12}
$$

Thus, dimensionless pi-numbers do not equal similarity principles, i.e., scaling factors are not necessarily invariant numbers.

4 ALLOMETRY OF THE MAMMALIAN CIRCULATORY SYSTEM

Allometric relations of anatomic structure and physiological functions are useful for identifying similarities of the circulatory function of different mammalian species [28, 29]. From a comparative physiological point of view, it is difficult to select the appropriate variables or parameters to describe precisely the circulatory function under a prescribed physiological condition. For this reason, circulatory allometric relations are generally obtained at resting

TABLE 1   Allometric relations of some hemodynamic parameters. $Y = aM^b$, M in kg. References are given in the text.

Parameter		Y	a	b
Heart rate	s^{-1}	f_h	3.60	-0.27
Stroke volume	ml	V_s	0.66	1.05
Pulse velocity	cm/s	c	446.0	0.0
Arterial pressure	dynes/cm^2	p	1.17×10^5	0.033
Radius of aorta	cm	r	0.205	0.36
Length of aorta	cm	L	17.5	0.31
Metabolic rate	ergs/sec	E_{MR}	3.41×10^7	0.734
Heart weight	kg	M_h	0.0066	0.98

heart rates. The obvious factors that are important in determining function are heart rate and size, cardiac efficiency and contractility, stroke volume, and blood pressure. Some examples of circulatory allometry are given in Table 1, and can be found in other sources [5, 6, 15, 16, 18, 28, 29, 39, 40, 41].

In mammals, the ratio of heart weight to body mass is an invariant with the heart accounting for about 0.6% of body mass. In allometric form [1, 10], this is

$$M_h = 6.6 \times 10^{-3} M^{0.98} \tag{13}$$

where the heart weight M_h and M are both in grams. With M_h in g and M in kg, this has been given [16] as

$$M_h = 2.61 \, M^{1.10}. \tag{14}$$

It is readily apparent that experimental conditions and ecological factors can influence the empirical constants a and b. The above exponents for the heart weight, however, do not differ significantly from the theoretical exponent of 1.0 ($M_h = M^{1.0}$). The deviations arise from statistical fits of regressions to experimental data. It is also readily apparent that if a variable scales as $M^{1.0}$, then the allometric equation can be made invariant by taking a ratio with M in the denominator, i.e., normalizing with body mass.

The stroke volume (V_s) is also an invariant when normalized to heart weight or to body mass,

$$V_s = 0.66 \, M^{1.05} \text{ ml} \quad or \quad = 0.74 \, M^{1.03} \text{ ml}. \tag{15}$$

It has long been considered an important hemodynamic quantity in assessing ventricular function. Its product, with blood pressure, bears a direct relation to the energy expenditure of the heart, or the external work, E_W,

$$E_W = pV_s. \tag{16}$$

This is the work performed by the heart in order to perfuse the vasculature during each contraction or, in other words, the work necessary to overcome the arterial load during each ejection. Blood pressures are generally invariant with respect to body mass in mammals (of course, there are exceptions [2, 35, 36], such as the giraffe [8]). This also indicates that the heart is a pressure source and maintaining a constant blood pressure is of utmost importance. The process of blood pressure control is complex, and important roles are played by baroreceptors, the renin-angiotensin system, and the autonomic nervous system, just to name a few subsystems. Allometrically, the mean arterial pressure is expressed as

$$p = 1.17 \times 10^5 \, M^{.033} \text{ dynes/cm}^2 = 87.8 \, M^{.033} \text{ mmHg} . \tag{17}$$

The exponent is slightly but not statistically significantly different from 0 $(p = aM^0)$. Thus, the external work is given by

$$E_W = 0.87 \times 10^5 \, M^{1.03} \text{ ergs} = 0.0087 M^{1.03} \text{ J} . \tag{18}$$

A larger ventricle generates a greater amount of external work. The quantity of blood that is ejected per beat (stroke volume), however, is a constant fraction of the amount contained in the heart (end-diastolic volume). Thus, ejection fraction, as it is termed, is an invariant among mammals,

$$F_{ej} = \frac{V_s}{V_{ed}} = 0.6 \text{ to } 0.7 . \tag{19}$$

In a failing heart, the ejection fraction can decrease substantially (to 0.2 say), as a result of a reduced stroke volume and an enlarged heart size.

The smaller the mammal, the smaller is its heart weight, but the faster its heart rate (f_h) [7],

$$f_h = 4.02 \, M^{-0.25} \text{ s}^{-1} . \tag{20}$$

Smaller mammals have shorter life spans, since the total number of heart beats in a mammal's lifetime is invariant. Within an individual mammal, rapid (and random) heart rhythms beyond normal often result in cardiac arrhythmias, such as ventricular tachycardia. On the other hand, it is interesting to note here that "cardiac slowing," which reduces heart rate, can actually have the beneficial consequence of increasing longitivity.

Cardiac output, deemed by Hales [12] as an important quantity describing ventricular function, is given as the product of stroke volume and heart rate, or the amount of blood pumped out of the ventricle per minute,

$$Q_{CO} = V_s f_h = (0.74 \, M^{1.03})(4.02 \, M^{-0.25}) \frac{60}{1000} = 0.178 \, M^{0.78} \text{ l/min} . \tag{21}$$

This is closely related to metabolic rate, since the heart supplies oxygen and nutrients for metabolism. Table 2 gives a comparison of cardiac output in

TABLE 2 Cardiac output of some mammalian hearts based on the allometric equation $Q_{CO} = 0.178\, M^{.78}$ l/min.

Species	Body mass (kg)	Cardiac output (l/min)
Elephant	2000	67
Horse	400	19
Man	70	5
Dog	20	1.8
Rabbit	3.5	0.5
Mouse	0.25	0.06
Tree Shrew	0.005	0.003

several species. Deviations from this equation have been found in very small mammals [45]. Since blood pressure is invariant, cardiac output is limited by the total peripheral resistance to blood flow of the mammalian systemic arterial tree, which is obtained as

$$Rs = \frac{p}{Q_{CO}} = 2.8 \times 10^6 M^{-0.747} \text{ dyn s cm}^{-5} . \tag{22}$$

Thus, the peripheral resistance follows the $-3/4$ power of mass [44], and is inversely proportional to the metabolic rate $(+3/4)$. This relation can be strongly altered under local conditions, such as vasoconstriction or vasodilation. This derived allometric equation can be compared to that reported by Gunther and Guerra [9] who have constructed an equation conforming more closely to the 2/3 power:

$$Rs = 3.35 \times 10^6 M^{-.68} \text{ dynes s cm}^{-5} . \tag{23}$$

5   THE ENERGETICS OF THE MAMMALIAN HEART

The heart as a muscular pump requires energy. The energy requirement of its constituent muscle fibers and the useful work it can generate are of considerable interest [22, 23, 38, 42]. They define the mechanical efficiency of the cardiac pump. In hemodynamic terms, the efficiency of the heart is defined as the ratio of external mechanical work to myocardial oxygen consumption,

$$e = \frac{E_W}{E_{MVO2}} = Qc \times O_{2a-v} , \tag{24}$$

where O_{2a-v} is the arterio-venous oxygen saturation and Qc, the coronary blood flow. The efficiency of the heart is an invariant among mammalian species [28] and is about 20%. The relation dictates the energy requirement and pumping performance of the heart.

E_W is also termed stroke work and is represented as the area under the left ventricular pressure-volume (P-V) diagram during each heart beat. The external mechanical work generated by the heart per unit body weight or heart weight is constant for the mammalian species [22, 23], i.e.,

$$\frac{E_W}{M} = \text{constant} = \frac{pV_s}{M}.\tag{25}$$

This result is also of considerable physiological importance, since it states that the cardiac external work intensity is invariant among mammals. Species differences in cardiac energetics, however, have been reported [33]. For man, taking $V_s = 75$ ml, $p = 100$ mmHg, $M = 70$ kg, and $M_h = 370$ g, the external work is about 1 J and the coefficient is about 2.7 J/kg. In terms of heart weight, this is

$$E_W = 2.7 \text{ J/kg} \quad or \quad E_W = 1/70 \text{ J/kg}\tag{26}$$

in terms of body mass. For a 2100-kg elephant, its left ventricle is estimated to generate about 30 J for each heart beat. Examination of the dimensions gives:

$$\frac{E_W}{M} = [M]^0[L]^2[T]^{-2},\tag{27}$$

which, although constant among mammals, is not dimensionless. Therefore, it is not an invariant number.

Cardiac output (Eq. (21)) is a direct function of metabolic rate, following the 3/4 power,

$$Q_{CO} \propto E_{MR} \propto M^{0.75}.\tag{28}$$

Normalizing cardiac output or metabolic rate with respect to body mass gives

$$\frac{Q_{CO}}{M} \propto M^{-0.25} \propto f_h \propto E_{MTR},\tag{29}$$

so that the metabolic turn-over rate (E_{MTR}) is a function of heart rate (Table 3). Combining Eqs. (25) and (29) gives

$$\frac{E_W/M}{E_{MTR}/f_h} = \text{constant} = K_h.\tag{30}$$

The dimensions of K_h is $[M]^0[L]^0[T]^0 = $ dimensionless constant. The value of K_h is shown in Table 3. It is an invariant number and serves to define the mammalian cardiac energetics in relation to the whole-body metabolism.

6   VASCULAR PERFUSION AND ARTERIAL PULSE TRANSMISSION CHARACTERISTICS

The arterial system exhibits geometric and elastic nonuniformities. Geometric taper of the aorta is associated with increased elastic modulus away from the

TABLE 3 Physiological parameters of three mammals showing the relation of metabolic turn-over rate to heart rate.

	Body Mass M (kg)	Metabolic Rate E_{MR} (J/s)	Metabolic Turn-over Rate E_{MTR} (J/s/kg)	Heart Rate f_h (min^{-1})	(E_{MTR}/f_h) (J/kg)	$(E_W/M)/$ (E_{MTR}/f_h) (dimensionless)
Man	70	82	1.17	70	1.0	0.014
Dog	20	32	1.6	90	1.07	0.013
Guinea Pig	0.7	2.6	3.7	240	0.93	0.015

heart. Vascular branching occurs where target organ perfusion is necessary. Arterial wall thickness to lumen radius ratios are invariant at corresponding anatomic sites in mammalian species. The ratio of aortic length to its diameter, is also an invariant [15, 26]. In addition, the sizes of terminal arterioles and capillaries, as well as red blood cells, are also virtually invariant among mammalian species regardless of body size. These represent structural invariants, giving rise to global vascular perfusion characteristics that are amazingly similar.

Similar pressure and flow waveforms are recorded in aortas of different mammalian species [26, 37] (Figure 1). This suggests that corresponding pulse transmission characteristics may also be similar. Nonuniformities in geometry and elasticity, as well as viscous damping, give rise to varying impedances to blood flow along the arterial tree. Pressure and flow pulses are, therefore, modified as they travel away from the heart and encounter mismatching of these impedances. Impedance to pulsatile flow is like resistance to steady flow and can be viewed as complex resistance that varies with frequency. Impedance is calculated as the complex ratio of pressure to flow for each harmonic, or multiples of heart rate. When the impedance is determined at the ascending aorta, or the entrance to the arterial tree, it is termed input impedance. Vascular input impedance (Z_{in}) can be used to characterize the global properties of the arterial system.

When the characteristic impedance of the proximal aorta (Z_o) is matched to the input impedance of the arterial tree, i.e., $Z_{in} = Z_o$, maximum transmission is present and reflection of the propagating pulse does not occur. Under this "matched impedances" condition, the pulsatile energy is totally transmitted to organ vascular beds. In normal physiological conditions, however, there is some mismatching of the impedances. This causes the reflection of the propagating pressure and flow pulses. The fraction of the propagating pulse that is reflected is given by the reflection coefficient [25], related to the impedances as

$$\Gamma = \frac{Z_{in} - Z_o}{Z_{in} + Z_o}. \tag{31}$$

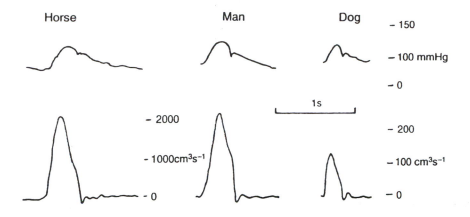

FIGURE 1   Simultaneously measured ascending aortic blood pressure and flow waveforms in three species of mammals.

The magnitude of the reflection coefficient at a normal resting heart rate is about 0.4, similar for many mammalian species (Table 4).

Pulse propagation characteristics [31] can be quantified with a propagation constant,

$$\gamma = \alpha + j\beta, \tag{32}$$

where α is the attenuation coefficient, describing pulse damping due to viscous losses, and β, the phase constant, denoting the relative amount of phase shift or pulse transmission time delay due to finite pulse propagation velocity, c. In the mammalian aorta, the pulse wave velocity is invariant, as seen from the allometric relation

$$c = \frac{\omega}{\beta} = 446\,M^0 \text{cm s}^{-1}, \tag{33}$$

where $\omega = 2\pi f_h$, f_h = heart rate (s^{-1}).

TABLE 4   Data for different mammals for analysis of arterial pulse transmission characteristics. $Z_o/R_s = 0.1$ is used in the calculation of Γ and γ.

	Body Mass M (Kg)	Heart Rate f_h (min^{-1})	Phase Velocity c (cm/s)	System Length l (cm)	Reflection Coefficient Γ	Γ (expt)	Propagation Constant $\times l$ γl
Horse	400	36	400	110	0.36	0.42	1.13
Man	70	70	500	65	0.38	0.45	1.06
Dog	20	90	400	45	0.39	0.42	1.01
Rabbit	3	210	450	25	0.41	0.48	0.93

To compare gross features of the arterial trees of different mammals, modeling approach can be particularly useful. The lumped modified windkessel model of the systemic arterial system (Figure 2) has been shown to represent well the features of the input impedance of the systemic arterial tree. For this representation, the input impedance is

$$Z_{in} = Z_o + \frac{R_s}{1 + j\omega C R_s} \tag{34}$$

dominated by Z_o; R_s, the systemic peripheral resistance as shown before; and C, the total systemic arterial compliance, representing the elastic storage properties of the arteries:

$$C = 0.18 \, 10^{-4} \, M^{0.95} \text{g cm}^4 \text{ s}^2 \,. \tag{35}$$

It is clear that the peripheral resistance decreases, while compliance increases with mammalian body size. Thus, the dynamic features of blood pressure and flow pulse transmission can be scaled through this kind of modeling. The ratio of Z_o/R_s corresponds to the ratio of pulsatile energy loss due to oscillatory flow to the energy dissipated due to steady flow (to overcome R_s), has been reported to be between 5–10%, and is an invariant for the mammalian arterial circulation [26, 28].

Some of the pulse transmission characteristics for horse, man, dog, and rabbit are summarized in Table 4. The ratio of pulse propagation wavelength, λ, for the fundamental harmonic, to the length of the aorta, l, equals about 6 and is independent of the body mass of the mammal. The product of λl is about 1, is again independent of the mammalian body mass, and confirms

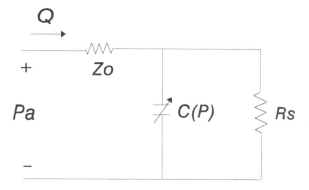

FIGURE 2   A linear lumped model of the mammalian arterial system. Z_o represents aortic characteristic impedance, C is arterial compliance and R_s is the peripheral vascular resistance. Pa and Q are aortic pressure and flow, respectively. A nonlinear model results when C is considered to be pressure-dependent, $C = C(P)$.

that the propagation characteristics along mammalian aortas are similar. The global reflection coefficient is also practically invariant. This occurs in spite of vast differences in the heart rate, systemic peripheral resistance, total systemic arterial compliance, and aortic characteristic impedance, that are associated with different body sizes.

These observed phenomena concerning pulse transmission, pulse wave velocity, and input impedance, as discussed above, must all be attributed to a common mechanism [30]. The architecture of the branching arterial junctions is such that only a portion of the pulse wave generated by the ventricle reaches the capillaries. Another part is reflected by the peripheral vessels, principally in the arteriolar beds. Reflected waves encounter mismatched branching sites on their return trip to the ventricle. As a result, a negligible fraction of the reflected pulse wave actually reaches the heart, with the exception of the fundamental frequency component for which the wavelength is longer than the effective length of the vascular system.

Another important feature of the optimal design of the mammalian arterial tree network is that there is minimal loss of pulsatile energy due to vascular branching [32]. The vascular junctions are practically impedance matched. In other words, the characteristic impedance of the mother vessel is closely matched to the branching daughter vessels. This implies that the geometric and elastic properties of the daughter vessels match that of the mother vessel. As such, pulse transmission at a vascular branching junction is met with minimal local reflection. This results in the facilitation of vascular perfusion with minimal energy loss en route to organ vascular beds.

7   THE NATURAL DESIGN CHARACTERISTICS

The cardiovascular systems of mammals exhibit amazingly similarity in both structural and functional design characteristics. A larger mammal has a larger heart size and is scaled with a greater number of fundamental building blocks, the sarcomeres or muscle cell units. These sarcomeres change their lengths to provide tension for developing an invariant magnitude of blood pressure necessary to propel blood to perfuse the vascular system. The cardiac efficiency and mechanical work intensity are both invariant features of the cardiac pump and only altered to meet changing vascular demands, such as activity. The mammalian cardiac energetics bears a constant relation to the whole body metabolism and this relation is maintained by changing heart rate to meet the demands of aerobic metabolism.

The geometrically tapered and branching design of the vascular system have nonuniform elastic properties. A larger mammal has greater aortic compliance to accommodate a larger stroke volume. The peripheral resistance is greater and closer to the heart as in smaller mammals and is scaled to maintain an inverse relationship to cardiac output. Invariant pulse transmission features are imbedded in the similar pulse pressure and flow waveforms observed at corresponding anatomical sites. The precision of natural design is even more amazing at vascular branching junctions, where branching vessel impedances are practically matched to ensure pulse wave transmission at utmost efficiency with minimal wave reflection and energy losses.

The optimality of the natural design characteristics of the mammalian cardiovascular system are such that many features are preserved to be invariant across species while similarities are governed by scaling laws.

REFERENCES

[1] Adolph, E. F. "Quantitative Relations in the Physiological Constitutions of Mammals." *Science* **109** (1949): 579.

[2] Bonner, J. T. *The Evolution of Complexity by Means of Natural Selection.* Princeton, NJ: Princeton University Press, 1983.

[3] Buckingham, E. "On Physically Similar Systems; Illustrations of the Use of Dimensional Equations." *Phys. Rev.* **4** (1915): 345.

[4] Burton, A. C. "Relation of Structure to Function of the Tissues of Walls of Blood Vessels." *Physiol. Rev.* **34** (1954): 619–642.

[5] Calder, W. A., III. "Scaling of Physiological Processes in Homeothermic Animals." *Ann. Rev. Physiol.* **43** (1981): 301.

[6] Calder, W. A., III. *Size, Function, and Life History.* New York: Dover, 1996.

[7] Clark, A. J. *Comparative Physiology of the Heart.* New York: Macmillan, 1927.

[8] Goetz, R. H., J. V. Warren, O. H. Gauer, J. L. Patterson, Jr., J. T. Doyle, E. N. Keen, and M. McGregor. "Circulation of the Giraffe." *Circ. Res.* **8** (1960): 1049–1058.

[9] Gunther, B. "Allometric Ratios, Invariant Numbers, and the Theory of Biological Similarity." *Physiol. Rev.* **55** (1975): 659.

[10] Gunther, B., and L. DeLa Barra. "Physiometry of the Mammalian Circulatory System." *Acta Physiol. Lat.-Am.* **16** (1966): 32.

[11] Gunther, B., and L. DeLa Barra. "Theories of Biological Similarities, Non-Dimensional Parameters and Invariant Numbers." *Bull. Math. Biophys.* **28** (1966): 9–102.

[12] Hales, S. *Statical Essays Containing Haemostaticks.* London, 1733.

[13] Harvey, W. *De Motu Cordis.* London, 1628. Reprinted. New York: Dover, 1995.

[14] Harvey, W. "Anatomical Exercises Concerning the Motion of the Heart and Blood in Living Creatures." Reprinted English translation. New York: Dover, 1995.

[15] Holt, J. P., E. A. Rhode, W. W. Holt, and H. Kines. "Geometric Similarity of Aorta, Venae Cavae, and Certain of Their Branches in Mammals." *Amer. J. Physiol.* **241** (1981): R100.

[16] Holt, J. P., E. A. Rhode, and H. Kines. "Ventricular Volumes and Body Weights in Mammals." *Amer. J. Physiol.* **215** (1968): 704.

[17] Huxley, J. S. *Problems of Relative Growth.* London: Methuen, 1932.

[18] Juznič, G., and H. Klensch. "Vergleichende Physiologische Untersuchunger über das Verhalten der Indices für Energieaufwand und Leistung des Herzens." *Arch. ges Physiol.* **280** (1964): 3845.

[19] Kenner, T. "Flow and Pressure in Arteries." In *Biomechanics*, edited by Y. C. Fung, N. Perroue, and M. Anliker. New York: Prentice-Hall, 1972.

[20] Kleiber, M. "Body Size and Metabolic Rate" *Physiol. Rev.* **27** (1947): 511–541.

[21] Lambert, R., and G. Teissier. "Théorie de la Similitude Biologique." *Ann. Physiol. Physiocochem. Biol.* **3** (1927): 212.

[22] Li, J. K-J. "A New Similarity Principle for Cardiac Energetics." *Bull. Math. Biol.* **45** (1983): 1005–1011.

[23] Li, J. K-J. "Hemodynamic Significance of Metabolic Turnover Rate." *J. Theor. Biol.* **103** (1983): 333–338.

[24] Li, J. K-J. "Comparative Cardiac Mechanics: Laplace's Law." *J. Theor. Biol.* **118** (1986): 339–343.

[25] Li, J. K-J. "Time Domain Resolution of Forward and Reflected Waves in the Aorta." *IEEE Trans. Biomed. Eng.* **BME-33** (1986): 783–785.

[26] Li, J. K-J. *Arterial System Dynamics.* New York: New York University Press, 1987.

[27] Li, J. K-J. "Laminar and Turbulent Flow in the Mammalian Aorta: Reynolds Number." *J. Theor. Biol.* **135** (1988): 409–414.

[28] Li, J. K-J. *Comparative Cardiovascular Dynamics of Mammals.* Boca Raton, FL: CRC Press, 1996.

[29] Li, J. K-J. "A New Approach to the Analysis of Cardiovascular Function: Allometry." In *Analysis and Assessment of Cardiovascular Function*, edited by G. Drzewiecki and J. K-J. Li, 13–29. New York: Springer-Verlag, 1998.

[30] Li, J. K-J., and A. Noordergraaf. "Similar Pressure Pulse Propagation and Reflection Characteristics in Aortas of Mammals." *Amer. J. Physiol.* **261** (1991): R519–R521.

[31] Li, J. K-J., J. Melbin, R. A. Riffle, and A. Noordergraaf. "Pulse Wave Propagation." *Circ. Res.* **49** (1981): 442–452.

[32] Li, J. K-J., J. Melbin, and A. Noordergraaf. "Directional Disparity of Pulse Wave Reflections in Dog Arteries." *Amer. J. Physiol.* **247** (1984): H95–99.

[33] Loiselle, D. S., and C. L. Gibbs. "Species Differences in Cardiac Energetics." *Amer. J. Physiol.* **237** (1979): H90–H98.

[34] Martin, R. R., and H. Haines. "'Application of Laplace's Law to Mammalian Hearts." *Comp. Biochem. Physiol.* **34** (1970): 959.

[35] McMahon, T. A. "Size and Shape in Biology." *Science* **179** (1973): 1201–1204.

[36] McMahon, T. A., and J. T. Bonner. "On Size and Life." New York: Scientific American Library, 1983.

[37] Noordergraaf, A., J. K-J. Li, and K. B. Campbell. "Mammalian Hemodynamics: A New Similarity Principle." *J. Theor. Biol.* **79** (1979): 485.

[38] Robard, S., F. Williams, and C. Williams. "The Spherical Dynamics of the Heart." *Am. Heart J.* **57** (1959): 348–360.

[39] Stahl, W. R. "Similarity Analysis of Biological Systems." *Persp. Biol. Med.* **6** (1963): 291.

[40] Stahl, W. R. "The Analysis of Biological Similarity." *Adv. Biol. Med. Phys.* **9** (1963): 356.

[41] Stahl, W. R. "Organ Weights in Primates and Other Mammals." *Science* **150** (1965): 1039–1042.

[42] Starling, E. H., and M. B. Visscher. "The Regulation of the Energy Output of the Heart." *J. Physiol.* **62** (1926): 243–261.

[43] Thompson, D. W. *On Growth and Form.* Cambridge, MA: Cambridge University Press, 1917.

[44] West, G. B., J. H. Brown, and B. J. Enquist. "A General Model for the Origin of Allometric Scaling Laws in Biology." *Science* **276** (1997): 122–126.

[45] White, L., H. Haines, and T. Adams. "Cardiac Output Related to Body Weights in Small Mammals." *Comp. Biochem. Physiol.* **27** (1968): 559–565.

[46] Woods, R. H. "A Few Applications of a Physical Theorem to Membranes in the Human Body in a State of Tension." *J. Anat. Physiol.* **26** (1892): 362–370.

Vascular System of the Human Heart: Some Branching and Scaling Issues

Mair Zamir

1 INTRODUCTION

The vascular system of the human heart, the "coronary network," is of particular interest because this network of vessels is responsible for blood supply to the heart for its own metabolism, and because failure of this supply is by far the most frequent cause of heart failure [2, 16, 21, 35, 45]. While the latter is usually mediated by a fairly complex vascular disease process, the ultimate effect remains simple: the heart fails because its vascular system fails to provide it with sufficient blood supply.

The coronary network is the most dense vascular network within the body, with the exception perhaps of the vascular systems of the lungs and kidneys, but these two systems have a predominately "processing" rather than metabolic function. The coronary network has a purely metabolic function, namely that of providing the heart muscle with sufficient blood supply for its own nourishment. Because of the intense energy requirements of this muscle, it is of particular interest to examine the way in which the coronary network meets the task of supplying it with blood and the way in which the supplying network is constructed.

The branching structure of blood vessels has been studied widely in terms of the efficiency of vascular networks as fluid flow systems. A vascular structure

Scaling in Biology, edited by J. H. Brown and G. B. West.
Oxford University Press, 2000.   **129**

which has been studied in particular is that of an open tree in which a single vessel undergoes a bifurcation into two daughter vessels, then each of these in turn undergoes a bifurcation and so on. In an idealized "uniform" and "complete" tree, all the bifurcations are symmetrical; that is, the two daughter branches at each bifurcation are of equal diameters, and branching extends equally far in all parts of the tree. These are basic fractal properties which have often been ascribed to vascular trees by theoretical studies [20, 34].

Arterial trees in the cardiovascular system, however, are found to be incomplete and highly nonuniform, though the dichotomy of arterial bifurcations is found to prevail widely. Trifurcations and other multiple-branch junctions are rare, thus the study of vascular trees is in large part a study of arterial bifurcations [38].

Optimality principles governing the geometry of arterial bifurcations have been formulated on the basis of minimizing pertinent properties of the bifurcation as a fluid flow conduit [24, 25, 33]. The result is a set of predictions regarding the diameters and angles of the two branches at an arterial bifurcation. These predictions have been tested against measured data and the results, on the whole, have been supportive but with considerable scatter, particularly in the case of branching angles [17, 22, 26, 42, 43, 44, 46, 47, 49]. Much of the data used in the past have been taken randomly from within one or more arterial trees, and in some cases within different organisms. Fully "mapped" data in which the hierarchic structure of a single arterial tree is determined and in which each bifurcation and the set of measurements taken from it are identified by the hierarchic position of the bifurcation on the tree structure, have only recently become available [19, 41]. These make it possible, for the first time, to examine the branching characteristics along different cross sections of an arterial tree, thereby providing more regional tests of branching laws within the tree structure.

Theoretical branching laws play an important role in the formulation of scaling laws in that they provide the link between scaling laws and biological function. Scaling laws for vascular trees are usually based on the assumption that the tree structure is uniform and complete [20, 34]. Comparison of measured data with these laws, therefore, provides a forum in which the relevant scaling laws can be discussed and in which the influence of function in a particular arterial tree can be assessed. In the coronary network, function plays a particularly important role, first because of the intensity and range of energy requirements of the heart muscle which the coronary vessels must provide for, and second because of the way these vessels are embedded within the myocardium and are subject to its violent contractions.

2   CORONARY NETWORK

Blood supply to the heart for the purpose of its own metabolism reaches it, ordinarily, via two branches of the ascending aorta, the first two branches

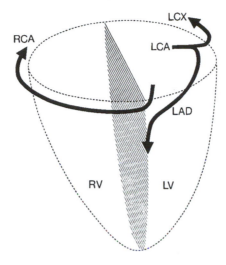

FIGURE 1   Schematic diagram of the way in which the left and right main coronary arteries (LCA,RCA) circle the heart and form the main "skeleton" of the coronary network. Adapted from Zamir [40]. Reprinted with permission from Springer-Verlag.

which the ascending aorta gives rise to as it just emerges from the left ventricle. The two vessels circle the left and right sides of the heart in the manner of a belt (Figure 1), giving rise to many branches in the process, and ultimately to a massive vascular network that reaches every part of the heart. The anatomy of these two vessels, generally known as the left and right main coronary arteries, and their major branches, have been described widely in the past, but mostly in relation to the anatomy of the heart [3, 13, 23]. The strategic design of the coronary network as a fluid conveying system and the manner in which it deals with the special needs of the heart has only recently been addressed [39, 40, 45]. Some of the issues involved are summarized below.

The first and most important issue is whether the coronary network has the form of an open tree structure or an interconnected mesh. Earlier thoughts on this subject favored the latter but considerable controversy ensued as observations led to contradictory and inconclusive results [3, 4, 5, 12, 14, 18, 29, 32]. Thinking then progressed to the notion that the underlying structure of the coronary network is that of an open tree but superimposed on it is a system of "collateral" vessels that provide alternate routes for blood flow in the event of an obstruction. The existence, topology, and functional significance of these vessels, however, has been surrounded by as much controversy as did the earlier concept, and some of this controversy persists today [1, 6, 7, 8, 9, 10, 11, 15, 28, 31]. Our work has led us to conclude that the coronary network has a decidedly open tree structure and that any collateral vasculature within it arises as an angiogenic response to ischemia on an *ad hoc* basis rather than as part of a permanent design [40].

The second important issue in the coronary circulation is how the left and right coronary arteries divide the myocardium between them (Figure 1). This is important since the left ventricle, which produces much of the pumping power of the heart, is supplied primarily by the left coronary artery with some assistance from the right coronary artery, but the extent of the latter is highly variable from one heart to another. The heart is said to be "left-dominant" when the left ventricle is supplied exclusively by the left coronary artery, "balanced" when the right coronary artery gives rise to the posterior descending branch which makes an important contribution to blood supply to the left ventricle, and "right-dominant" when the right coronary artery goes much beyond this contribution to supply a larger portion of the left ventricle. Approximately 70% of human hearts are right-dominant, 20% are balanced, and only 10% are left-dominant [23]. Thus in 90% of human hearts the left ventricle is supplied by both the left and right main coronary arteries, which affirms that this feature of the coronary network is the result of design rather than random anatomical variation.

An example of a right coronary artery from a right-dominant human heart [41] is shown in Figure 2 to illustrate its highly nonuniform and incomplete structure. The underlying tree structure is shown schematically in Figure 3 where we also indicate how the "levels" of the tree are defined in the present work.

3   BRANCHING AND SCALING LAWS

Branching laws have their origin in Murray's so-called cube law [24], which suggests that the flow rate q in a blood vessel varies as the cube of the diameter d of the vessel or, conversely, from the point of view of the design of a vascular tree, that the diameter d should vary as $q^{1/3}$. The law is based on a compromise between the rate of energy expenditure required to drive the flow and the metabolic rate of energy expenditure required for maintaining the volume of blood needed to fill the vessel [37]. It is assumed that the first of these is proportional to d^{-4} as in Poiseuille flow, and the second is proportional to the volume of the vessel and hence to d^2.

Applied to an arterial bifurcation where the diameters of the parent vessel and two branches are denoted by d_0, d_1, d_2, respectively, and where conservation of mass requires that flow rate in the parent vessel must equal the sum of flow rates in the two branches, the law provides an important relationship between the diameters of the three vessels at an arterial bifurcation, namely

$$d_0^3 = d_1^3 + d_2^3 . \tag{1}$$

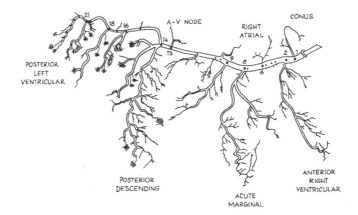

FIGURE 2   Some of the main branches which the right coronary artery gives rise to as it circles the heart (adapted from Zamir [41]). Numbers along the main vessel identify junctions as they arise sequentially along the course of the vessel. The numbers are related to the "levels" of the tree as illustrated schematically in Figure 3 and are useful for mapping branches arising at these junctions.

This law has been generalized by several authors in order to examine the consequences of other power laws and in order to deal with the scatter found in measured data [27, 30, 36]. The more general form can be written as

$$d_0^k = d_1^k + d_2^k \,, \tag{2}$$

where k is the power in the assumed relation between vessel diameter and flow rate, that is $q \propto d^k$.

Scaling laws, because of their concern with the global properties of a tree as a whole, have usually dealt with mostly symmetrical bifurcations where $d_1 = d_2 = d$ [34]. Branching laws, on the other hand, are actually concerned with the local asymmetry of arterial bifurcations. A measure of that asymmetry is the bifurcation index $\alpha = d_2/d_1$. If d_2 is always chosen to designate the smaller of the two diameters, the value of α lies conveniently between 0 and 1.0 for all possible degrees of asymmetry.

In terms of the power k and of the bifurcation index α, the diameter ratios at an arterial bifurcation become

$$\frac{d_1}{d_0} = \frac{1}{(1+\alpha^k)^{1/k}}, \quad \frac{d_2}{d_0} = \frac{\alpha}{(1+\alpha^k)^{1/k}}. \tag{3}$$

For a symmetrical bifurcation ($\alpha = 1.0$) these become

$$\frac{d_1}{d_0} = \frac{d_2}{d_0} = 2^{-1/k} \,, \tag{4}$$

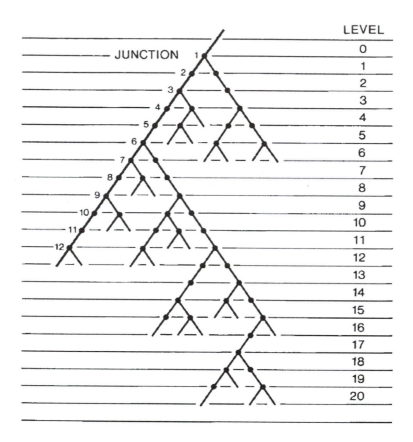

FIGURE 3   A tree structure shown schematically to illustrate the nonuniformity and incompleteness usually encountered in arterial trees. The figure also defines the "levels" of the tree as they are used in the present work. The junction numbers along the main trunk correspond to those in Figure 2 and are shown here to illustrate how they relate to the rest of the tree structure.

which in combination with the cube law ($k = 3$) give

$$\frac{d_1}{d_0} = \frac{d_2}{d_0} = 2^{-1/3} \tag{5}$$

and in combination with the square law ($k = 2$) give

$$\frac{d_1}{d_0} = \frac{d_2}{d_0} = 2^{-1/2}. \tag{6}$$

Thus, if a vessel undergoes n successive bifurcations which have the same degree of asymmetry $\alpha = d_2/d_1$, and if at each bifurcation we follow the

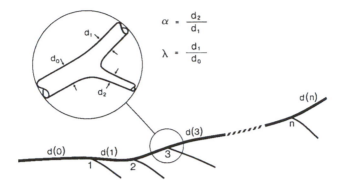

FIGURE 4 Notation for diameter ratios at an arterial bifurcation. Adapted from Zamir [39]. Reprinted with permission from The Rockefeller University Press.

branch with the larger diameter, namely d_1, then the diameter of the vessel will decrease according to (Eq. (3))

$$\frac{d(n)}{d(0)} = \frac{1}{(1+\alpha^k)^{n/k}},\tag{7}$$

where $d(0), d(n)$ denote the diameters of the vessel at the start and after n bifurcations, respectively, as illustrated in Figure 4.

It is clear that if α is small, the diameter of the vessel will not diminish as rapidly as it does when α is near 1.0. This situation is depicted in Figure 5, based on the cube law, compared with data from the right coronary artery.

The figure indicates clearly that many bifurcations are nonsymmetrical, but many others, particularly at higher levels of the tree or at smaller branches, have values of α near 1.0. The figure also affirms the difference between delivering and distributing vessels, which has been noted previously [39]. The top four curves which begin with smaller values of α represent the diameter gradients of four major distributing vessels: the right main coronary artery and three of its major branches at junctions 2, 8, and 14 (Figure 2), while the other curves represent diameter gradients of smaller delivering vessels.

Scaling laws, which must usually be based on symmetrical bifurcations, would correspond to the bottom curve in Figure 5, for which $\alpha = 1.0$. If the square law is used instead of the cube law, the curve would be modified somewhat to that marked by a dotted line. Both curves are closer to the delivering category of vessels with values of α near 1.0 than to the distributing group with values of α near zero. The difference between the cube and square laws

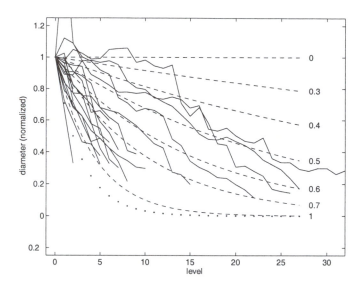

FIGURE 5 Change in normalized diameter along the right coronary artery and some of its main branches. The dashed lines in the background follow the cube law (Eqs. (3) and (7)), with values of α as indicated at the right. The dotted curve at the bottom corresponds to the square law and symmetrical bifurcations, that is $\alpha = 1.0$.

is fairly small compared with the much larger difference between symmetrical ($\alpha = 1.0$) and nonsymmetrical ($\alpha < 1.0$) bifurcations.

4   AREA EXPANSIONS

An important property of arterial bifurcations and hence of arterial trees is the area ratio β which represents a measure of expansion in the cross-sectional area available to the flow at each bifurcation and hence in the tree as a whole

$$\beta = \frac{d_1^2 + d_2^2}{d_0^2}. \tag{8}$$

An increase in cross-sectional area is associated with a decrease in flow velocity, and vice versa; hence, values of β at different levels of the arterial tree have functional significance.

In terms of the power k and bifurcation index α the area ratio is, in general, given by

$$\beta = \frac{1 + \alpha^2}{(1 + \alpha^k)^{2/k}}. \tag{9}$$

For a symmetrical bifurcation this becomes

$$\beta = 2^{1-2/k}, \tag{10}$$

which in combination with the cube law ($k = 3$) gives $\beta = 2^{1/3}$, and in combination with the square law ($k = 2$) it gives $\beta = 1$.

Comparison with data taken from the right coronary artery of Figure 2 is shown in Figure 6(a). It is again observed that nonsymmetrical bifurcations are highly prevalent along the full range of α. Despite this, however, comparison of branching and scaling laws with data and with each other is meaningful since the difference between the cube and square laws is small compared with the range of scatter of the data.

Also, and perhaps most important, the data indicates that symmetrical and nonsymmetrical bifurcations exhibit essentially the same range of area ratios. Thus branching and scaling laws based on symmetrical bifurcations are relevant despite the nonsymmetrical nature of bifurcations in the cardiovascular system.

Finally, it has been suggested that the square law may prevail only at the first few levels of the arterial tree, changing over to the cube law thereafter [34, 48]. The data shown in Figure 6(a) comes from all levels of the coronary tree and it is not clear here whether there is a change from one law to the other. For this purpose we show in Figures 6(b)–(f) the same data divided according to branching levels of the tree. There is some trend toward the square law in Figure 6(b), which corresponds to the first 10 levels of the tree, but the trend is fairly weak and far from conclusive. At other levels of the tree (Figures 6(c)–(f)) and in the tree as a whole (Figure 6(a)) regression lines fall insipidly between the cube and square laws and, as remarked earlier, the difference between them is very small compared with the scatter in the data.

An important functional difference between the square and cube laws is that the first implies constant velocity of the flow as it progresses from central to peripheral levels of the arterial tree, while the second implies a diminishing velocity in that direction. Thus the square law is untenable on physiological grounds since flow in the capillaries is required to move much more slowly than flow in the aorta. The suggestion that the square law may prevail at some central levels of the arterial tree and then give way to the cube law [34, 48] resolves this difficulty on theoretical grounds, but more data is required to establish the actual existence of this transition in the arterial tree. The present results provide no conclusive evidence for a transition, but it must be remembered that these results derive entirely from the coronary network which may not be representative of the arterial tree as a whole.

ACKNOWLEDGMENTS

This work was supported by the Natural Sciences and Engineering Research Council of Canada.

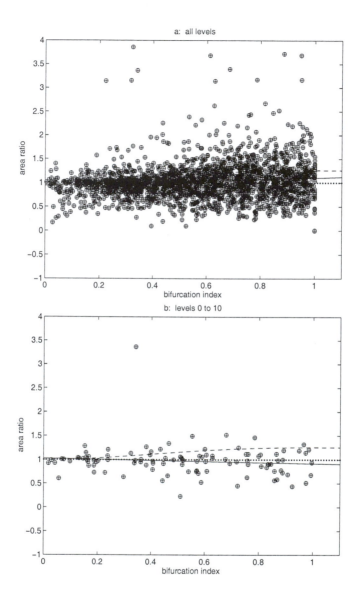

FIGURE 6   Bifurcation index (α) and area ratio (β) as measured at different levels of the tree structure of the right coronary artery, as indicated at the top of each figure, where α is a measure of local asymmetry at arterial bifurcations and β is a measure of change in cross-sectional area, as defined in the text. Dashed and dotted lines in the background represent the relation between α and β according to the cube and square laws, respectively, while the solid line represents simple regression of the measured data points.

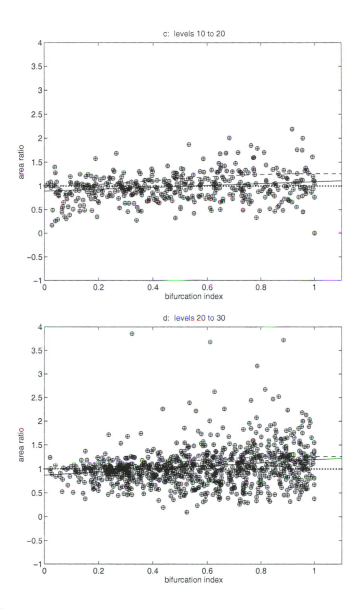

FIGURE 6   Continued.

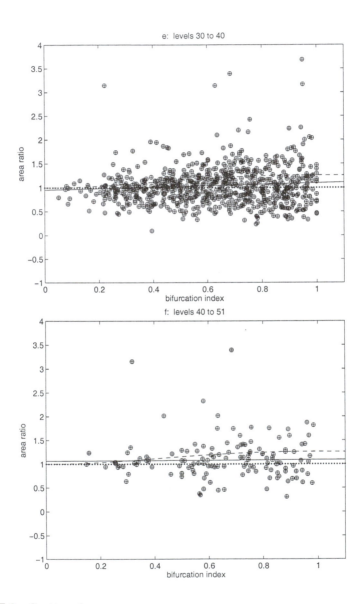

FIGURE 6   Continued.

REFERENCES

[1] Anastasiou-Nana, M., J. N. Nanas, R. B. Sutton, and T. J. Tsagaris. "Left Main Coronary Artery Occlusion." *Cardiology* **72** (1985): 208–213.

[2] Baroldi, G. "Diseases of the Coronary Artery." In *Cardiovascular Pathology*, edited by M. D. Silver. Baltimore, MD: University Park Press, 1983.

[3] Baroldi, G., and G. Scomazzoni. *Coronary Circulation in the Normal and Pathologic Heart.* Washington, DC: Armed Forces Institute of Pathology, 1967.

[4] Baroldi, G., O. Mantero, and G. Scomazzoni. "The Collaterals of the Coronary Arteries in Normal and Pathologic Hearts." *Circul. Res.* **4** (1956): 223–229.

[5] Blumgart, H. L., P. M. Zoll, A. S. Freedberg, and D. R. Gilligan. "The Experimental Production of Intercoronary Arterial Anastomoses and Their Functional Significance." *Circulation* **1** (1950): 10–27.

[6] Boxt, L. M., and D. C. Levin. "A Primer of Coronary Angiography." *Cardiovascular Medicine* **August** (1985): 21–28.

[7] Brazzamano, S., J. M. Fedor, J. C. Rembert, and J. C. Greenfield, Jr. "Increase in Myocardial Collateral Blood Flow During Repeated Brief Episodes of Ischemia in the Awake Dog." *Basic Res. Cardiol.* **79** (1984): 448–453.

[8] Cohen, M. V. "The Functional Value of Coronary Collaterals in Myocardial Ischemia and Therapeutic Approach to Enhance Collateral Flow." *Am. Heart J.* **95** (1978): 396–404.

[9] DeWood, M. A., W. F. Stifter, C. S. Simpson, J. Spores, G. S. Eugster, T. P. Judge, and M. L. Hinnen. "Coronary Arteriographic Findings Soon After Non-q-Wave Myocardial Infarction." *N. Eng. J. Med.* **315** (1986): 417–423.

[10] Elayda McArthur, A., V. S. Mathur, R. J. Hall, G. A. Massumi, E. Garcia, and C. M. de Castro. "Collateral Circulation in Coronary Artery Disease." *Am. J. Cardiol.* **55(1)** (1985): 58–60.

[11] Eng, C., and E. S. Kirk. "Flow Into Ischemic Myocardium and Across Coronary Collateral Vessels Is Modulated by a Waterfall Mechanism." *Circul. Res.* **56** (1985): 10–17.

[12] Fulton, W. F. M. *The Coronary Arteries.* Springfield, IL: C. C. Thomas, 1965.

[13] Gensini, G. G. *Coronary Arteriography.* New York: Futura Mount Kisco, 1975.

[14] Gensini, G. G., and B. C. B. Da Costa. "The Coronary Collateral Circulation in Living Man." *Am. J. Cardiol.* **24** (1969): 393–400.

[15] Gottwik, M. G., S. Puschmann, B. Wusten, C. Nienaber, K.-D. Muller, M. Hofmann, and W. Schaper. "Myocardial Protection by Collateral Vessels During Experimental Coronary Ligation: A Prospective Study in a Canine Two-Infarction Model." *Basic Res. Cardiol.* **79** (1984): 337–343.

[16] Gregg, D. E. *Coronary Circulation in Health and Disease.* Philadelphia, PA: Lea & Febiger, 1950.

[17] Hutchins, G. M., M. M. Miner, and J. K. Boitnott. "Vessel Caliber and Branch-Angle of Human Coronary Artery Branch-Points." *Circul. Res.* **38** (1976): 572–576.

[18] James, T. N. "The Delivery and Distribution of Coronary Collateral Circulation." *Chest* **58** (1970): 183–203.

[19] Kassab, G. S., C. A. Rider, N. J. Tang, Y. C. Fung, and C. M. Bloor. "Morphometry of Pig Coronary Arterial Trees." *Am. J. Physiol.* **265** (1993): H350–H365.

[20] Mandelbrot, B. B. *Fractals: Form, Chance, and Dimensions.* San Francisco, CA: Freeman, 1977.

[21] Marcus, M. L. *The Coronary Circulation in Health and Disease.* New York: McGraw-Hill, 1983.

[22] Mayrovitz, H. N., and J. Roy. "Microvascular Blood Flow: Evidence Indicating a Cubic Dependence on Arteriolar Diameter." *Am. J. Physiol.* **14** (1983): H1031–H1038.

[23] McAlpine, W. A. *Heart and Coronary Arteries.* New York: Springer-Verlag, 1975.

[24] Murray, C. D. "The Physiological Principle of Minimum Work. I. The Vascular System and the Cost of Blood Volume." *PNAS* **12** (1926): 207–214.

[25] Murray, C. D. "The Physiological Principle of Minimum Work Applied to the Angle of Branching of Arteries." *J. Gen. Physiol.* **9** (1926): 835–841.

[26] Rodbard, S. "Vascular Caliber." *Cardiology* **60** (1975): 4–49.

[27] Roy, A. G., and M. J. Woldenberg. "A Generalization of the Optimal Models of Arterial Branching." *Bull. Math. Biol.* **44** (1982): 349–360.

[28] Schaper, W. *The Collateral Circulation of the Heart.* Amsterdam: North Holland, 1971.

[29] Schlesinger, M. J. "An Injection Plus Dissection Study of Coronary Artery Occlusions and Anastomoses." *Am. Heart. J.* **15** (1938): 528–568.

[30] Sherman, T. F. "On Connecting Large Vessels to Small. The Meaning of Murray's Law." *J. Gen. Physiol.* **78** (1981): 431–453.

[31] Sjoquist, P.-O., G. Duker, and O. Almgren. "Distribution of the Collateral Blood Flow at the Lateral Border of the Ischemic Myocardium After Acute Coronary Occlusion in the Pig and the Dog." *Basic Res. Cardiol.* **79** (1984): 164–175.

[32] Spalteholz, W. "Die Coronararterien des Herzen." Verhandlungen der Anatomischen Gesellschaft (Supplement to Anatomischer Anzeiger) 21st Meeting at Wurzburg, 141–153. Jena: Verlag von Gustav Fischer, 1907.

[33] Thompson, D'A. W. *On Growth and Form.* Cambridge, UK: Cambridge University Press, 1942.

[34] West, G. B., J. H. Brown, and B. J. Enquist. "A General Model for the Origin of Allometric Scaling Laws in Biology." *Science* **276** (1997): 122–126.

[35] Willerson, J. T., L. D. Hillis, and L. M. Buja. *Ischemic Heart Disease. Clinical and Pathological Aspects*. New York: Raven Press, 1982.

[36] Woldenberg, M. J., and K. Horsfield. "Finding the Optimal Length for Three Branches at a Junction." *J. Theor. Biol.* **104** (1983): 301–318.

[37] Zamir, M. "Shear Forces and Blood Vessel Radii in the Cardiovascular System." *J. Gen. Physiol.* **69** (1977): 449–461.

[38] Zamir, M. "Nonsymmetrical Bifurcations in Arterial Branching. *J. Gen. Physiol.* **72** (1978): 837–845.

[39] Zamir, M. "Distributing and Delivering Vessels of the Human Heart." *J. Gen. Physiol.* **91** (1988): 725–735.

[40] Zamir, M. "Flow Strategy and Functional Design of the Coronary Network." In *Coronary Circulation*, edited by F. Kajiya, G. A. Klassen, J. A. E. Spaan, and J. I. E. Hoffman. Tokyo: Springer-Verlag, 1990.

[41] Zamir, M. "Tree Structure and Branching Characteristics of the Right Coronary Artery in a Right-Dominant Human Heart." *Can. J. Cardiol.* **12** (1996): 593–599.

[42] Zamir, M., and N. Brown. "Arterial Branching in Various Parts of the Cardiovascular System." *Am. J. Anat.* **163** (1982): 295–307.

[43] Zamir, M., and H. Chee. "Branching Characteristics of Human Coronary Arteries." *Can. J. Physiol. Pharmacol.* **64** (1986): 661–668.

[44] Zamir, M., and J. A. Medeiros. "Arterial Branching in Man and Monkey." *J. Gen. Physiol.* **79** (1982): 353–360.

[45] Zamir, M., and M. D. Silver. "Morpho-Functional Anatomy of the Human Coronary Arteries with Reference to Myocardial Ischemia." *Can. J. Cardiol.* **1** (1985): 363–372.

[46] Zamir, M., J. A. Medeiros, and T. K. Cunningham. "Arterial Bifurcations in the Human Retina." *J. Gen. Physiol.* **74** (1979): 537–548.

[47] Zamir, M., S. Phipps, B. L. Langille, and T. H. Wonnacott. "Branching Characteristics of Coronary Arteries in Rats." *Can. J. Physiol. Pharmacol.* **62** (1984): 1453–1459.

[48] Zamir, M., P. Sinclair, and T. H. Wonnacott. "Relation Between Diameter and Flow in Major Branches of the Arch of the Aorta. *J. Biomechanics* **25** (1992): 1303–1310.

[49] Zamir, M., S. M. Wrigley, and B. L. Langille. "Arterial Bifurcations in the Cardiovascular System of a Rat." *J. Gen. Physiol.* **81** (1983): 325–335.

Constrained Constructive Optimization of Arterial Tree Models

Wolfgang Schreiner
Rudolf Karch
Friederike Neumann
Martin Neumann

1   INTRODUCTION

The arterial systems in mammals are essentially networks of binary branching trees, made up of vascular segments. Successive branching gives rise to self-similar structures, and arterial models have indeed been established on the basis of self-similarity. It is, therefore, of particular interest to present another modeling method here, which is also capable of generating arterial models with several thousands of branchings but is based on optimization principles rather than fractal generators. A highly fruitful comparison can be envisaged.

Basically the arterial system of any mammal fulfills the task of conveying blood from the heart to all parts of the body and delivering it to the tissue. Due to the vital importance of circulation, the necessity of its continuous non-interrupted operation, and the fact that maintaining the circulation consumes substantial energy, we can well assume that optimization of the binary branching arterial networks has taken place during phylogenesis.

The heart itself is the most complex part of the circulatory system. It provides the pumping action necessary to maintain the whole body's circulation and is at the same time also an organ depending on the support by that very circulation. This fact implies the dual role of the myocardial muscle, involving a complex relationship between contractility and perfusion. As a con-

Scaling in Biology, edited by J. H. Brown and G. B. West.
Oxford University Press, 2000.   **145**

sequence, numerous efforts have been made over the last decades to scrutinize the hemodynamic mechanisms involved and to provide quantitative models for them [6, 7, 8, 9, 10, 15, 38].

For any quantitative treatment of arterial hemodynamics by means of computer simulation, the establishment of an appropriate model is the first necessary step. Several approaches are possible:

1. *Compartment models*: Large groups of vessels (arteries, microvessels, veins) are lumped together into compartments, each of which is then characterized globally, e.g., by a global resistance and pressure-volume relation [2, 14].
2. *Anatomical models*: Specific parts of arterial trees are modeled in close relationship to anatomical parameters obtained by quantitative measurements [21], possibly also including lesions (stenoses).
3. *Optimization models*: Physiological principles regarding the functional optimality of the circulatory system are formulated mathematically and serve as a basis for generating optimized structures on the computer. A recent approach is "Constrained Constructive Optimization" (CCO) capable of generating very realistic models of arterial trees involving several thousand vascular segments [22, 23].
4. *Fractal models*: Based on a generating law, self-similar models are constructed over successive orders of bifurcations, the generating laws either being constant or stochastic in nature. The choice of generators can either be based on the statistical evaluation of experimental data (e.g., morphometric results obtained from corrosion casts) [1, 5, 32] or they can be derived from optimization principles [33, 34].

In this chapter we focus on the CCO algorithm, refer to comparisons between models and measurements, outline several modalities under which CCO can be performed and, finally, relate CCO models to fractal concepts generally relevant for scaling in biology.

2   METHOD OF CONSTRAINED CONSTRUCTIVE OPTIMIZATION (CCO)

The procedure starts from a geometrically defined perfusion area, representing a part of tissue to be supplied with blood, see Figure 1. The arterial tree model will be composed of a binary branching network of straight cylindrical tubes, perfused according to Poisseuille's law [8]. Each terminal segment is supposed to supply the same flow at the same pressure into its respective microcirculatory black box (see the shaded areas in Figure 1).

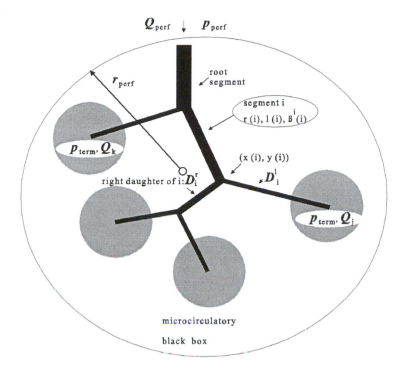

FIGURE 1 Schematic of CCO, showing perfusion through the root segment and blood delivery by terminal segments at four randomly chosen locations within a given perfusion area.

2.1 SCALING

Obviously, flow through any bifurcation must fulfill the continuity constraint

$$Q_0 = Q_1 + Q_2\,, \tag{1}$$

where the subscripts 0, 1, and 2 denote the parent and the larger and the smaller daughter segments of the bifurcation, respectively. Presetting the terminal flows ("terminal flow constraint") implies, due to the continuity condition [8], that flows in all segments of the binary branching tree of given topology (connective structure) are determined.

If, in addition to predefined flows, radii of parent and daughter segments at each bifurcation are forced to obey a "bifurcation law"

$$r_0^\gamma = r_1^\gamma + r_2^\gamma\,, \tag{2}$$

all ratios of radii become determined throughout the tree. The appropriate choice of γ has been thoroughly discussed in the literature [38, 39], and theoretical arguments as well as experimental measurements indicate that only values in the range $2 \leq \gamma \leq 3$ are physiologically reasonable, CCO, being capable of handling arbitrary values, is run for several choices, see Section 2.5 below.

If, finally, the whole tree is to conduct a certain total flow over a given pressure difference between perfusion pressure and terminal pressure, absolute values result for the root radius and (through the defined ratios) for all radii.

It is noteworthy that the above "scaling" procedure is feasible for any dichotomously branching tree, regardless of its connective or geometrical structure [33].

2.2 GEOMETRIC OPTIMIZATION

For a given tree, scaled as described above, a target function is evaluated according to

$$
T = \sum_{i=1}^{N_{\text{tot}}} \ell_i \, r_i^{\lambda} \rightarrow \min \,,
\tag{3}
$$

which globally characterizes the "optimality of the tree," e.g., by being proportional to the total volume for $\lambda = 2$. Any geometrical displacement of a bifurcation, in spite of unchanged connective structure, generally changes the value of the target function since lengths ℓ_i change and radii have to be adjusted so as to maintain the constraints (which is always possible, as explained above). On this basis, a gradient method can be used to optimize the geometrical position of each bifurcation in such a tree until convergence is obtained.

2.3 STRUCTURAL OPTIMIZATION AND GROWTH

Suppose a new terminal segment is to be connected to an existing CCO tree. To which of the existing segments will this connection be made? The answer lies in a procedure called "connection search": Tentative connections are performed to all (reasonable) segment candidates, and the best one (the optimum one in terms of the target function) is finally adopted to become permanent. Each tentative connection is geometrically optimized, the value of the target function recorded, and the connection dissolved again. Since the "connection search" represents structural optimization, geometrical optimization is nested within structural optimization, see Figure 2.

Repeating this process of growth—encompassing the operations of scaling, geometric optimization, and connection search—enables the CCO algorithm, similar to the mathematical principle of induction, to generate highly detailed tree structures with large numbers of segments according to optimization principles.

Geometric Optimization is nested **structural**   **Result:**
for a given structure within **optimization**   **optimum**
                                                          connection

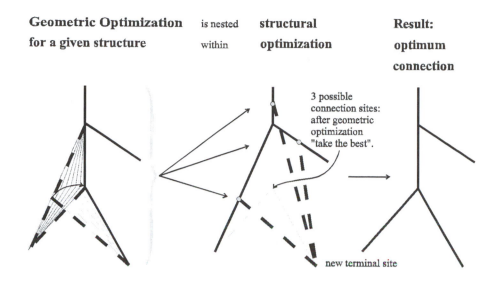

3 possible
connection sites:
after geometric
optimization
"take the best".

new terminal site

FIGURE 2   Growth of a CCO model via structural optimization. The optimum topology is found by test-wise, transient connections, for each of which full geometric optimization is performed. The best candidate is finally selected.

2.4   COMPARISONS WITH EXPERIMENTAL DATA

Models generated by CCO have been evaluated with regard to several predicted physiological and morphometric properties such as branching angles, pressure profile, branching asymmetry, and the relationship between segment radii and bifurcation level [3, 11, 19, 20, 22, 24, 26, 27, 28]. In general, satisfactory agreement with measurements reported in the literature was found. This is remarkable since no anatomic or morphometric information is plugged into the CCO algorithm. The geometry and structure of CCO models are generated solely on the basis of optimization principles embedded within appropriate boundary conditions.

2.5   DIFFERENT MODALITIES OF CONSTRAINED CONSTRUCTIVE OPTIMIZATION

As a most simple reference case, CCO can be applied to generate an optimized tree within a circular (two-dimensional) perfusion area, with the geometry and connective structure being optimized according to minimum intravascular blood volume, see Figure 3.

The CCO algorithm allows for several modalities regarding the way optimized trees are generated. The first type of modality is the use of different pseudo-random number-sequences (PRNSs) for generating the locations of

FIGURE 3   CCO tree with 4,000 terminal segments generated in a two-dimensional perfusion area, optimized for minimum intravascular volume at bifurcation exponent $\gamma = 3$; no minimum symmetry required (see below).

the terminal segments. Each PRNS induces a different topological structure although the tree is optimized according to the same target function and meets the very same boundary conditions of pressures, flows, and the bifurcation law. Figure 4 shows two representatives obviously differing in "connective structure," although the "character" of these tree models is very similar. Variations between CCO models, which are only due to different PRNSs, can be correlated to the anatomical variability of (cardio) vascular trees in different individuals of the same species [18, 28, 36]. It was surprising to find that significantly different anatomical structures do not necessarily induce drastic differences in properties intimately related to blood flow. This agrees with results reported by Zamir [38].

Another simple type of modality for CCO concerns the shape of the perfusion volume. It is obvious from Figure 5 that different shapes of the perfusion areas induce different branching patterns ("anatomies"). Quantitative evaluations have shown that not only the shape of tissue to be perfused, but that the location of the feeding vessel also influences the degree of optimality which can be achieved: for some shapes the optimized tree needs slightly more intravascular volume than for other shapes, despite the same total flow being

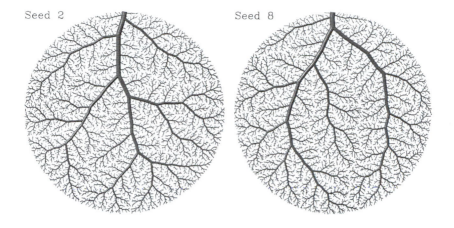

Seed 2   Seed 8

FIGURE 4   CCO trees generated with different pseudo-random number-sequences ("seeds"); 4,000 terminal segments; optimization for minimum intravascular volume; $\gamma = 3$.

distributed over equal areas. Moreover it is not only plausible but has also been quantitatively demonstrated that morphometric parameters (e.g., measures of bifurcation symmetry) depend on the shape of the perfusion area as well as on the site of inlet [26]. It should be noted that the present implementation of CCO requires the perfusion volume to be either entirely convex or exhibit one nonconvex zone at the most. In that case (see Figure 5, upper left panel) the feeding vessel (root segment) must enter at that very site.

Another parameter of CCO that can be varied is the bifurcation exponent, γ, governing the shrinkage of radii from parent to daughter segments (cf. Eq. (2)). In the literature, the choice of $\gamma = 3$ is considered reasonable by many authors, although smaller values (e.g., $\gamma = 2.55$) are also considered optimum regarding pulsewave transmission in distensible tree models [4, 16, 29, 32]. Note that $\gamma = 2$ implies that the cross-sectional area of a parent vessel equals the sum of the cross sections of both daughters, which further implies equal mean flow velocities throughout all successive generations of segments (from large to small segments). Values $\gamma > 2$ imply that the total cross-sectional area increases from parents to daughters and, hence, flow velocity decreases on the average. Figure 6 shows two CCO trees, each generated for a different value of γ. It is obvious that small values of γ (left panel) imply large cross sections of the big vessels as compared to small ones. This relation is inversed as γ increases (right panel). Thus, different values of γ have a distinct impact on the distribution of flow velocity and the pressure profile from the root to the terminal segments. As opposed to that, γ has but a low influence on the connective structure developing under CCO (the large vessels in both panels of Figure 6 exhibit fairly similar anatomical courses).

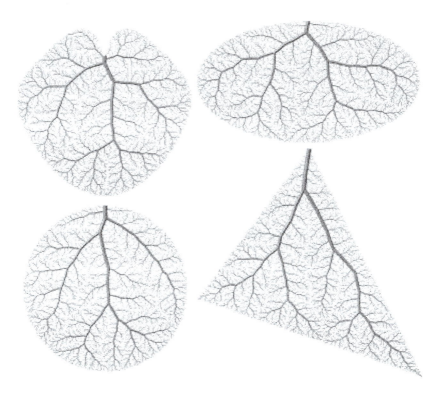

FIGURE 5   CCO trees spreading over perfusion areas with different shapes. Parameters as in Figures 3 and 4.

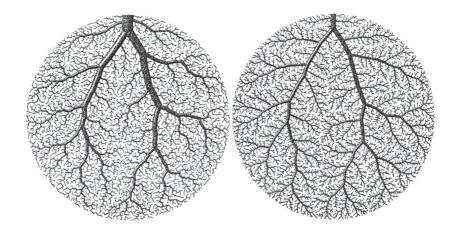

FIGURE 6   CCO trees optimized under different bifurcation laws: $\gamma = 2.1$ (left panel) and $\gamma = 2.3$ (right panel); target function parameter $\lambda = 2$ in both trees.

Another interesting effect of different choices of γ relates to the shear stress between blood and the vessel walls. It can be mathematically derived that equal shear stress in each parent and its daughter segments (which in fact implies equal shear stress all over the tree) is only possible with $\gamma = 3$ (see, Zamir [37]). This is, however, only a necessary but not a sufficient condition. What is needed in addition, is a flow splitting at bifurcations exactly adapted to segment radii so as to make $Q_i/r_i^3 = \text{const}$. This can in fact only be achieved by appropriately balancing the resistances of the subtrees distal to both daughters. If this is not provided for, even with $\gamma = 3$ the shear stress will generally be different in parent and daughter segments. Accordingly, Figure 7 shows the shear stress averaged within Strahler orders [1, 30] of CCO trees, each curve corresponding to one specific Strahler order. A set of 21 CCO trees was generated, each with a different value of $\gamma = 2.0, 2.05, 2.10, \ldots, 3.0$, and shear stress was averaged as outlined above. The result was that for low values of γ there is a large spread of shear stress between Strahler orders within one and the same tree (left part of the plot in Figure 7). Low Strahler orders, in particular the terminal segments with Strahler order 0, experience large shear stress while the large segments (large Strahler orders) experience low shear stress.

This is plausible since for $\gamma = 2$ the average cross-sectional area is constant and so is average flow velocity, giving rise to low shear stress in the large vessels as compared to the small ones. As γ increases, the spread of shear stress within trees diminishes and passes through a minimum around $\gamma = 3$. Increasing $\gamma > 3$ reshuffles large values of shear stress from the smaller into the larger segments and thus the shear-wise ranking between large and small segments is reversed. Concomitantly, spread again increases toward $\gamma = 4$.

This trend is again plausible, since the total cross section of large segments decreases whereas that of small segments increases on the average as γ increases. This result also demonstrates that the choice of $\gamma = 3$ is not only a necessary condition for uniform shear stress in case the flows are adjusted to these ends, but that $\gamma = 3$ also marks the region where the shear stress ranking of segments is reversed and the spread minimized.

Another modality of CCO that can easily be varied is the target function to be minimized during geometric and structural optimization. For generating the reference tree (Figure 3) intravascular volume was minimized, which by many authors is considered a reasonable choice since blood is a costly substance to maintain [12, 16, 17, 31]. However, we may as well minimize total intravascular surface, total vascular length or even total vascular hypervolume [25], simply by inserting different exponents λ into Eq. (3). As shown in Figure 8, the resulting structures are totally different in visual appearance. For $\lambda = 0$ segment radius does not play any role in optimization, and, as a consequence, large vessels are allowed to develop just as curvy as small ones (upper left panel of Figure 8). The larger the exponent λ of the radius, the more optimization weight is put on large caliber vessels which, as a result, become more and more streamlined (lower right panel in Figure 8).

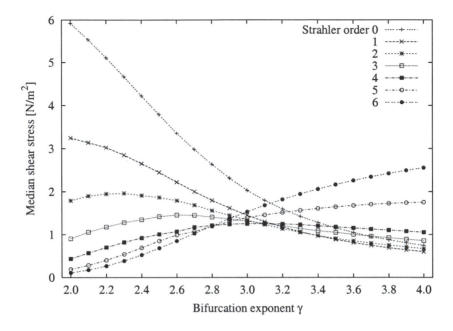

FIGURE 7   Spread of shear stress dependent on the bifurcation law. x-axis: Bifurcation exponent γ, see Eq. (2). For each value of γ a separate CCO tree was generated. y-axis: Shear stress. Each data point represents the shear stress averaged over all segments belonging to a specific Strahler order (see figure legend) in the respective tree.

A quantity directly related to curviness is the pathlength to reach a given geometric destination within the perfusion area through the vascular tree model, see the vessel paths shown in solid fill in Figure 8. Naturally this pathlength decreases as curviness is reduced and detours diminish. In physiological terms a straight path relates to a short distance access for any biochemical substance (e.g., a messenger) to be carried by blood into the tissue. However, some tradeoff between a short distance for access and the intravascular volume has to be made; a slightly larger volume is necessary to permit a shorter distance for access, as shown in Figure 9.

As a further illustration of CCO modalities, we focus on the relation between bifurcation symmetry and the shrinkage of segment radii.

In his morphometric analysis of human coronary arteries, Zamir [37] classifies vessels as "distributing" if they carry blood across large distances, on their way loosing comparatively little flow into side branches. These side branches are rather small, while the "distributing" vessels themselves decrease only slightly in radius across such (very asymmetric) bifurcations. Plotting segment radii (normalized with respect to the root segment, r_0) against the

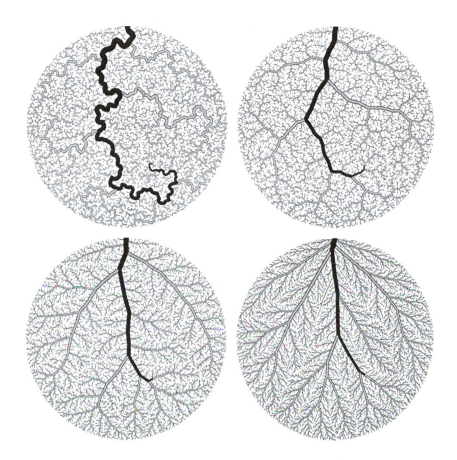

FIGURE 8 Dependence of structure on optimization target. Minimum total segment length (upper left), minimum total surface (upper right), minimum volume (lower left), and minimum hypervolume (lower right panel) with values of $\lambda = 0, 1, 2$, and 3. In each tree the path to the very same, deliberately chosen geometrical location is outlined in solid fill.

bifurcation level (n) thus yields a curve with a modest decrease, which can be quantified by the "branching rate"

$$b^n = r_n/r_0. \tag{4}$$

b is found by a least squares fit to measured radii along a path over n successive bifurcations. However, there are also vessels within the coronary vasculature which bifurcate much more symmetrically into side branches, and even the larger daughter thereby looses significantly in radius. These vessels

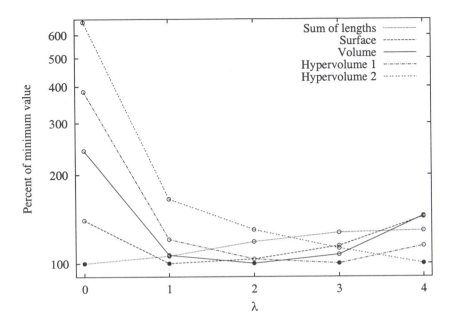

FIGURE 9   Relative magnitude of global characteristics depending on the optimization target. x-axis: Optimization target parameter λ, see Eq. (3). y-axis: Percentage of global quantity, relative to the respective minimum value. The four characteristics, total segment length, surface, volume, and hypervolume ($\lambda = 0, 1, 2, 3$) were calculated, while optimization was based on the quantity defined by the value of λ shown on the x-axis. Evidently, each global quantity must assume its minimum value when being an optimization target at the same time.

are called "delivering" and yield lower values of b upon the same kind of analysis, see the two panels in Figure 10 reproduced from Zamir's work.

Within the CCO algorithm, there is a very straightforward concept of provoking the generation of either "distributing" or "delivering" vessels. Within the connection search, each tentative connection is not only evaluated according to the optimization target but also with regard to the "individual branching rate" of the new bifurcation generated. This is achieved by simply imposing a lower limit on the degree of symmetry via $r_2/r_1 \geq \xi^{low}$. Those tentative connections which fail to pass the criterion are precluded from becoming permanent. The impact on structure is striking, see Figure 11 with $\xi^{low} = 0.4$, as compared to Figure 3 with $\xi^{low} = 0.0$.

The "branching rate" for this CCO tree (in comparison to the tree in Figure 3) shows a much more rapid decline in radius when proceeding from the root toward the periphery along the "main path" (i.e., always entering the larger daughter at bifurcations). Curbing asymmetry tunes the CCO model

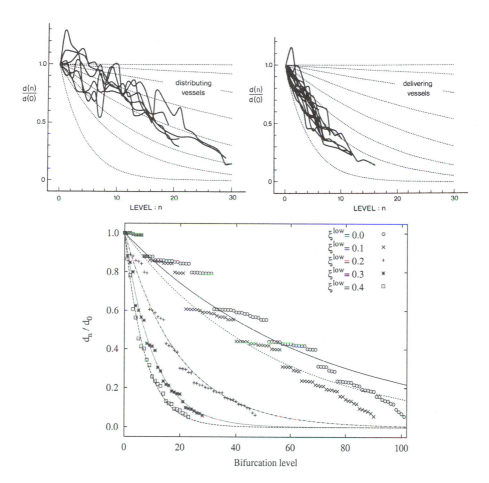

FIGURE 10 Shrinkage of radii along delivering and distributing types of vessels. *Top*: Experimental data of Zamir [37], showing the decline in segment diameter relative to a reference diameter (*y*-axis) as a function of the bifurcation level (*x*-axis). Distributing vessels (top left) carry blood across larger distances and radii shrink more slowly. Delivering vessels (top right) accomplish the short distance supply, and radii shrink more rapidly. (Printed with permission from The Rockefeller University Press.) *Bottom*: The CCO model reveals the intrinsic link between bifurcation symmetry and these transport characteristics. Increasing the minimum of bifurcation symmetry (ξ^{low}) gradually changes tree structure from "distributing" (circles, $\xi^{low} = 0.0$, $b = 0.99$) to "delivering" (squares, $\xi^{low} = 0.4$, $b = 0.87$).

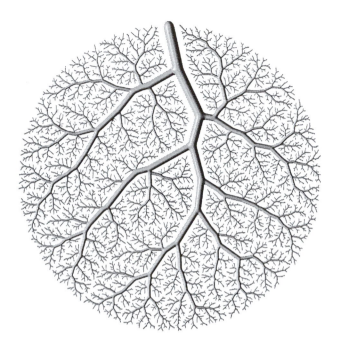

FIGURE 11   CCO tree generated while enforcing a minimum symmetry for new branchings. During the connection search only those bifurcations for which $r_2/r_1 \geq \xi^{low} = 0.4$ are allowed to become permanent. As a consequence, regions directly adjacent to large vessels are supplied through detours with gradually declining radii rather than by very small terminals directly branching off from the large vessel. Different degrees of minimum symmetry can be enforced, giving rise to graded changes in structure.

into a delivering type of tree, and results nicely compare to the experimental measurements, see Figure 10.

3   SUMMARY: KEY FEATURES OF CCO

Some features of CCO relate closely to physiological conditions and biological development and will be summarized in the following:

- At each stage of development of a CCO model tree, appropriate physiological function is provided since the very same boundary conditions of pressure and flows are fulfilled. This model feature relates to the fact that each living system must develop (ontogenetically) in a way that not only the fully differentiated organism but also each intermediate stage of development is viable.

- New terminal sites in CCO models are predominantly added in regions with perfusion densities lower than average, corresponding to a demand-induced angiogenesis *in vivo*.
- Segments generated in early phases of CCO tree development are transmitted to become main vessels in later stages. This obviously parallels embryologic development.
- Site, type, and direction of growth in each step is influenced by the structure developed in previous steps. This feature of CCO models again relates closely to embryologic development, where differentiation is ruled by the surrounding structure that is already existing.
- Topology and geometry of CCO models could therefore be influenced by defining zones of successive growth, which would guide the process of structural development.

4   GOALS AND AIMS

One key idea of CCO relates to the fact that models for anatomical structures are generated via optimization rather than through the input of anatomical knowledge or quantitative morphometric data. Still, the resulting models resemble real vascular trees to a high degree, including quantitative morphometric parameters. This offers the possibility for testing different optimization targets against each other so as to obtain a kind of indirect evidence on "what nature really optimized during phylogenesis."

Besides these issues of basic research, CCO lends itself to produce high-quality, detailed models of vascular structures, serving as a basis for computer simulations of complex hemodynamic phenomena within myocardial perfusion, which cannot sufficiently be tackled within the frame of more coarse models. Examples of complex and important questions regarding myocardial flow impediment, calling for advanced quantitative treatment of coronary circulation, are the consequence of

- local ischemia,
- local or global loss of contractility,
- modified coronary perfusion pressure waveforms due to intra-aortic balloon-pumping or left-ventricular assist devices, and
- mural tension in the ventricle and external (pericardial) compression due to cardiac assist devices or resuscitation.

Finally, as with all realistic vascular models, CCO models can serve as a basis for quantitatively predicting the hemodynamic consequences of vascular-surgical interventions, such as bypassing, balloon-dilations, and stent implantation.

5   SYNOPSIS OF CCO PROSPECTS WITH ISSUES DISCUSSED AT THE SFI MEETING

Since the nature of scaling laws was the central topic of the Santa Fe Institute (SFI) workshop, self-similar structures and descriptive parameters were extensively discussed. CCO vascular tree models repeatedly bifurcate, develop a remarkable degree of detail (several thousand terminal segments in one model), and hence suggest a quantitative evaluation also in terms of fractal parameters. However, three major differences remain:

1. There is no unique generator for self-similar structures, not even a stochastic generator. In contrast to fractal models, the geometric arrangement of segments plays the major role in model "generation," besides optimization itself.
2. Distributions of segments and terminals generated in CCO models may be evaluated with respect to scaling and fractal properties.
3. Even when furnished with several thousand terminal sites, CCO models have to be truncated fairly early in comparison to real fractal models. The bias in the computation of any fractal parameters induced by this "boundary effect" is considerable. This is true to some extent also for real arterial trees, which terminate in capillaries and exhibit fractal properties within some finite range.

An interesting extension of the CCO method to three dimensions (see Figure 5) was recently performed by Karch et al. [13], which for the first time offers the possibility for studying the difference in fractal properties between CCO models in two and three dimensions, respectively. While several morphometric parameters have already been evaluated for CCO models in order to assess their adequacy [24, 26, 33], it became clear at the SFI workshop that plunging into the fractal aspects of CCO models will be a major point of interest in the future. In particular, it is evident that the steric effect of vessel intersection—which represents an important (negative) selection criterion during the connection search in CCO—is drastically different in three dimensions. Compared with two dimensions, many more degrees of freedom for nonintersecting geometries are available. It will be interesting how this fact influences the values of fractal parameters deduced from CCO models.

Another key issue addressed and the SFI workshop was whether the constraints chosen for CCO so far were indeed general enough. In particular, G. B. West raised the question of whether the constant choice of γ throughout a tree is really adequate. It was well acknowledged that even with the present version of CCO different model trees can readily be generated for different values of γ (see Figure 6), but on top of that, a γ-variation within a tree would be most desirable and make the models even more realistic. After all, several experimental results indicate such a variation [7] *in vivo*, and also theoretical derivations explain "that, why and how" such a variation is to be

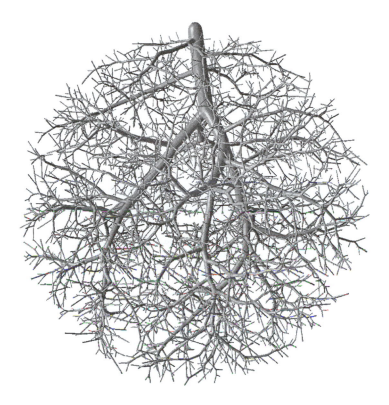

FIGURE 12   Three-dimensional CCO tree generated by the generalized algorithm due to Karch [13]; 4,000 terminal sites; diameter of perfused sphere is chosen to make the perfusion volume equal $100\,\text{cm}^3$; all other parameters equal to those in Figure 3.

expected [32]. Hence, the CCO algorithm will be evaluated by the Vienna Group regarding the possibility of such an extension within the near future. In the meantime some of this work is already in progress.

6 SUMMARY

We describe the method of "Constrained Constructive Optimization" (CCO) which is capable of generating arterial tree models from first principles by an iterative algorithm based on the optimization of global target functions. The search for a global optimum over all possible topologies of trees is replaced by a stepwise optimization, in which only the most recent step of growth is globally optimized. At each stage of development, the model fulfills physiologically reasonable boundary conditions for pressures and flows. The procedure, in essence, is derived from the growth of real arterial networks and allows us to generate complex arterial tree structures, which exceed any previously known optimized tree models in detail. The self-similar type of arterial structures generated allows essential characteristics of whole subtrees to be stored along with those segments representing the entrance into the respective subtree. In order to reduce the computational expense necessary for models of several thousand segments we implemented optimized traversing wetlands based on recursive definitions of quantities. Constrained optimization for continuous variables (bifurcation coordinates) is used to find the geometric optimum of single bifurcations while, simultaneously, the structural change in every step of growth is guided by discrete optimization according to the same target function. Quantitative evaluations of segment radii, branching angles and the pressure conditions within CCO models have confirmed their physiological and anatomical adequacy. Even though no anatomical knowledge or data are put into the models, the structures resulting from CCO closely resemble real arterial trees. Possible applications and future developments can be envisaged in several aspects:

- Comparison of different optimization targets allows for an "indirect search for the appropriate target." "Appropriate" means that the evolution and ontogony of arterial trees in mammals leads to very similar structures. Besides the targets used up to now (surface, volume, etc.), other aspects, such as impedance matching, may be considered either within the target or within the boundary conditions.
- Since CCO models offer full numerical access to very realistic structures, they lend themselves to the study of their scaling phenomena and fractal properties. "Proper scaling" may serve as an additional criterion in the search for the optimum combination of "target & boundaries" for CCO.
- The difference in scaling properties between two- and three-dimensional systems can be evaluated, which is in fact the topic of a current study.
- The properties of fractal generators derived from functional optimization principles may be compared with the "generators that can be statistically derived from CCO models," which have been optimized under the additional requirement that segments must be arranged in space.
- The structural substrate of CCO models can be used to compute hemodynamic properties and behavior by numerical simulation techniques. From these studies we can expect to gain much understanding for complex phenomena such as the redistribution of blood between the inner and outer layers of the heart muscle following a loss of contractility due to underperfusion.

REFERENCES

[1] Bassingthwaighte, J. B., R. B. King, and S. A. Roger. "Fractal Nature of Regional Myocardial Blood Flow Heterogeneity." *Circ. Res.* **65** (1989): 578–590.

[2] Bruinsma, P., T. Arts, J. Dankelman, and J. A. E. Spaan. "Model of the Coronary Circulation Based on Pressure Dependence of Coronary Resistance and Compliance." *Basic Resh. Cardiol.* **83** (1988): 510–524.

[3] Chilian, W. M., S. M. Layne, and S. H. Nellis. "Microvascular Pressure Profiles in the Left and Right Coronary Circulations." In *Coronary Circulation: Basic Mechanism and Clinical Relevance*, edited by F. Kajiya, G. A. Klassen, and J. A. E. Spaan, 173–190. Tokyo, Berlin, Heidelberg, New York: Springer-Verlag, 1990.

[4] Collins, M. W., and X. Y. Xu. "A Predictive Scheme for Flow in Arterial Bifurcations: Comparison with Laboratory Measurements." In *Biomechanical Transport Processes*, edited by F. Mosora, 125–133. New York: Plenum Press, 1990.

[5] Dawant, B., M. Levin, and A. S. Popel. "Effect of Dispersion of Vessel Diameters and Lengths in Stochastic Networks I. Modeling of Microcirculatory Flow." *Microvascular Resh.* **31** (1985): 203–222.

[6] Dawson, T. H. *Engineering Design of the Cardiovascular System of Mammals*. Englewood Cliffs, NJ: Prentice Hall, 1991.

[7] Feigl, E. O. "Coronary Circulation." In *Textbook of Physiology*, edited by H. D. Patton, A. F. Fuchs, B. Hille, A. M. Scher, and R. Steiner, 21st ed., vol. 2, chap. 48, 933–951. Philadelphia, PA: Saunders, 1989.

[8] Fung, Y. C. *Biodynamics: Circulation*. New York, Berlin, Heidelberg, Tokyo: Springer-Verlag, 1984.

[9] Hoffman, J. I. E. "Transmyocardial Perfusion." *Progress in Cardiovascular Diseases* **29** (1987): 429–464.

[10] Jones, C. J. H., L. Kuo, M. J. Davis, and W. M. Chilian. "Distribution and Control of Coronary Microvascular Resistance." *Interactive Phenomena in Cardiac System* **346** (1993): 181–188.

[11] Joyner, W. L., and M. J. Davis. "Pressure Profile Along the Microvascular Network and Its Control." *Federal Proceedings* **46** (1987): 266–269.

[12] Kamiya, A., and T. Togawa. "Optimal Branching Structure of the Vascular Tree." *Bull. Math. Biophys.* **34** (1972): 431–438.

[13] Karch, R., F. Neumann, M. Neumann, and W. Schreiner. "A Three-Dimensional Model for Arterial Tree Representation, Generated by Constrained Constructive Optimization." *Comp. Biol. & Med.* **29** (1999): 19–38.

[14] Karlsson, M. "Modelling and Simulation of the Human Arterial Tree—A Combined Lumped-Parameter and Transmission Line Element Approach." In *Computer Simulations in Biomedicine*, edited by H. Power and R. T. Hart, 11–18. Southampton, UK: Computational Mechanics Publication, 1995.

[15] Klocke, F. J., R. E. Mates, J. M. Canty, and A. K. Ellis. "Coronary Pressure-Flow Relationships, Controversial Issues and Probable Implications." *Circ. Res.* **56** (1985): 310–323.

[16] LaBarbera, M. "Principles of Design of Fluid Transport Systems in Zoology." *Science* **249** (1990): 992–999.

[17] Lefevre, J. "Teleonomical Optimization of a Fractal Model of the Pulmonary Arterial Bed." *J. Theor. Biol.* **102** (1983): 225–248.

[18] McAlpine, W. A. *Heart and Coronary Arteries. An Anatomical Atlas for Clinical Diagnosis, Radiological Investigation and Surgical Treatment.* Berlin: Springer-Verlag, 1975.

[19] Nellis, S. H., A. J. Liedtke, and L. Whitesell. "Small Coronary Vessel Pressure and Diameter in an Intact Beating Rabbit Heart Using Fixed-Position and Free-Motion Techniques." *Circ. Res.* **49** (1981): 342–353.

[20] Neumann, F., W. Schreiner, and M. Neumann. "Constrained Constructive Optimization of Binary Branching Arterial Tree Models." In *Computer Aided Optimum Design of Structures IV: Structural Optimization,* edited by S. Hernandesz, M. El-Sayed, and C. A. Brebbia, 181–188. Glasgow: Bell & Brain, 1995.

[21] Rooz, E., T. F. Wiesner, and R. M. Nerem. "Epicardial Coronary Blood Flow Including the Presence of Stenoses and Aorto-Coronary Bypasses I: Model and Numerical Method." *J. Biomech.* **107** (1985): 361–367.

[22] Schreiner, W. "Computer Generation of Complex Arterial Tree Models." *J. Biomed. Eng.* **15** (1993): 148–150.

[23] Schreiner, W., and P. F. Buxbaum. "Computer-Optimization of Vascular Trees." *IEEE Trans. Biomed. Eng.* **40** (1993): 482–491.

[24] Schreiner, W., M. Neumann, F. Neumann, S. M. Rödler, A. End, P. F. Buxbaum, M. R. Müller, and P. Spieckermann. "The Branching Angles in Computer-Generated Optimized Models of Arterial Trees." *J. Gen. Physiol.* **103** (1994): 975–989.

[25] Schreiner, W., F. Neumann, M. Neumann, A. End, S. M. Rödler, and S. H. Aharinejad. "The Influence of Optimization Target Selection on the Structure of Arterial Tree Models Generated by Constrained Constructive Optimization." *J. Gen. Physiol.* **106** (1995): 583–599.

[26] Schreiner, W., F. Neumann, M. Neumann, A. End, and M. R. Müller. "Structural Quantification and Bifurcation Symmetry in Arterial Tree Models Generated by Constrained Constructive Optimization." *J. Theor. Biol.* **180** (1996): 161–174.

[27] Schreiner, W., F. Neumann, M. Neumann, R. Karch, A. End, and S. M. Rödler. "Limited Bifurcation Asymmetry in Coronary Arterial Tree Models Generated by Constrained Constructive Optimization." *J. Gen. Physiol.* **109** (1997): 129–140.

[28] Schreiner, W., F. Neumann, M. Neumann, A. End, and S. M. Rödler. "Anatomical Variability and Functional Ability of Vascular Trees Modeled by Constrained Constructive Optimization." *J. Theor. Biol.* **187** (1997): 147–158.

[29] Sherman, T. F. "On Connecting Large Vessels to Small: The Meaning of Murray's Law." *J. Gen. Physiol.* **78** (1981): 431–453.

[30] Strahler, A. N. "Revisions of Horton's Quantitative Factors in Erosional Terrain." *Trans. Am. Geophys. Union* **34** (1953): 345–365.

[31] Thompson, D. W. *On Growth and Form*, 2d ed. Cambridge, MA: Cambridge University Press, 1942.

[32] van Bavel, E., J. A. E. Spaan. "Branching Patterns in the Porcine Coronary Arterial Tree: Estimation of Flow Heterogeneity." *Circ. Res.* **71** (1992): 1200–1212.

[33] West, B. J., and A. L. Goldberger. "Physiology in Fractal Dimensions." *Amer. Sci.* **75** (1987): 354–364.

[34] West, G. B., J. H. Brown, and B. J. Enquist. "A General Model for the Origin of Allometric Scaling Laws in Biology." *Science* **276** (1997): 122–126.

[35] Zamir, M., and N. Brown. "Arterial Branching in Various Parts of the Cardiovascular System." *Am. J. Anatomy* **163** (1982): 295–307.

[36] Zamir, M., and D. C. Bigelow. "Cost of Departure from Optimality in Arterial Branching." *J. Theor. Biol.* **109** (1984): 401–409.

[37] Zamir, M. "Distributing and Delivering Vessels of the Human Heart." *J. Gen. Physiol.* **91** (1988): 725–735.

[38] Zamir, M. "Optimality Principles in Arterial Branching." *J. Theor. Biol.* **62** (1976): 227–251.

[39] Zamir, M., and H. Chee. "Branching Characteristics of Human Coronary Arteries." *Canadian J. Physiol. & Pharmacology* **64** (1986): 661–668.

Quarter-Power Allometric Scaling in Vascular Plants: Functional Basis and Ecological Consequences

Brian J. Enquist
Geoffrey B. West
James H. Brown

1 INTRODUCTION: PLANT STRUCTURE AND FUNCTION

Although there are approximately 230,000 species [78], of vascular plants, all share essentially the same anatomical and physiological design. Vascular plants span well over 12 orders of magnitude in body size. A single *Sequoia* functions across this entire size range as it develops from seedling to adult tree. There is a rich empirical literature that provides many details of the anatomy and physiology of plants of differing sizes, growth forms, taxonomic groups, and environmental settings (e.g., Shinozaki et al. [99]; Yoda et al. [132]; Whittaker and Woodwell [128]; Shidei and Kira [98]; Cannell [11]; Niklas [74]). Nevertheless, there are few mechanistic models that attempt to link the relationships between whole plant architectural geometry, microscopic anatomy of the vascular system, and the physiological process of fluid flow, either within the different parts of a single plant or among plants that differ in size (see Niklas [74]; Tyree and Ewers [111]; Dewar et al. [21]). The only widely cited general model that integrates the salient features as a complete system for distributing resources from rootlet to leaf is the simplistic and often criticized "pipe model" [99]. Furthermore, it remains to be seen if anatomical and physiological processes can be "scaled up" to understand how structure and

Scaling in Biology, edited by J. H. Brown and G. B. West.
Oxford University Press, 2000.   **167**

function of individual plants affect larger-scale ecological and evolutionary patterns [22, 55].

1.1 MAJOR RESEARCH THEMES

Four major themes have motivated present empirical and theoretical studies of plant form and function.

1.1.1 Pipe Model.

The first theme stems from Leonardo da Vinci's original observation of tree construction (see Richter [89]; Zimmermann [137]; Horn this volume). da Vinci observed that "all the branches of a tree at every stage of its height, when put together, are equal in thickness to the trunk below them." Assuming that branch architecture directly reflects the hydraulic architecture leads to the "pipe model," in which a unit of leaf area is supplied by a given unit area of conducting tissue [43, 99]. The pipe model has been used as a basis for understanding the structural and functional design of trees (see Waring et al. [116]; Berninger and Nikinmaa [5]; Horn this volume) Although several authors have pointed out serious problems with its assumptions and predictions (e.g., Tyree and Ewers [111]), the pipe model is still the most widely cited model for whole-plant structure [35].

1.1.2 Hydraulic Architecture.

The second and somewhat related theme stems from research on the anatomy and physiology of plant vascular systems. Here the focus has been primarily on individual vascular elements, especially the xylem. The aim has been to understand the physical and biological processes which govern the flow of fluid within the vascular system and how they have influenced the evolution of xylem anatomy [13, 14, 44, 45, 68, 69, 102, 109, 111, 113, 135, 136, 137, 138]. An important question focuses on how trees more than 100 m tall are able to transport water and nutrients to such impressive heights (see Zimmermann [137]; Canny [14]). Some studies have considered functional explanations for differences in anatomy and physiology of the vascular system among plants living in unique environments such as deserts or swamps, or having unique growth architectures such as vines and hemiepiphytes [15, 82, 112, 137].

1.1.3 Resistance-Capacitance Models.

A third theme has been the detailed modeling of fluid flow along a water potential gradient from the soil through the plant to the atmosphere. These "resistance capacitance" models are designed to show how anatomical, physiological, and physical attributes of plants and their environment influence the water potential gradient and the rate of fluid transport throughout an individual. Such models often invoke many complex attributes of anatomy and physiology to account for known patterns of vascular resistance (e.g., van den Honert [115]; Jones [49]; Smith et al. [101]; Tyree and Sperry [114]; see also Jones [50]; Schulte and Costa [95]). This approach tends to ignore the complex dynamics of fluid flow through the microcapillary vascular tubes and chooses instead to characterize bulk flow in terms of analogies to electrical circuits.

1.1.4 Plant Architecture and Form.
A fourth theme has been the application of mathematical and biomechanical principles to understand plant architecture, with only indirect reference to the flow of fluid through the vascular system. This research has noted that although there is an enormous diversity of vascular plants, most can be classified as sharing a limited number of branched architectural forms [38]. Some of this research has interpreted plant branching patterns in terms of phyllotaxic schemes, Fibonacci series, and fractal geometry [3, 29, 41, 53, 70, 88]. Other work has focused on the architectural design of plants to optimize interception of sunlight, water, or nutrient resources by leaves, branches, or roots [8, 24, 32, 33, 42, 64, 70, 76, 79, 127], to resist buckling due to wind and gravity [36, 51, 60, 62, 73, 77], and to obey other biomechanical principles [71].

1.2 PLANT ALLOMETRY

Largely missing from studies of plant structure and function is an explicit consideration of the role body size plays in the structural and functional design of the vascular system and in mediating the relationship between the biomechanical constraints and the resource requirements of individual plants and resource availability in their environment (see Niklas [73]). Traditionally, the effect of plant size has not been a major focus of botanical studies (for notable exceptions see Sinnott [100]; Pearsall [83]; Murray [66]; Turrell [108]; and Niklas [74]). In contrast, however, studies of allometry had a major influence on animal anatomy, physiology, and ecology (e.g., Peters [84]; Schmidt-Nielsen [94]; Calder [10]; and Brown [9]). Since Huxley [48] defined the allometric equation many structural and functional variables of organisms (Y) have been shown to scale as power functions of body mass (M):

$$Y = Y_0 M^b , \tag{1}$$

where Y_o is a constant that varies with the variable and kind of organism, and b is another constant that typically assumes values that are multiples of $1/4$.

Applications of allometry to plant biology have consisted primarily of applying biomechanical principles to explain the scaling of structural and functional features within plants of varying size (McMahon [60]; Niklas [71]; Niklas [74]; but see Thomas [105, 106]) or of developing empirical relationships among size-related variables for application to agriculture, forestry, and ecosystem ecology (e.g., Waring et al. [116]; Shidei and Kira [98]; and Cannell [11]). Few investigators have applied allometry to more mechanistic studies of resource uptake, growth, and ecology (see Niklas [74]). One exception is the study of the explicitly allometric thinning law, describing the relationship between population density and plant size in ecological systems [57, 117, 125, 131], but this has traditionally been interpreted to be an outcome of simple geometric packing of volumes rather than a consequence of size-dependent rates of resource use (see also Dewar [20]).

Furthermore, studies of plant form and function generally do not include an explicit integration and treatment of fractal geometry. Several investigators have called attention to the obvious fractal-like nature of plant architecture [3, 29, 59, 65], and measured the fractal dimensions of plant structures (e.g., Morse et al. [65]; Tatsumi et al. [104]; Fitter and Strickland [30]; Nielson et al. [67]; Bernston and Stoll [6]; and Eshel [27]). However, there have been few attempts to relate fractal architectures of plants to their function or to explore their ecological ramifications.

2   A GENERAL MODEL FOR THE ORIGIN OF ALLOMETRIC SCALING LAWS IN BIOLOGY

In this chapter we present an overview of a complete "rootlet to leaf" quantitative model [122] of plant vascular systems. This model predicts the allometric scaling of many structural and functional attributes, both among branches of varying size within a single plant and between plants that differ in size. In doing so, it provides a mechanistic framework that links morphological pattern and physiological process in vascular plants. This approach incorporates many anatomical and physiological features of plants. In order to explain essential features of vascular systems, the model makes a number of simplifying assumptions but does not violate what is known about the fundamental aspects of plant anatomy and physiology. The model predicts that total resource use or metabolic rate scales as $M^{3/4}$ and explains why the maximum height of trees is limited to approximately 100 m. Furthermore, it shows how the distribution of resources through fractal-like networks within a plant ramifies to limit population density, energy, and ecosystem flux of plants across diverse ecosystems.

The plant model is an extension of a zeroth-order general model for biological resource distribution systems [121, 122]. It is based on the transport of essential materials through branching vascular networks. Despite the many idiosyncratic features observed across different kinds of organisms, most, if not all, living systems appear to obey a common set of design principles. The general model predicts that these "vascular" systems are fractal-like networks which: (i) branch to supply resources to all parts of a three-dimensional body, (ii) minimize the energy required to distribute materials, and (iii) have terminal elements (i.e. terminal vessels, capillaries) that do not vary as body size changes. As a result of these simple principles of functional design, organisms exhibit a common set of allometric scaling relationships.

3   THE PLANT MODEL

We start with a simple zeroth-order model of resource distribution within a plant. It is motivated by the basic facts of plant biology that the uptake of essential resources occurs in specialized vascular tissues and that these materials are distributed throughout the plant in a liquid medium under negative pressure by a vascular system. We model the transport of fluid through xylem vessels of an angiosperm tree, from the base of the trunk to the petioles. The model should also apply, with only minor modification, to the specifics of transport through tracheids, through phloem, and within roots. A more detailed treatment of the model is given in West et al. [122].

In addition to the three assumptions mentioned above, the plant model assumes that the xylem network system is composed of multiple microtubular vascular elements aligned in parallel and running continuously from rootlet to leaf (Figure 1). For simplicity, all tubes are assumed to be of equal length, and their diameters are constant within a branch segment but allowed to vary between segments, thereby allowing for possible tapering of tubes from trunk to petiole. This variation is shown to be critical in circumventing the problem of hydrodynamic resistance increasing with tube length. Thickness and structure of tube walls are ignored, as are lateral connections between parallel tubes. Lastly, we allow the ratio of conducting to nonconducting tissue to vary with tree height, thereby resolving a possible conflict between hydrodynamic and mechanical constraints.

Based on these assumptions, we show that the architecture of a tree must be a self-similar fractal with specific scaling exponents. These predictions are well supported by the data. We are aware that some of our assumptions are oversimplified abstractions of a more complex anatomy and physiology, see Section 3.3 below we consider some of the consequences of relaxing them. For the moment, however, they provide a quantitative zeroth-order model of the entire plant network, which makes testable predictions and can be used as a point of departure for more detailed investigations.

3.1   TWO CHALLENGES IN THE DIVERSIFICATION IN PLANT SIZE

Numerous studies have documented that increase in plant size often leads to greater access to limiting resources and increased reproductive output (e.g., Harper [40]). However, in order for vascular plants to diversify in size, two important problems needed to be overcome. Increase in size also brings (i) additional need for mechanical support and (ii) increased vascular hydrodynamic resistance with increasing transport distance between soil and leaves. The allometric solutions to these problems will be shown to dictate how branch and xylem vessel radii change with changes in plant size and branching level.

The model can be described as a hierarchical branching network running from the trunk (level 0) to the petioles (level N). An arbitrary level in the plant branching network is denoted by k. We characterize the architecture of

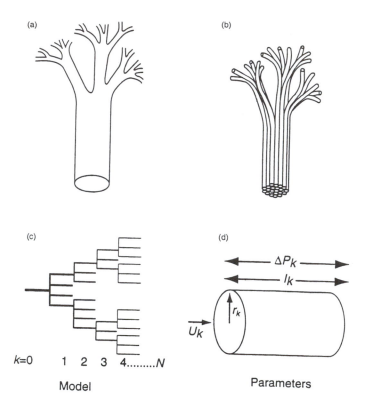

(a) (b) (c) (d)

k=0   1 2 3   4.........N

Model   Parameters

FIGURE 1   (a) Macroscopic architecture of part of a plant branching network; (b) pipe model representation of the microscopic architecture of such a network illustrating a tightly bundled, plant vascular system comprised of diverging vessel elements; (c) topological representation of such a network, where k specifies the order of the level, beginning with the trunk, or main stem, ($k = 0$) and ending with the petiole ($k = N$); (d) parameters of a typical branch at the kth level include the pressure differential δP_k, length l_k, radius r_k, and fluid velocity u_k.

the branching network using three parameters to define the relationship of daughter to parent branches (Figure 1). These are defined through the ratios of branch radii, $\beta_k \equiv r_{k+1}/r_k \equiv n^{-a/2}$, vessel tube radii $\bar{\beta}_k \equiv a_{k+1}/a_k \equiv n^{-\bar{a}/2}$, and branch lengths $\gamma_k \equiv l_{k+1}/l_k$. The branching ratio, n, the number of daughter branches derived from one parent branch, is typically 2, and is assumed to be independent of k. Elsewhere, we have shown [121] that for the network to be volume filling, $\gamma = n^{-1/3}$ independent of k. Below, we show that if biomechanical constraints are uniform throughout the tree, then β_k and a are also independent of k, proving that the network is a self-similar fractal. If, in addition, we assume that $\bar{a}$ is independent of k, so that possible tapering of tubes is uniform, it can be shown that the scaling of branch radii and xylem vessel radii is

$$\frac{r_k}{r_N} = n^{(N-k)a/2}, \quad \frac{a_k}{a_N} = \left(\frac{r_k}{r_N}\right)^{\bar{a}/a}; \tag{2}$$

and the scaling of branch lengths is

$$\frac{l_k}{l_N} = \left(\frac{r_k}{r_N}\right)^{2/3a}, \tag{3}$$

where r_N and l_N are the petiole radius and length, respectively, and N is the total number of branchings. Given that the total number of tubes is preserved at each branching, then multiple scaling laws can be derived. For example, the number of terminal branches is given by

$$n_k^L = \left(\frac{r_k}{r_N}\right)^{2/a}, \tag{4}$$

the area of conductive tissue by

$$A_k^{CT} = A_N^{CT}\left(\frac{r_k}{r_N}\right)^{2(1+\bar{a})/a}, \tag{5}$$

and the proportion of conductive tissue by

$$f_k \equiv \frac{A_k^{CT}}{A_k^{TOT}} = n_N\left(\frac{a_N^2}{r_N^2}\right)\left(\frac{r_k}{r_N}\right)^{2(1+\bar{a}-a)/a}. \tag{6}$$

Thus, the total cross-sectional area of all daughter branches, nA_{k+1}, at any level k, is given by $nA_{k+1}^{TOT}/A_k^{TOT} = n\beta_k^2 = n^{1-a}$. When $a = 1$ this reduces to unity and the branching is "area preserving," namely the total cross-sectional area at any given level, $nA_{k+1}^{TOT} = A_k^{TOT}$ (Richter [89]; Horn this volume). The pipe model where all tubes have the same constant diameter (i.e. $\bar{a} = 0$), are tightly bundled, and that there is no nonconducting tissue is a simple case of this. In reality, however, tubes are not tightly packed in sapwood and there may be heartwood which provides additional mechanical stability (Figure 2).

Assuming that branching is volume filling ($\gamma = n^{-1/3}$), the above scaling relations can be parameterized in terms of just two exponents, a and $\bar{a}$. Here we show how a and $\bar{a}$ are determined from two critical constraints: a from biomechanical stability, and $\bar{a}$ from minimization of hydrodynamic resistance.

3.1.1 Mechanical Constraint.

As plants increase in size they must be able to support themselves against buckling due to gravitational forces and wind. Using biomechanical principles, the design of trunks and branches to resist buckling leads to some optimum relationship between the length and radius of branches: $l_k = (E/\rho)r_k^\alpha$, where E is Young's elastic modulus and ρ is the

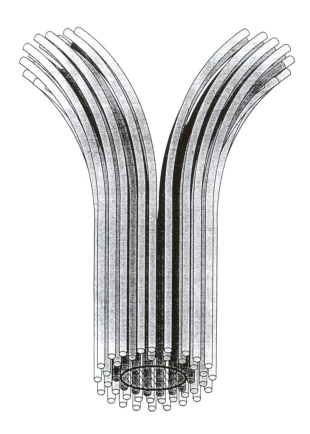

FIGURE 2   Symbolic representation of the vascular structure of a "realistic" branch, showing the division between conducting and nonconducting tissue; this is to be contrasted to the tightly bundled vascular system of the pipe model, shown in Figure 1(a), in which all tissue is conducting.

wood density of the branch. Previous studies have shown that if the condition of mechanical stability is the same for all branches, then α is constant, independent of k. In this case a and β_k are also constant, giving a branching architecture that is a self-similar fractal. Analyses based on scale-invariant solutions to the bending moment equations for beams (elastic similarity) give $\alpha = 2/3$ [36, 51, 62, 71, 74, 77]. This constraint, which is most important for the trunk and large branches, agrees well with data for these segments (McMahon and Kronauer [62]; Bertram [7]; Niklas [77]; Horn this volume). Comparison with Eq. (2) and (3) give $a = 2/(3/\alpha)$. If this holds uniformly for all branches, α is constant, independent of k. In this case a and β_k are also constant, which when coupled with the volume-filling constraint leads to $a = 1$ and is precisely the condition for area-preserving branching (Horn this volume). In other words, biomechanical stability leads to branches behaving as if they were tightly packed, vascular bundles of constant-diameter tubes. This was the simple model that we originally used to motivate area-preserving branching in plants [121]. Here, however, we relax those assumptions by incorporating nonconducting heartwood, allowing tubes to vary in diameter, and to be loosely packed in the sapwood.

Note that the result $a = 1$ implies that the leaf area distal to the kth branch $A_k^L = C_L r_k^2$, where $C_L \equiv A_L/r_N^2$, is invariant and A_L is the area of a leaf. In addition, the number of branches of a given size $N_k = n^N(r_N/r_k)^{2/a}$, or $N_k \propto r_k^{-2}$. If reproductive tissues are supplied by vascular elements in the same way as leaves, they should exhibit a similar scaling behavior. All of these predictions are in good agreement with data [72, 74, 99, 103, 106].

3.1.2 Hydrodynamic Constraint.

Increases in plant size also brings about an increase in the path distance over which resources must be transported. The hydrodynamic resistance of a given tube, Z_k^i, is given by the classic Poiseuille formula which governs flow through pipes, $Z_k^i = 8\eta l_k/a_k^4$, where η is the fluid viscosity [137]. Note that any slight change in tube diameter leads to a disproportionate change in total tube resistance because of the fourth-power dependence on a_k. If the tube diameter does not change, then hydrodynamic resistance increases linearly with transport distance, l_k, independent of any mechanical constraint. Furthermore, because path lengths from the soil to the leaves and branch meristems differ, resources would tend to be delivered at higher rates over the shorter paths, limiting resource supply to apical meristems and terminal shoots. This linear increase in resistance would seemingly mitigate against diversification of plant size (see also Raven and Handley [87] and Raven [86]).

The total resistance of a single tube running from the trunk to the petiole can be summed across all vascular elements to give

$$Z_i = \left[\frac{1 - n^{(1/3-2\bar{a})(N+1)}}{1 - n^{(1/3-2\bar{a})}}\right] Z_N = \left[\frac{1 - \{(n^{1/3} - 1)1_N\}^{(1-6\bar{a})}}{1 - n^{(1/3-2\bar{a})}}\right] Z_N, \quad (7)$$

where Z_N is the resistance of the petiole. Thus, the behavior of the resistance of the total system, Z_i, depends critically on the exponent that governs xylem vessel tapering with branching level: namely, whether $\bar{a}$ is greater than, less than, or equal to $1/6$. If $\bar{a}$ is less than $1/6$ the total resistance of the network increases as the path length from soil to petiole increases with size. However, if $\bar{a} > 1/6$, then this equation has the remarkable property that the total tube resistance is constant, independent of both the number of branchings, N, and the total path length, l_T, of the branch. Therefore, Z_i is invariant with size for all plants. It is exactly what is needed to solve the problem of ensuring that all leaves have comparable rates of resource supply independent of total branch length.

Since large $\bar{a}$ corresponds to steeper tapering, this would eventually lead either to unrealistically large tube radii in the trunk or unrealistically small ones in the petiole. We assume that the physical properties of water in differing environments set limits on maximum and minimum sizes of xylem vessels [15]. To avoid such extreme tapering, $\bar{a}$ should therefore be as close as possible to the minimum value consistent with Eq. (7), namely 1/6. As shown below, an extension of this argument leads to an expression for the maximum height of trees.

3.2 CONSEQUENCES: ALLOMETRIC RELATIONSHIPS AND MAXIMUM TREE HEIGHT

3.2.1 *Allometric Relations.* Having determined how the branch radii, branch lengths, and tube radii change with branching level, allometric scaling relationships for several physiological and anatomical traits can be shown to follow. For example, the conductivity ($K_k \equiv l_k/Z_k$) of a branch segment is given by

$$K_k \equiv K_N \left(\frac{r_k}{r_N} \right)^{2(1+2\bar{a})/a}, \tag{8}$$

where Z_k is the total resistance in the branch segment. Similarly, the leaf-specific conductivity (the conductivity per unit leaf area) of a branch segment is given by

$$L_k \equiv \frac{K_k}{n_k^L a_L} = L_k = L_N \left(\frac{r_k}{r_N} \right)^{4\bar{a}/a}, \tag{9}$$

where a_L and n_N^L is the number of leaves distal to branch k and the area of a leaf, respectively. With $\bar{a} = 1/6$ and $a = 1$, this predicts $K_k \propto r_k^{8/3}$, $L_k \propto r_k^{2/3}$, and $a_k/a_N = (r_k/r_N)^{1/6}$. Taking the radius of a vessel in the petiole, a_N, as 10 μm and the number of tubes in a petiole, n_N, as 200 give for the normalization, $K_N \approx 7 \times 10^{-10}$ m^{4}sec^{-1} MPa^{-1}. These relations can also be expressed as a function of the conducting tissue area: $K_k \propto (A_k^{CT})^{(1+2\bar{a})/(1+\bar{a})} \propto (A_k^{CT})^{8/7}$ and $L_k \propto (A_k^{CT})^{(2\bar{a})/(1+\bar{a})} \propto (A_k^{CT})^{2/7}$. Notice that these exponents do not depend explicitly on a. All of these predictions are in good agreement with data [52, 82, 130]. For comparison, the pipe model, where $a = 1$ and $\bar{a} = 0$, gives $K_k \propto r_k^2$ and $L_k \propto r_k^0$. Note also that, since n_N and a_L are invariant, this implies that $L_k \propto a_k^4$, independent of a and $\bar{a}$.

It can further be shown that $n^N \propto M^{3/(1+3a)} \propto M^{3/4}$ when $a = 1$. In addition, the total number of terminal branches, or leaves, is predicted to scale as $n_0^L \propto r_0^2 \propto M^{3/4}$ and can be combined to give $l_k \propto M^{(1-k/N)/(a+3)} \propto M^{(1-k/N)/4}$, and $r_k \propto M^{(1-k/N)3a/2(a+3)} \propto M^{3(1-k/N)/8}$. Notice that the length and radius of a kth-level branch scales more slowly with M than does the trunk: $l_0 \propto M^{1/4}$ and $r_0 \propto M^{3/8}$. Since the total height of a given tree, h, is equivalent to the length of a tube from trunk to leaf: $h = l_T \approx l_0/(1 - \gamma)$, this gives $h \propto M^{1/4}$. The tube radius scales as $a_k/a_N = (r_k/r_N)^{1/6} \propto M^{3\bar{a}(1-k/N)/2(a+3)} \propto M^{(1-k/N)/16}$, so that $a_0/a_N = n^{N/12} \propto M^{1/16}$. Taking $N = 18$, and $n = 2$ this gives $a_0/a_N \approx 2.8$, so that, if the radius of a vessel in a petiole $a_N \approx 10\mu m$, then the radius of a vessel in the trunk $a_0 \approx 30\mu m$ [137]. Furthermore, even over 12 orders of magnitude variation in mass, a_0 is predicted to change by only about 60%. So the small variation in tube radius predicted by the model and observed empirically (see Zimmermann [135, 137]; Ewers and Zimmermann [28]) not only overcomes the problem of resistance increasing with tube length but also is the basis for many observed scaling relationships.

3.2.2 Metabolic Rate.

Driven by a pressure gradient, ΔP, which is independent of plant size, the volume rate of fluid flow through a single tube is $\dot{Q}_i = \Delta P / Z_i$. We have shown above that when $\bar{a} > 1/6$, Z_i is also independent of plant size, scaling as M^0. Thus, $\dot{Q}_i \propto M^0$ so that the volume flow rate through a single vascular tube from trunk to petiole is predicted to be the same for all plants. The total volume flow rate through all vascular tubes is $\dot{Q}_0 \propto M^{3/(1+3a)} = M^{3/4}$, when $a = 1$. The model therefore predicts that whole-plant metabolic rate scales as $M^{3/4}$, which is the same relationship observed in animals. For comparison note that in the naïve pipe model where $\bar{a} = 0$ and $a = 1$, then $Z_i \propto n^{N/3} \propto M^{1/4}$, which would give $\dot{Q}_0 \propto M^{1/2}$.

We tested this prediction by analyzing data from the literature (see Enquist et al. [26]). Several studies report the total rate of fluid transport in the xylem ($\dot{Q}_o$) as a function of stem diameter D, (e.g., Huber [44]; Sakurantani [93]; Schulze et al. [96]). Figure 3 shows that this relationship can be described as $\dot{Q}_o \propto D^{1.778}$. Other studies report relationships between stem diameter and above-ground dry mass (e.g., White [125]); averaging these gives $D \propto M^{0.412}$ ($n = 8$, $SD = 0.356$). So, substituting gives $\dot{Q}_o \propto M^{0.732}$. These relationships are nearly indistinguishable from those predicted from our model: $\dot{Q}_o \propto D^2, D \propto M^{3/8}$, and $\dot{Q}_o \propto M^{3/4}$. Small deviations from the predicted exponents can have many sources, including using wet or dry mass (see Table 1) and measuring techniques that slightly overestimate the diameters of large trees [80].

Whole-plant xylem transport provides a measure, not only of nutrient and water use, but also of gross photosynthesis and therefore of metabolic rate. Because of stoichiometric and physiological constraints, the allometric scaling exponents for water, nutrient, and photosynthate fluxes must be equivalent. Thus, rates of transpiration or xylem transport are appropriate, although generally overlooked, indices of whole-plant metabolism. Both the theoretical model and the empirical evidence indicate that whole-plant metabolic rates scale as $M^{3/4}$, so that mass- or tissue-specific rates scale as $M^{-1/4}$. This agrees with the qualitative observation that size-specific growth rates are generally highest in annuals and small herbs and lowest in large trees [37, 107]. More quantitatively, Whittaker and Woodwell [128] have found that whole-plant rates of twig (P_L) and leaf production, and wood and bark production (P_B), in six species of temperate trees, scale as $P_L \propto D^{1.653}$ and $P_B \propto D^{1.807}$. Using the scaling of stem mass in these species ($D \propto M^{0.438}$), the rates of new tissue production are $P_L \propto M^{0.724}$ and $P_B \propto M^{0.791}$. Using total above

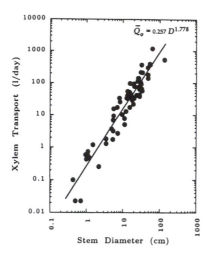

FIGURE 3   Relationship between rate of whole-plant xylem transport and basal stem diameter for 69 individuals and 37 species, including herbaceous plants, shrubs, tree seedlings, and mature evergreen and deciduous 37 ($r^2 = 0.913$, $n = 69$, $p < 0.0001$, 95% CI: 1.644 to 1.912).

ground biomass gives $P_L \propto M^{0.774}$ and $P_B \propto M^{0.846}$ respectively. These values are essentially indistinguishable from the predicted value of $M^{3/4}$.

3.2.3   Pressure Gradient.   The pressure drop across a kth-level branch is $\Delta P_k = \dot{Q}_i Z_k^i$, so the ratio of pressure gradients between adjacent branch levels is $(\Delta P_{k+1}/l_{k+1})/(\Delta P_k/l_k) = (l_k/l_{k+1})^{6\bar{a}} = l_k/l_{k+1}$, when $\bar{a} = 1/6$. The pressure gradient is therefore steeper in smaller branches than in larger ones; in particular, the ratio between trunk and petiole is predicted to be $(\Delta P_0/l_0)/(\Delta P_N/l_N) \propto M^{-1/4}$. So, for a tree with trunk length 4 m and petiole length 4 cm, this ratio is $\sim 1/100$. This also predicts that the ratio of pressure gradients should be smaller in larger trees and that the pressure drop, ΔP_k, should be the same for all branch segments, independent of level. Regardless of the mechanism generating ΔP, these predictions are in good agreement with data [137]. In contrast, the naïve pipe model predicts that the pressure gradient rather than the pressure difference is size independent.

3.2.4   Removing Branch Segments.   A particularly sensitive test of the model is its quantitative prediction of how total plant resistance changes as progressively larger branch segments are removed. Suppose that only branches up to the kth-level remain, then the total number of tubes in the trunk is still n_0; however, they now terminate at the kth level rather than the Nth. The ratio of resistances at various levels $R_k \equiv Z_k^{\mathrm{TOT}}/Z^{\mathrm{TOT}}$ can be expressed as

$$R_k = \frac{(r_{k+1}/r_0)^p - 1}{(r_{N+1}/r_0)^p - 1} \tag{10}$$

where $p \equiv 2(1 - 6\bar{a})/3a$ and $r_{N+1} \equiv r_N n^{-a/2}$. Note that $R_N = 1$ and $R_{-1} = 0$, the latter corresponding to the limit where all conducting tissue has been removed. Figure 4 shows empirical data for removing branch segments with various values for $\bar{a}$. The theoretical prediction with $\bar{a} = 1/6$ agrees very well with empirical data (from Yang and Tyree [129]), and provides a much better fit than the classic pipe model, where $\bar{a} = 0$.

3.2.5 Fluid Velocity.

Since tubes taper, fluid velocity, u_k, must increase in smaller branches. Allometrically, u_k scales as

$$M^{-(1-k/N)a/2(a+3)} \propto M^{-(1-k/N)/8},$$

when $a = 1$. For the trunk this gives $u_0 \propto M^{-1/8}$. So, over a range of eight orders of magnitude in mass, corresponding roughly to a 50-cm sapling relative to a 50-m tree, the fluid velocity in the trunk is predicted to decrease by a factor of ~ 10. Again, the above predictions are supported by the data [50, 137]. In contrast, the pipe model predicts flow velocity to be independent of plant size.

3.2.6 Conducting Tissue and the Maximum Height of Trees.

From above, the area of conducting tissue in a branch at the kth level scales as $A_k^{CT} \propto r_k^{2(1+\bar{a})/a} \propto A_k^{\mathrm{TOT}\,7/6}$, where A_k^{TOT} is the total cross-sectional area of a branch. Alternatively, it can be shown that for the total number of leaves of a given branch, n_k^L, $A_k^{CT} \propto (n_k^L)^{(1+\bar{a})} \propto (n_k^L)^{7/6}$, which depends only on $\bar{a}$. The proportion of conductive tissue relative to total branch cross-sectional area is given by: $f_k = (n_N a_N^2/r_N^2)(r_k/r_N)^{2(1+\bar{a}-a)/a} \propto r_k^{1/3}$. All of these results are well supported by the data [56, 82, 91]. Since $f_k \leq 1$, this last formula leads to a limitation on the maximum height and radius of a tree. Maximum tree height is predicted to occur when the hydrodynamic resistance is no longer independent of branch length. This can be shown as

$$r_0^{\mathrm{MAX}} = r_N \left(\frac{r_N^2}{a_N^2 n_N}\right)^{a/2(1+\bar{a}-a)} \quad , \quad l_0^{\mathrm{MAX}} = l_N \left(\frac{r_N^2}{a_N^2 n_N}\right)^{1/3(1+\bar{a}-a)} . \tag{11}$$

Since maximum height is equivalent to the total tube length, $h^{\mathrm{MAX}} = l_0^{\mathrm{MAX}}/(1 - n^{-1/3})$, these give $r_0^{\mathrm{MAX}} = r_N^7/a_N^6 n_N^3$ and $l_0^{\mathrm{MAX}} = l_N r_N^4/a_N^4 n_N^2$. The maximum size is therefore extremely sensitive to the parameters of the petiole, which include the size of the petiole, r_N, the total number of vessels, n_N, and their sizes, a_N. We assume that the physical properties of water, environmental conditions, and anatomical constraints ultimately determine the size range of xylem vessels (e.g., Zimmermann [137]; Carlquist [15]; Tyree and Ewers [111]), which ultimately limits canopy height across ecological communities. As an example, with $a = 1$ and $\bar{a} = 1/6$, take $r_N = 0.5$ mm, $a_N = 10$ μm, and $n_N = 200$, then these give $r_0^{MAX} \approx 1$ m and $h^{MAX} \approx 40$ m. If instead, we change the diameter of tubes in the petiole so $a_N = 8$ μm but the other parameters remain the same, then $r_0^{MAX} \approx 4$ m and $h^{MAX} \approx 100$ m. Thus, these formulae cannot be used to accurately calculate maximum size. On the other hand, the argument does show why h^{MAX} is of the order of 100 m rather than 1 m or 1000 m. It is of note that if $\bar{a}$ were increased, so that tapering of xylem tubes were much steeper, then the maximum height of trees would be greatly reduced. Therefore, if plants are selected to maximize height, $\bar{a}$ must approach its minimum value, $1/6$, consistent with total tube resistance being a constant (see Figure 4). Thus, the model provides an explanation from fundamental principles why the size of trees is limited and how that size is related to basic parameters, which depend on both mechanical and hydrodynamic constraints.

3.2.7 Fractal Dimensions of Plant Architecture. Our model predicts a fractal-like hierarchical branching network with specific scaling properties. Elsewhere (see West et al. this volume) we have shown that this general framework can also be used to calculate fractal dimensions of plant networks. In particular, the model predicts a fractal dimension of 1.5 for plant roots and shoots that fill a three-dimensional volume while shoots or roots that instead fill two-dimensional space should have a dimension of 1.33. Both of these predictions appear to be well supported by experimental data (see Aono and Tosiyasu [3]; Morse et al. [65]; Fitter and Strickland [30]; Nielson et al. [67]).

3.3 DEVIATIONS FROM PREDICTED SCALING EXPONENTS

Variation in any of our assumptions (space-filling, minimization of resistance, uniform biomechanical constraints) or in limiting aspects of ecological environments will lead to calculable deviations from the predicted exponents. We assume that biomechanical constraints lead to $l_k \propto r_k^{\alpha}$ with α independent of k. Resistance to elastic buckling, however, which gives $\alpha = 2/3$ leading to $a=1$ and area-preserving branching, may not apply throughout the plant, especially in the "butt swell" at the base of the trunk and in the smallest branches (e.g., Bertram [7]). This can be easily incorporated into the model as a variation in α for either small or large k, and will lead to calculable corrections to corresponding scaling laws.

Furthermore, the above ground portion of grasses, palms, saplings, and shrubs with relatively few branches, clearly violate the "area filling" assumption. A more likely geometry is $\gamma = 1/2$ rather than $1/3$. Since their branches are predominantly conducting tissue they maintain area-preserving branching, where $a = 1$ and therefore $l_k \propto r_k$, which is in agreement with observation [77]. Thus, throughout ontogeny as a sapling grows to adult size γ changes, so that, $\gamma = 1/2 \rightarrow 1/3$. Thus, a will vary from $a = 2/(2\,\alpha) \rightarrow 2/(3\,\alpha)$. This will lead to calculable deviations in several allometric exponents. In short, variation in any of our assumptions, or in certain aspects of the environment, will lead to predictable deviations from the theoretical model.

4 SUMMARY OF THE MODEL

The above treatment should be viewed as a zeroth-order model. It represents a variant of our general model for linear branching resource networks, which incorporates salient features of plants. It makes several simplifying assumptions, and incorporates only those essentials of plant anatomy and physiology necessary to derive a complete integrated characterization of the architecture, biomechanics, and hydrodynamics of vascular plant design. It can serve as a starting point for more elaborate models that incorporate special features of particular kinds of plants growing in different environments.

Our model accurately predicts many attributes of vascular plants and provides a more realistic characterization of plant structure and function than previous models such as the pipe model (see Table 1). The pipe model does not explicitly include biomechanical constraints, or allow for the presence of nonconducting tissue. More critically, it does not incorporate the paramount problem of total hydrodynamic resistance increasing with increasing path length from root to leaf. Thus, it fails to predict the observed scaling exponents for conductivity and leaf-specific conductivity, observed sapwood/heartwood ratios, and how total vascular resistances of remaining branch segments should change as terminal branches are experimentally removed.

Our model provides a basis for which to understand many fundamental features of plant structure and function. First, it predicts several anatomical and physiological scaling relationships, which compare favorably with empirical values (see Table 2). The close correspondence between predicted and observed scaling exponents demonstrates the power of this single model to provide a quantitative integrated explanation for many features of vascular anatomy and physiology as well as whole-plant architecture. Because it also predicts several scaling relationships that have not yet been measured, the model is subject to rigorous tests. Second, the model predicts the magnitudes of certain variables, including: conductivity of different branch segments, surface area of leaf supplied by each tube, pressure gradient differential between leaf and trunk, the ratio of conducting to nonconducting tissue, and maximum trunk radius and total height of a tree. These predictions correspond well with observed values and illustrate how the design of resource distribution networks ultimately constrains anatomy and physiology; they follow from an interplay between geometrical, hydrodynamical, and biomechanical principles. Perhaps the most important insight is that "quarter power" scaling, which has been widely commented on in animals, is predicted theoretically and demonstrated empirically in vascular plants.

More importantly, the model shows how plants can overcome the potentially devastating effect of resistance increasing with tube length so as to insure comparable xylem flow to all leaves of the plant. This prediction is supported by measuring changes in vessel radii and branch resistance [28, 137]. Such regulation can potentially be accomplished in two ways: (1) by

TABLE 1   Predicted values of scaling exponents for some physiological and anatomical variables of plant vascular systems as a function of total plant mass, M, and branch radius, r_k. For the latter case, predictions are compared with measured values in the last column. Except for the case of branch length, where it is given as ±0.036, confidence levels are not quoted by the authors cited. ND indicates no data yet available, and LSC stands for leaf-specific conductivity. Since there is very little data on allometric scaling with mass, no measured values for these exponents are quoted.

PLANT MASS		Variable	BRANCH RADIUS		
Exponent	Symbol		Symbol	Exponent	
$\frac{3}{4}=0.75$	n_0^L	Number of Leaves	n_k^L	$2=2.00$	2.007^*
$\frac{3}{4}=0.75$	N_0	Number of Branches	N_k	$-2=-2.00$	$-2.00^\dagger$
$\frac{3}{4}=0.75$	n_0	Number of Tubes	n_k	$2=2.00$	ND
$\frac{1}{4}=0.25$	l_0	Branch Length	l_k	$\frac{2}{3}=0.67$	$0.652^\ddagger$
$\frac{3}{8}=0.375$	r_0	Branch Radius			
$\frac{7}{8}=0.875$	A_0^{CT}	Area of Conductive Tissue	A_k^{CT}	$\frac{7}{3}=2.33$	$2.13^\S$
$\frac{1}{16}=0.0625$	a_0	Tube Radius	a_k	$\frac{1}{6}=0.167$	ND
$1=1.00$	K_0	Conductivity	K_k	$\frac{8}{3}=2.67$	$2.63^\P$
$\frac{1}{4}=0.25$	L_0	LSC	L_k	$\frac{2}{3}=0.67$	$0.727^\P$
		Fluid Flow Rate	$\dot{Q}_k$	$2=2.00$	ND
$\frac{3}{4}=0.75$	$\dot{Q}_0$	Metabolic Rate			
$-\frac{1}{4}=-0.25$	$\Delta P_0/l_0$	Pressure Gradient	$\Delta P_k/l_k$	$-\frac{2}{3}=-0.67$	ND
$-\frac{1}{8}=-0.125$	u_0	Fluid Velocity	u_k	$-\frac{1}{3}=-0.33$	ND
$-\frac{3}{4}=-0.75$	Z_0	Branch Resistance	Z_k	$-\frac{3}{4}=-0.75$	ND
$\frac{1}{4}=0.25$	h	Tree Height			
$\frac{3}{4}=0.75$		Reproductive Biomass		$2=2.00$	
$\frac{25}{24}=1.0415$		Total Fluid Volume			

*Niklas [77].
†Shinozaki et al. [99].
‡Bertram [7], see also Niklas [77], and Horn this volume.
§Patino et al. [82].
¶Yang and Tyree [129].

TABLE 2   Comparison of predictions made by the pipe model and the general allometric model presented here for several physiological and anatomical variables. Notice, that both models diverge on several variables. Because the pipe model does not explicitly include biomechanics or hydrodynamics it makes a series of faulty predictions (see Table 1).

Variable	Pipe Model	General Allometric Model
Leaf Area	$A_k^L \propto r_k^2$	$A_k^L \propto r_k^2$
Reproductive Biomass	$M_{REP} \propto r_k^2$	$M_{REP} \propto r_k^2$
Number of Branches	$N_k \propto r^{-2}$	$N_k \propto r^{-2}$
Fluid Velocity	$u_0 \propto M^0$	$u_0 \propto M^{-1/8}$
Proportion of Cond. Tissue	$f_k \propto r_k^1$	$f_k \propto r_k^{1/3}$
Pressure Gradient	$\dfrac{(\Delta P_0/l_0)}{(\Delta P_N/l_N)} \propto M^0$	$\dfrac{(\Delta P_0/l_0)}{(\Delta P_N/l_N)} \propto M^{-1/4}$
Total Fluid Flow/ Metabolic Rate	$\dot{Q}_0 \propto r_0^{4/3}$, $\dot{Q}_0 \propto M^{1/2}$	$\dot{Q}_0 \propto r_0^2$, $\dot{Q} \propto M^{3/4}$
Conductivity	$K_k \propto r_k^2$	$K_k \propto r_k^{8/3}$
Conductivity	$K_k \propto (A_k^{CT})_k^0$	$K_k \propto (A_k^{CT})_k^{8/7}$
Leaf-Specific Conductivity	$L_k \propto (A_k^{CT})_k^0$	$L_k \propto (A_k^{CT})_k^0$
Leaf-Specific Conductivity	t $L_k \propto r_k^0$	$L_k \propto r_k^{2/3}$

moderate tapering of the tubes over the generations of branching or (2) by placing constrictions in the tubes. Real plants appear to use both mechanisms. There are several studies documenting that vessel radii taper from trunk to leaf, and the magnitude of this variation is consistent with our model (e.g., Zimmermann [137]). Furthermore, there is evidence for increasing range in xylem tube size with increasing plant size in the fossil record—supporting the prediction of increasing xylem tube diameter with increases in branch diameter [68, 69].

It is still, however, unclear whether the presence of constrictions plays a major role in dictating whole-plant hydraulic architecture. In 1983, Martin Zimmermann proposed the hydraulic segmentation hypothesis of plant architecture. He hypothesized that strong selection for restricting embolisms to peripheral branches has led to observed patterns of vascular resistance [137]. During times of extreme negative water potentials, a xylem conduit is prone to cavitation, and the resulting air embolism permanently shuts off fluid flow. Zimmermann hypothesized that constriction of xylem tubes at branching junctions served to isolate these damaging embolisms and were responsible for observed patterns of hydraulic architecture (i.e., conductivity, pressure differentials, etc.). Xylem in expendable peripheral branches and petioles were therefore more subject to catastrophic failure, whereas vessels in the main trunk and larger branches were protected from damaging embolisms.

There is evidence for such branch constrictions in several species of vascular plants [2, 25, 54, 137]. If these constrictions are of approximately the same size, z, at *all* branch junctions, they would contribute Nz to the total tube resistance. The resulting effect on total plant hydrodynamic resistance is almost identical to the result with tapering tubes, i.e. $\bar{a} = 1/6$, and leading to an additional resistance which grows logarithmically with length. Nevertheless, recent studies have indicated that these constrictions appear to play a very minor role in the total hydrodynamic resistance of xylem network [110, 130]. While our work proposes a different mechanism for whole-plant hydrodynamic resistance than that proposed by Zimmermann, our model shows how several anatomical, physiological, and architectural characteristics of vascular plants can be shown to be functionally linked by the biological demands of resource transport through fractal-like branching networks. Furthermore, selection for isolation of embolisms to peripheral branches may occur in the framework of our model so long as the number of constrictions are: (1) proportional to the number of branchings; and (2) the overall hydrodynamic resistance, due to constrictions, does not increase too rapidly with plant size.

Competition for limited resources resulting in increased size has apparently led to a design of trees that maximizes height, with compensating tapering of vascular tubes. Our model gives the novel prediction that in a given environment, with a fixed pressure difference between air and soil, *all xylem tubes of all plants conduct water and nutrients at approximately the same rate*. This counterintuitive result provides the basis for the "energetic equivalence rule" (see next section), where, across diverse ecosystems, the total use of resources is independent of plant size.

5 ECOLOGICAL CONSEQUENCES OF ALLOMETRIC SCALING

5.1 SCALING OF PLANT POPULATION DENSITY

The general plant model offers a mechanistic hypothesis for many observed anatomical and physiological allometries at the level of the individual. Here we show that it also offers a basis from which to construct mechanistic connections between these organismal processes and their ecological consequences. The biological and physical principles imposed upon vascular networks powerfully limit rates of whole-plant resource use which, in turn, ramifies to constrain biological organization of populations, ecological communities, and ecosystems. One example is the relationship between population density and body size in ecological communities [18, 26].

When the average dry mass of the average plant ($\bar{M}$) in mature populations is plotted against the maximum plant density ($N_{\max}$) there is a distinct upper boundary which traditionally has been characterized by a power law with an exponent $\approx -3/2$ [34, 39, 126, 131]. This pattern, known as the "$-3/2$

thinning law," has been proposed to hold for plants in both single and mixed species stands and over a size range spanning 12 orders of magnitude from unicellular algae to the tallest trees (Gorham [34]; White [126]; see also Agusti et al. [1]). The fact that as plants grow they fill a three-dimensional volume (a linear distance, L^3) and cover an exclusive area (a linear distance, L^2) has suggested a simple geometric explanation for the "$-3/2$" density/mass relationship (see Yoda et al. [131]; Miyanishi et al. [63]; White [125]; Norberg [80]). The constraints of packing geometric shapes into a finite area leads to a geometric limit between density and mass.

Initially, the $-3/2$ thinning relationship was cited one of the most general principles of plant ecology (e.g., Harper [39]). Recently, however, the theoretical and empirical bases for the density-mass boundary have been questioned and hotly debated [20, 31, 47, 57, 80, 81, 85, 117, 118, 119, 120, 133, 134]. The $-3/2$ exponent, derived from purely geometric considerations, is difficult to reconcile with known mechanisms of plant growth, resource uptake, and competition. Furthermore, increasingly precise data suggest that the interspecific boundary is closer to $-4/3$ (Weller [118]; see also Lonsdale [57]), indicating that population density scales as $M^{-3/4}$, which is the same exponent reported in animals [9, 18, 19, 61]. Because metabolic rates of animals scale as $M^{3/4}$, similar relationships in plants suggest that both share a common scaling law which reflects how resource requirements of individual organisms affect competition and spacing among individuals within ecological communities.

Here we develop a simple extension of our general model by following the ecological ramifications of allometric scaling of resource use in ecological communities. We assume that: (1) sessile plans compete for spatially limiting resources; (2) their allometric rate of resource use scales as $M^{3/4}$; and (3) they grow until they are limited by resources, so that the rate of resource use by plants approximates the rate of resource supply, R. The maximum number of individuals, N_{max}, that can be supported per unit area is related to the average rate of resource use per individual, $\bar{Q}$, by

$$R = N_{max}\bar{Q} \propto N_{max}\bar{M}^{-3/4}. \tag{12}$$

At equilibrium, when the rate of resource use approximates rates of resource supply, R is constant, giving

$$N_{max} \propto \bar{M}^{-3/4}. \tag{13}$$

This ecological extension of the general model, based on resource use by individual plants, therefore predicts a mass-density scaling exponent of $-4/3$ rather than $-3/2$ as predicted by the geometric model [131]. We analyzed data from the literature relating $\bar{M}$ and N_{max} (Figure 5) and found an exponent, -1.341, which is nearly identical to the predicted -1.333 and has statistical confidence intervals that include $-4/3$ but not $-3/2$. This supports recent studies which also suggest that the thinning exponent is closer to $-4/3$ than $-3/2$ (see Weller [117]; Lonsdale [57]; Franco and Kelly [31]). Other sources in the literature express total above-ground plant biomass per unit area (M_{tot}) as a function of maximum population density (see Weller [57, 85, 117, 118]). Using Eq. (13) our model predicts the scaling of total plant biomass per unit area:

$$M_{tot} = N_{max}\bar{M} \propto N_{max}N_{max}^{-4/3} \propto N_{max}^{-1/3} . \tag{14}$$

The geometric model [131] predicts an exponent of $-1/2$. In a previous analysis of data on interspecific populations, Lonsdale [57] found an exponent of -0.379, which is closer to $-1/3$ than to $-1/2$. Using a different data set, we performed a similar analysis (Figure 6) and found that maximum values of biomass per unit area, N_{max}, scales to the -0.325 power, which is statistically indistinguishable from the $-1/3$ predicted by our extension of the general model.

5.2   INVARIANCE OF ENERGY USE IN ECOSYSTEMS

Because the rate of resource use per unit area, Q_{tot} is the product of population density, N_{max}, and the average rate of resource use per individual, $\bar{Q}$, we have

$$Q_{tot} = N_{max}\bar{Q} \propto \bar{M}^{-3/4}\bar{M}^{3/4} \propto \bar{M}^0 . \tag{15}$$

Therefore, the model gives the nonintuitive prediction that the total resource use, Q_{tot}, or productivity of plants in ecosystems per unit area, is invariant with respect to body size. We tested this prediction by calculating Q_{tot} from the data used to compile Figures 3 and 5. As shown in Figure 7, the flux per surface area scales as $M^{0.0135}$. This empirical value does not differ statistically from the size invariance, M^0, predicted by the model. The relationship holds across 12 orders of magnitude variation in plant size. The variation around the regression line is expected to reflect variation in resource supply and therefore in productivity among ecosystems ranging from grasslands and tundra to temperate and tropical forests (e.g., Rosenzweig [92]). Our allometric-based estimation of maximum xylem surface flux in terrestrial ecosystems, ($\bar{x} = 16.3$ l m^{-2} d^{-1}, sd = 16.44) is within an order of magnitude of maximum values reported for terrestrial ecosystem flux rates (data summarized in Jones [50]). Together, these results indicate that a common allometric framework not only can explain anatomical and physiological parameters of individuals, but also can account for larger-scale patterns across ecological communities and ecosystems.

Most of the dissatisfaction with the original geometric formulation of the thinning law stems from its lack of a mechanism to predict empirically measured thinning trajectories, and the resulting variation in sizes and densities of plants among different environments (see White [126]; Ellison [23]; Weller [117, 118]; Norberg [80]; Lonsdale [57]). For example, as plant populations thin, individuals attain a certain maximum size that is characteristic of local environmental conditions. Not only are individuals larger in forests than in grasslands, but also each community typically contains multiple codominant species of nearly identical size [16, 107]. Our model does not predict thinning trajectories. It does, however, predict that the rate of resource use per unit area varies among plant communities with differences in resource supply but not with plant size. This result stems directly from the invariance of flux rate per individual xylem tube across plants of differing size predicted by our general model. Thus, ecosystems composed of plants of contrasting sizes and life forms, such as adjacent forests, grasslands, and agricultural fields, can have identical productivity (see Rosenzweig [92]; Harper [40]; Schulze et al. [97]; see also Dewar et al. [21]; K. Gross personal conversation). This does not appear

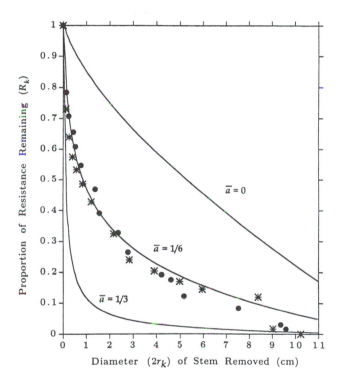

FIGURE 4 Effect of removing branch segments: the proportion of total resistance remaining, R_k, as a function of the diameter of the removed stem, $2r_k$. The data points, taken from [129], represent two different *Acer saccharum* trees. The solid lines are derived from our model. With $\bar{a} = 1/6$, as predicted from our model, the agreement is excellent. By contrast, with $\bar{a} = 0$, corresponding to no tapering of tubes as in the pipe model, and with $\bar{a} = 1/3$, the agreement is poor. Note that the curves terminate when only the trunk remains. If extrapolated all would converge at the trunk diameter, ~ 14.5 cm, where $2r_k = 2r_0$ and no conducting tissue remains so that $R_k = R_1 = 0$.

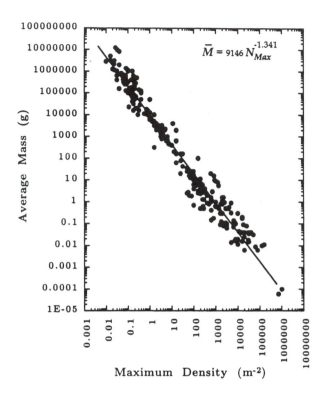

FIGURE 5 Relationship between average plant mass and maximum population density for 254 populations ($r^2 = 0.963$, $n=254$, $p < 0.0001$, 95% CI: -1.373 to -1.308). The exponent is statistically indistinguishable from $-4/3$ indicating that maximum population density scales as $M^{-3/4}$.

to have been appreciated despite much work by plant ecologists on relationships between population density and plant size and between individual plant performance and ecosystem productivity.

We have shown that metabolic rates of plants scale indistinguishably from the predicted $M^{3/4}$, and that a model which incorporates this scaling can account for the average sizes and maximum densities of plants observed in different communities. Traditionally, plant ecologists have implicitly treated the sizes of individuals as if they were determined by population density: plotting mass as the dependent variable in depicting thinning relationships. Animal physiologists and ecologists have done just the opposite: plotting density and other variables as functions of body mass (but see Hughes and Griffiths [46]; and Armstrong [4]). The theoretical and empirical advantage of the latter approach is that variables are expressed in terms of standard allometric

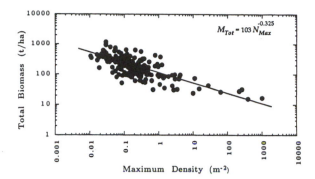

FIGURE 6   Relationship between total plant biomass (roots, shoots, and leaves) and maximal population density [11] ($r^2 = 0.534$, $n = 189$, $p < 0.0001$, 95% CI: -0.281 to -0.368). The fitted equation has an exponent statistically indistinguishable from $N_{\mathrm{Max}}^{-1/3}$ predicted from our model.

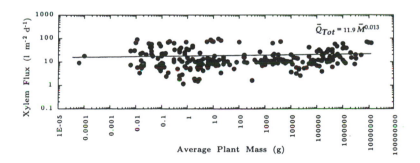

FIGURE 7   Relationship between total xylem flux and average size of the dominant plants in diverse ecosystems ($r^2 = 0.009$, $n = 254$, $p = 0.127$, 95% CI: -0.003 to 0.031). The stoichiometric equivalence of rates of water, nutrient, and photosynthate flux within individual plants (see text, and references therein) explains why actual evapotranspiration can be used to estimate productivity [92]. Since the allometric equation has an exponent that is statistically indistinguishable from zero, ecosystem productivity is independent of plant size.

equations which highlight the universality of the 3/4-power scaling of resource use and the related quarter-power scaling of other structural, functional, and ecological attributes [10, 84, 94]. It has long been known that life history variables, such as growth rate, life span, and age to first reproduction, scale with body size (e.g., Yoda et al. [132]; Grime and Hunt [37]; Tilman [107]; Chapin [16]; Charnov [17]). Despite this variation, it is noteworthy that rates of resource use per unit area are independent of body size. This is observed empirically in animals [18, 19] and demonstrated here, both empirically and theoretically for plants.

6 CONCLUSIONS

A general theory of allometry appears to offer a theoretical framework by which to draw mechanistic connections among several major areas of botanical research. Quarter-power allometric scaling laws, well known in animals, also apply to many characteristics of plants. There are numerous allometric parallels, including: metabolic rate $\propto M^{3/4}$, radius of trunk or aorta $\propto M^{3/8}$, size of and fluid velocity in terminal vessels $\propto M^0$, population density $\propto M^{-3/4}$, and energy use per unit area M^0. We hypothesize that the allometric constraints of transporting and distributing resources through biological networks causes many attributes of whole-organism biology and ecology, including the major botanical research themes listed at the beginning of this chapter. Our general model also shows why plants are excellent study systems for connecting biological attributes of individuals to macroscopic structures and dynamics in ecological communities. Because the anatomical, physiological, and ecological basis of resource extraction and distribution are relatively well understood, it is becoming possible to predict how anatomical and physiological traits change with varying size and across differing selective environments. Because plants are sessile and easy to sample, there are databases available for many diverse ecological and physiological attributes of both individual plants and entire ecological communities. This provides the basis from which to build a general framework for how allometric constraints at the level of the individual lead to ecological and evolutionary consequences of body size. Furthermore, the fact that both animals and plants share many identical allometric exponents suggests a common influence of body size in nearly all aspects of biological structure, function, and diversity.

ACKNOWLEDGMENTS

We thank K. J. Niklas and M. Tyree for their constructive comments. BJE was supported by an NSF post-doctoral fellowship and NSF grant GER-9553623. JHB was supported by NSF grant DEB-931896, and GBW by DOE contract ERWE161. We also acknowledge the generous support of the Thaw Charitable Trust.

REFERENCES

[1] Agusti, S., C. M. Duarte, and J. Kalff. "Algal Size and the Maximum Density and Biomass of Phytoplankton." *Limnol. Oceanog.* **32** (1987): 983–986.

[2] Aloni, R. J., D., Alexander, and M. T. Tyree. "Natural and Experimentally Altered Hydraulic Architecture of Branch Junctions in *Acer saccharum* March. and *Quercus velutina* Lam. Trees." *Trees* **11** (1997): 255–264.

[3] Aono, M., and L. K. Tosiyasu. "Botanical Tree Image Generation." *I.E.E.E. Comp. Graph. & Appl.* **27** (1984): 10–34.

[4] Armstrong, J. D. "Self-Thinning in Juvenile Sea Trout and Other Salmonid Fishes Revisited." *J. Animal Ecol.* **66** (1997): 519–526.

[5] Berninger, F., and E. Nikinmaa. "Implications of Varying Pipe Model Relationships on Scots Pine Growth in Different Climates." *Functional Ecology* **11** (1997): 146–156.

[6] Bernston, G. M., and P. Stoll. "Correcting for Finite Spatial Scales of Self-Similarity when Calculating the Fractal Dimensions of Real-World Structures." *Proc. Roy. Soc. Lond. B* **264** (1997): 1531–1537.

[7] Bertram, J. E. A. "Size-Dependent Differential Scaling in Branches: The Mechanical Design of Trees Revisited." *Trees* **4** (1989): 241–253.

[8] Borchert, R., and N. A. Slade. "Bifurcation Ratios and the Adaptive Geometry of Trees." *Botanical Gazette* **142** (1981): 394–401.

[9] Brown, J. H. *Macroecology.* Chicago, IL: University Chicago Press, 1995.

[10] Calder, W. A. I. *Size, Function and Life History.* Cambridge, MA: Harvard University Press, 1984.

[11] Cannell, M. G. R. *World Forest Biomass and Primary Production Data.* New York: Academic Press, 1982.

[12] Canny, M. J. "The Transpiration Stream in the Leaf Apoplast: Water and Solutes." *Phil. Trans. Roy. Soc. London Series B* **341** (1993): 87–100.

[13] Canny, M. J. "Apoplastic Water and Solute Movement: New Rules for an Old Space." *Ann. Rev. Ecol. & System.* **46** (1995): 215–236.

[14] Canny, M. J. "Applications of the Compensating Pressure Theory of Water Transport." *Amer. J. Botany* **85** (1998): 897–909.

[15] Carlquist, S. *Comparative Wood Anatomy.* New York: Springer-Verlag, 1988.

[16] Chapin, F. S., III. "Functional Role of Growth Forms in Ecosystem and Global Processes." In *Scaling Physiological Processes: Leaf to Globe,* edited by J. R. Ehleringer and C. B. Field, 287–308. New York: Academic Press, 1993.

[17] Charnov, E. L. *Life History Invariants: Some Explorations of Symmetry in Evolutionary Ecology.* Oxford: Oxford University Press, 1993.

[18] Damuth, J. "Population Density and Body Size in Mammals." *Nature* **290** (1981): 699–700.

[19] Damuth, J. "Interspecific Allometry of Population Density in Mammals and Other Animals: The Independence of Body Mass and Population Energy-Use." *Biol. J. Linn. Soc.* **31** (1987): 193–246.

[20] Dewar, R. C. "A Mechanistic Analysis of Self-Thinning in Terms of Carbon Balance of Trees." *Ann. Botany* **71** (1993): 147–159.

[21] Dewar, R. C., B. E. Medlyn, and R. R. McMurtrie. "A Mechanistic Analysis of Light and Carbon Use Efficiencies." *Plant Cell & Envir.* **21** (1998): 573–588.

[22] Ehleringer, J. R., and C. B. Field. *Scaling Physiological Processes: Leaf to Globe.* New York: Academic Press, 1993.

[23] Ellison, A. M. "Density-Dependent Dynamics of *Salicornia europaea* Monocultures." *Ecology* **68** (1987): 737–741.

[24] Ellison, A. M., and K. J. Niklas. "Branching Patterns of *Salicornia europaea* (Chenopodiaceae) at Different Successional Stages: A Comparison of Theoretical and Real Plants." *Amer. J. Botany* **75** (1988): 501–512.

[25] Ellison, A. M., K. J. Niklas, and S. Shumway. "Xylem Vascular Anatomy and Water Transport of *Salicornia europaea.*" *Aquatic Botany* **45** (1998): 325–339.

[26] Enquist, B. J., J. H. Brown, and G. B. West. "Allometric Scaling of Plant Energetics and Population Density." *Nature* **395** (1998): 163–165.

[27] Eshel, A. "On the Fractal Dimensions of a Root System." *Plant, Cell & Envir.* **21** (1998): 247–251.

[28] Ewers, F. W., and M. H. Zimmermann. "The Hydraulic Architecture of Eastern Hemlock (*Tsuga canadensis*)." *Canad. J. Botany* **62** (1984): 940–946.

[29] Farnsworth, K. D., and K. J. Niklas. "Theories of Optimization, Form, and Function in Branching Architecture in Plants." *Func. Ecology* **9** (1995): 355–363.

[30] Fitter, A. H., and T. R. Strickland. "Fractal Characterization of Root System Architecture." *Funct. Ecology* **6** (1992): 632–635.

[31] Franco, M., and C. K. Kelly. "The Interspecific Mass-Density Relationship and Plant Geometry." *Proc. Natl. Acad. Sci.* **95** (1998): 7830–7835.

[32] Givnish, T. "On the Adaptive Significance of Leaf Height in Forest Herbs." *Amer. Natur.* **120** (1982): 353–381.

[33] Givnish, T. "Comparative Studies of Leaf Form: Assessing the Relative Roles of Selective Pressures and Phylogenetic Constraints." *New Phytologist* **106** (suppl.) (1987): 131–160.

[34] Gorham, E. "Shoot Height, Weight, and Standing Crop in Relation to Density in Monospecific Plant Stands." *Nature* **279** (1979): 148–150.

[35] Grace, J. "Plant Water Relations." In *Plant Ecology*, edited by M. J. Crawley, 28–50. Cambridge, UK: Blackwell Science, 1997.

[36] Greenhill, A. G. "Determination of the Greatest Height Consistent with Stability that a Vertical Pole or Mast Can Be Made, and the Greatest

Height to Which a Tree of Given Proportions Can Grow." *Proc. Cam. Phil. Soc.* **4** (1881): 65–73.

[37] Grime, J. P., and R. Hunt. "Relative Growth-Rate: Its Range and Adaptive Significance in Local Flora." *J. Ecology* **63** (1975): 393–422.

[38] Halle, F., R. A. A. Oldemann, and P. B. Tomlinson. *Tropical Trees and Forests: An Architectural Analysis.* Berlin: Springer-Verlag, 1978.

[39] Harper, J. "A Darwinian Approach to Plant Ecology." *J. Ecology* **55** (1967): 247–270.

[40] Harper, J. *Population Biology of Plants.* New York: Academic Press, 1977.

[41] Honda, H. "Description of the Form of the Tree-like Body: Effects of Branching Angle and the Branch Length on the Shape of the Tree-like Body." *J. Theor. Biol.* **31** (1971): 331–338.

[42] Horn, H. S. *The Adaptive Geometry of Trees.* Princeton, NJ: Princeton University Press, 1971.

[43] Huber, B. "Weitere Quantitative Untersuchungen uber sas Wasserleitungssystem der Pflanzen." *Jb. Wiss. Bot.* **67** (1928): 877–959.

[44] Huber, B. "Observation and Measurement of Plant Sap Streams." *Berichte Deutsche Botanische Gesellschaft* **50** (1932): 89–109.

[45] Huber, B., and E. Schmidt. "Weitere Thermo-elektrische Uuntersuchungen uber den Transpirationsstrom der Baume." *Tharandt Forst* Jb **87** (1936): 369–412.

[46] Hughes, R. N., and C. L. Griffiths. "Self-Thinning in Barnacles and Mussels: The Geometry of Packing." *Amer. Natur.* **132** (1988): 484–491.

[47] Hutchings, M. "Ecology's Law in Search of a Theory." *New Scientist* **98** (1983): 765–767.

[48] Huxley, J. S. *Problems of Relative Growth.* London: Methuea, 1932.

[49] Jones, H. G. "Modeling Diurnal Trends of Leaf Water Potential in Transpiring Wheat." *J. Appl. Ecol.* **15** (1978): 613–626.

[50] Jones, H. G. *Plants and Microclimate: A Quantitative Approach to Environmental Plant Physiology.* Cambridge, MA: Cambridge University Press, 1992.

[51] King, D. A., and O. L. Louks. "The Theory of Tree Bole and Branch Form." *Radiat. Env. Biophys.* **15** (1978): 141–165.

[52] Kuuluvainen, T., D. G. Sprugel, and J. R. Brooks. "Hydraulic Architecture and Structure of *Abies lasiocarpa* Seedlings in Three Alpine Meadows of Different Moisture Status in the Eastern Olympic Mountains, Washington, U.S.A." *Arctic & Alpine Res.* **28** (1996): 60–64.

[53] Leopold, L. B. "Trees and Streams: The Efficiency of Branching Patterns." *J. Theor. Biol.* **31** (1971): 339–354.

[54] Lev-Yadun, S., and R. Aloni. "Vascular Differentiation in Branch Junctions of Trees: Circular Patterns and Functional Significance." *Trees* **4** (1990): 49–54.

[55] Levin, S. "The Problem of Pattern and Scale in Ecology." *Ecology* **73** (1992): 1943–1967.

[56] Long, J. N., F. W. Smith, and D. R. M. Scott. "The Role of Douglas-Fir Stem Sapwood and Heartwood in the Mechanical and Physiological Support of Crowns and Development of Stem Form." *Canad. J. Forest Res.* **11** (1981): 459–464.

[57] Lonsdale, W. M. "The Self-Thinning Rule: Dead or Alive?" *Ecology* **71** (1990): 1373–1388.

[58] Machado, J., and M. T. Tyree. "Patterns of Hydraulic Architecture and Water Relations of Two Tropical Trees with Contrasting Leaf Phenologies: *Ochroma pyramidale* and *Pseudobombax septenatum*." *Tree Physiol.* **14** (1994): 219–240.

[59] Mandelbrot, B. B. *The Fractal Geometry of Nature*. New York: W. H. Freeman, 1977.

[60] McMahon, T. A. "Size and Shape in Biology." *Science* **179** (1973): 1201–1204.

[61] McMahon, T. A., and J. T. Bonner. *On Size and Life*. New York: Scientific American Library, 1983.

[62] McMahon, T. A., and R. E. Kronauer. "Tree Structures: Deducing the Principle of Mechanical Design." *J. Theor. Biol.* **59** (1976): 443–466.

[63] Miyanishi, K., A. R. Hoy, and P. B. Cavers. "A Generalized Law of Self-Thinning in Plant Populations." *J. Theor. Biol.* **78** (1979): 439–442.

[64] Morgan, J., and M. G. R. Cannell. "Support Costs of Different Branch Designs: Effects of Position, Number, Angle, and Deflection of Laterals." *Tree Physiology* **4** (1988): 303–313.

[65] Morse, D. R., J. H. Lawton, J. H. Dodson, and M. M. Williamson. "Fractal Dimension of Vegetation and the Distribution of Arthropod Body Lengths." *Nature* **314** (1985): 731–733.

[66] Murray, C. D. "A Relationship Between Circumference and Weight in Trees and Its Bearing on Branching Angles." *J. Gen. Phys.* **10** (1927): 725–739.

[67] Nielson, K. L., J. P. Lynch, and H. N. Weiss. "Fractal Geometry of Bean Root Systems: Correlations Between Spatial and Fractal Dimension." *Amer. J. Botany* **84** (1997): 26–33.

[68] Niklas, K. J. "Size-Related Changes in the Primary Xylem Anatomy of Some Early Tracheophytes." *Paleobiology* **10** (1984): 487–506.

[69] Niklas, K. J. "The Evolution of Tracheid Diameter in Early Vascular Plants and Its Implications on the Hydraulic Conductance of the Primary Xylem Strand." *Evolution* **39** (1985): 1110–1122.

[70] Niklas, K. J. "The Role of Phyllotactic Pattern as a Developmental Constraint on the Interception of Light by Leaf Surfaces." *Evolution* **40** (1988): 1–16.

[71] Niklas, K. J. *Plant Biomechanics: An Engineering Approach to Plant Form and Function*. Chicago, IL: University Chicago Press, 1992.

[72] Niklas, K. J. "The Allometry of Plant Reproductive Biomass and Stem Diameter." *Amer. J. Botany* **80** (1993): 461–467.

[73] Niklas, K. J. "Morphological Evolution Through Complex Domains of Fitness." *Proc. Natl. Acad. Sci.* **91** (1994): 6772–6779.

[74] Niklas, K. J. *Plant Allometry: The Scaling of Form and Process.* Chicago, IL: University Chicago Press, 1994.

[75] Niklas, K. J. "Size-Dependent Variations in Plant Growth Rates and the '3/4-Power Rule.'" *Amer. J. Botany* **81** (1994): 134–145.

[76] Niklas, K. J. "Effects of Hypothetical Developmental Barriers and Abrupt Changes on Adaptive Walks in a Computer Generated Domain for Early Vascular Land Plants." *Paleobiology* **23** (1997): 63–67.

[77] Niklas, K. J. "Size- and Age-Dependent Variation in the Properties of Sap- and Heartwood in Black Locust (*Robinia pseudoacacia* L.)" *Ann. Botany* **79** (1997): 473–478.

[78] Niklas, K. J. *The Evolutionary Biology of Plants.* Chicago, IL: University Chicago Press, 1997.

[79] Niklas, K. J., and V. Kerchner. "Mechanical and Photosynthetic Constraints on the Evolution of Plant Shape." *Paleobiology* **10** (1984): 79–101.

[80] Norberg, R. A. "Theory of Growth Geometry of Plants and Self-Thinning of Plant Populations: Geometric Similarity, Elastic Similarity, and Different Growth Modes of Plant Parts." *Amer. Natur.* **131** (1988): 220–256.

[81] Osawa, A., and S. Suigita. "The Self-Thinning Rule: Another Interpretation of Weller's Results." *Ecology* **70** (1989): 279–283.

[82] Patino, S., M. T. Tyree, and H. Herre. "Comparison of Hydraulic Architecture of Wood Plants of Differing Phylogeny and Growth Form with Special Reference to Free Standing and Hemi-epiphytic *Ficus* Species from Panama." *New Phytologist* **129** (1995): 125–134.

[83] Pearsall, W. H. "Growth Studies. VI. On the Relative Sizes of Growing Plant Organs." *Ann. Botany* **41** (1927): 549–556.

[84] Peters, R. H. *The Ecological Implications of Body Size.* Cambridge: Cambridge University Press, 1983.

[85] Petraitis, P. S. "Use of Averages vs. Total Biomass in Self-Thinning Relationships." *Ecology* **76** (1995): 656–658.

[86] Raven, J. A. "The Evolution of Vascular Plants in Relation to Quantitative Functioning of Dead Water Conducting Cells and Atomata." *Biol. Rev.* **68** (1993): 337–363.

[87] Raven, J. A., and L. L. Handley. "Transport Processes and Water Relations." *New Phytologist* (Suppl.) **106** (1987): 217–233.

[88] Rashevsky, N. "The Principle of Adequate Design." In *Foundations of Mathematical Biology*, edited by R. Rosen, vol. III, 143–175. New York: Academic Press, 1973.

[89] Richter, J. P. *The Notebooks of Leonardo da Vinci, 1452–1519.* New York: Dover, 1970.

[90] Robichaud, E., and I. R. Methven. "The Applicability of the Pipe Model Theory for the Prediction of Foliage Biomass in Trees from Natural, Untreated Black Spruce Stands." *Canad. J. Forest Res.* **22** (1992): 1118–1123.

[91] Rogers, R., and T. M. Hinckley. "Foliar Weight and Area Related to Current Sapwood Area in Oak." *Forest Sci.* **25** (1979): 298–303.

[92] Rosenzweig, M. L. "Net Primary Productivity of Terrestrial Communities: Prediction from Climatological Data." *Amer. Natur.* **102** (1968): 67–74.

[93] Sakurantani, T. "A Heat Balance Method for Measuring Water Flux in the Stem of Intact Plants." *J. Agr. Met.* **37** (1981): 9–17.

[94] Schmidt-Nielsen, K. *Scaling: Why Is Animal Size so Important?* Cambridge, MA: Cambridge University Press, 1984.

[95] Schulte, P. J., and D. G. Costa. "A Mathematical Description of Water Flow through Plant Tissues." *J. Theor. Biol.* **180** (1996): 61–70.

[96] Schulze, E. D., J. Cermak, R. Matyssek, M. Penka, R. Zimmermann, F. Vasicek, W. Gries, and J. Kucera. "Canopy Transpiration and Water Fluxes in the Xylem of the Trunk of *Larix* and *Picea* Trees—A Comparison of Xylem Flow, Porometer, and Cuvette Measurements." *Oecologia* **66** (1985): 475–483.

[97] Schulze, E. D., F. M. Kelliher, C. Korner, J. Lloyd, and R. Leuning. "Relationships Among Maximal Stomatal Conductance, Carbon Assimilation Rate, and Plant Nitrogen Nutrition: A Global Ecology Scaling Exercise." *Ann. Rev. Ecol. & System.* **25** (1994): 629–660.

[98] Shidei, T., and T. Kira. *Primary Productivity of Japanese Forests: Productivity of Terrestrial Communities (JIBP Synthesis)*. Tokyo: University of Tokyo Press, 1977.

[99] Shinozaki, K., K. Yoda, K. Hozumi, and T. Kira. "A Quantitative Analysis of Plant Form—The Pipe Model Theory: I. Basic Analysis." *Jap. J. Ecol.* **14** (1964): 97–105.

[100] Sinnott, E. W. "The Relation Between Body Size and Organ Size in Plants." *Amer. Natur.* **55** (1921): 385–403.

[101] Smith, J. A. C., P. J. Schulte, and P. S. Nobel. "Water Flow and Water Storage in *Agave deserti*: Osmotic Implications of Crassulacean Acid Metabolism." *Plant, Cell, & Environment* **10** (1987): 639–648.

[102] Sperry, J. S., N. N. Alder, and S. E. Eastlack. "The Effect of Reduced Hydraulic Conductance on Stomatal Conductance and Xylem Cavitation." *J. Exper. Botany* **44** (1993): 1075–1082.

[103] Stevens, G. C. "Lianas as Structural Parasites: The *Bursera simaruba* Example." *Ecology* **68** (1987): 77–81.

[104] Tatsumi, J. A., A. Yamauchi, and Y. Kono. "Fractal Analysis of Plant Root Systems." *Ann. Botany* **64** (1989): 499–503.

[105] Thomas, S. C. "Asymptotic Height as a Predictor of Growth and Allometric Characterizations in Malaysian Rain Forest Trees." *Amer. J. Botany* **83** (1996): 556–566.

[106] Thomas, S. C. "Reproductive Allometry in Malaysian Rain Forest Trees: Biomechanics Versus Optimal Allocation." *Evol. Ecology* **10** (1996): 517–530.

[107] Tilman, D. *Plant Strategies and the Dynamics and Structure of Plant Communities.* Princeton, NJ: Princeton University Press, 1988.

[108] Turrell, F. M. "Growth of the Photosynthetic Area of *Citrus.*" *Botanical Gazette* **122** (1961): 284–298.

[109] Tyree, M. T. "A Dynamic Model for Water Flow in a Single Tree: Evidence that Models Must Account for Hydraulic Architecture." *Tree Physiol.* **4** (1988): 195–217.

[110] Tyree, M. T., and J. D. Alexander. "Hydraulic Conductivity of Branch Junctions in Three Temperate Tree Species." *Trees* **7** (1993): 156–159.

[111] Tyree, M. T., and F. W. Ewers. "The Hydraulic Architecture of Trees and Other Woody Plants." *New Phytologist* **119** (1991): 345–360.

[112] Tyree, M. T., and F. W. Ewers. "Hydraulic Architecture of Woody Tropical Plants." In *Tropical Forest Plant Ecophysiology*, edited S. S. Mulkey, R. L. Chazdon, and A. P. Smith, 217–243. New York: Chapman and Hall, 1996.

[113] Tyree, M. T., M. E. D. Graham, K. E. Cooper, and L. J. Bazos. "The Hydraulic Architecture of *Thuja occidentalis.*" *Canad. J. Botany* **61** (1983): 2105–2111.

[114] Tyree, M. T., and J. S. Sperry. "Do Woody Plants Operate Near the Point of Catastrophic Xylem Disfunction Caused by Dynamic Water Stress? Answers from a Model." *Plant Physiology* **88** (1988): 574–580.

[115] Van den Honert, T. H. "Water Transport in Plants as a Catenary Process." *Discuss. Faraday Soc.* **3** (1948): 146–153.

[116] Waring, R. H., P. E. Schroeder, and R. Oren. "Application of the Pipe Model Theory to Predict Canopy Leaf Area." *Canad. J. Forest Res.* **12** (1982): 556–560.

[117] Weller, D. E. "A Re-Evaluation of the −3/2 Power Rule of Plant Self-Thinning." *Ecol. Monographs* **57** (1987): 23–43.

[118] Weller, D. E. "The Interspecific Size-Density Relationship Among Crowded Plant Stands and Its Implications for the −3/2 Power Rule of Self-Thinning." *Amer. Natur.* **133** (1989): 20–41.

[119] Weller, D. E. "Will the Real Self-Thinning Rule Please Stand Up?—A Reply to Osawa and Sugita." *Ecology* **71** (1990): 1204–1207.

[120] Weller, D. E. "The Self-Thinning Rule: Dead or Unsupported?—A Reply to Lonsdale." *Ecology* **72** (1991): 747–750.

[121] West, G. B., J. H. Brown, and B. J. Enquist. "A General Model for the Origin of Allometric Scaling Laws in Biology." *Science* **276** (1997): 122–126.

[122] West, G. B., J. H. Brown, and B. J. Enquist. "A General Model for the Structure and Allometry of Plant Vascular Systems." *Nature* (1999): submitted.

[123] Westoby, M. "Self-Thinning Driven by Leaf Area not by Weight." *Nature* **265** (1977): 330–331.

[124] Westoby, M. "The Self-Thinning Rule." *Adv. Ecol. Res.* **14** (1984): 167–225.

[125] White, J. "The Allometric Interpretation of the Self-Thinning Rule." *J. Theor. Biol.* **89** (1981): 475–500.

[126] White, J. "The Thinning Rule and Its Application to Mixtures of Plant Populations." In *Studies on Plant Demography: A Festschrift for John L. Harper*, edited by J. White, 291–309. New York: Academic Press, 1985.

[127] Whitney, G. G. "The Bifurcation Ratio as an Indicator of Adaptive Strategy in Woody Plant Species." *Bull. Torr. Bot. Club* **103** (1976): 67–72.

[128] Whittaker, R. H., and G. M. Woodwell. "Dimension and Production Relations of Trees and Shrubs in the Brookhaven Forest, New York." *J. Ecology* **57** (1968): 1–25.

[129] Yang, S., and M. T. Tyree. "Hydraulic Resistance in *Acer saccharum* Shoots and Its Influence on Leaf Water Potential and Transpiration." *Tree Physiology* **12** (1993): 231–242.

[130] Yang, S., and M. T. Tyree. "Hydraulic Architecture of *Acer saccharum* and *A. rubrum*: Comparison of Branches to Whole Trees and the Contribution of Leaves to Hydraulic Resistance." *J. Exper. Botany* **45** (1994): 179–186.

[131] Yoda, K., T. Kira, H. Ogawa, and K. Hozumi. "Self-Thinning in Overcrowded Pure Stands Under Cultivated and Natural Conditions." *J. Biol. Osaka City University* **14** (1963): 107–129.

[132] Yoda, K., K. Shinozaki, H. Ogawa, K. Hozumi, and T. Kira. "Estimation of the Total Amount of Respiration in Woody Organs of Trees and Forest Communities." *J. Biol. Osaka City University* **16** (1965): 15–26.

[133] Zeide, B. "Analysis of the 3/2 Power Law of Self-Thinning." *Forest Sci.* **33** (1987): 517–537.

[134] Zeide, B. "Self-Thinning and Stand Density." *Forest Science* **37** (1991): 517–523.

[135] Zimmermann, M. H. "Hydraulic Architecture of Some Diffuse-Porous Trees." *Canad. J. Botany* **56** (1978): 2286–2295.

[136] Zimmermann, M. H. "Structural Requirements for Optimal Water Conduction in Tree Stems." In *Tropical Trees as Living Systems*, edited by P. B. Tomlinson and M. H. Zimmermann. New York: Cambridge University Press, 1978.

[137] Zimmermann, M. H. *Xylem Structure and the Ascent of Sap.* Berlin: Springer-Verlag, 1983.

[138] Zimmermann, M. H., and C. L. Brown. *Trees. Structure and Function.* New York: Springer, 1971.

Twigs, Trees, and the Dynamics of Carbon in the Landscape

Henry S. Horn

Terrestrial plants in general and trees in particular grow by two mechanisms, the geometrical proliferation of buds that encapsulate the instructions for further proliferation, and the expansion of tissues in approximate proportion to the amount of tissue already present. Limits to this growth are imposed by the availability of resources that must be shared with neighbors or with competing parts of the plant itself. If all resources were divided equally within a plant, a fractal or nearly fractal structure would result. A few simple measurements on a seedling could predict its adult form. A few simple measurements on a single individual, or even on part of an individual, could be scaled up allometrically to predict functional properties of a whole community. Turning this argument around, some basic questions about economic and conservative mechanisms of growth, allocation, and competition can be posed and addressed by students who are willing to take and analyze a few simple measurements.

Biological scaling provides splendid opportunities to design quantitative exercises for courses in field biology, ecology, and environmental science. Almost any pair of characters can be measured on individuals of different sizes, and plotted against each other on log-log axes. The resulting allometry can be compared and contrasted between species or within species between environments (Bonner and Horn this volume and Harvey this volume). Measurements on individuals can be extended to estimate parameters that are functionally

Scaling in Biology, edited by J. H. Brown and G. B. West.
Oxford University Press, 2000. **199**

significant for populations (Cyr this volume), and for communities (Enquist et al. [11], also this volume). And allometric techniques can even scale results to whole regions (Brown et al. this volume, Harte this volume, and Calder this volume). Allometries and isometries can suggest ways of calculating maxima and minima to bound the true parameters between dicey extremes.

This chapter gives three examples, all of them works in progress rather than definitive results. The first example dissects a tree with a chain saw to reconstruct its history of growth in height and diameter. The standing taper of the tree and its developmental trajectory are then compared with economic resistance to buckling [13, 28, 29]. The second example examines Leonardo da Vinci's assertion that cross-sectional area is preserved whenever a tree branches [21]. This is an important starting point for West, Brown, and Enquist's model [48] (also this volume) that unifies the explanation of a variety of allometries from scales of physiological morphology to community structure, yet Tyree [45] argues that this assumption sometimes fails. The third example starts with a primitive computer program that draws fractal trees from measurements made on a single twig [19]. At the level of outright caricature, some of the computer-drawn trees share some of the characteristics of some species of real trees. But some ecologically important characteristics of the computer trees are relatively independent of the details of the fractal model. One of those characteristics affects the proportion of a tree's productivity that is fixed in long-term carbon storage. And the carbon stored in the above-ground wood is easy to measure if Leonardo's assertion holds. So my second and third examples span the scales from measurements of twigs to implications for global warming.

1 DISSECTED TREE EXAMPLE

McMahon [26] calculated the height above which a tree of a given diameter is likely to suffer elastic collapse. By balancing the torque exerted by a leaning column against the elastic force needed to set it upright, and calculating the minimum diameter needed to exert this force, he found that $D^2 Y/g/\rho/H^3 =$ constant, where D and H are respectively diameter and height, Y is Young's modulus of elasticity, g is the universal gravitational constant, ρ is the specific density of wood, and the constant is a sort of safety factor. He discovered that the ratio Y/ρ (elasticity/density) is approximately constant, i.e., that heavy woods tend to be stiff or springy and light woods tend to be floppy. So $H^3 \alpha D^2$, equivalently $H \alpha D^{2/3}$, or $\log H = 2/3 \log D + \text{constant}$. Furthermore, by a clever analogy between a tree and a fractally divided ax-head-shaped beam, McMahon and Kronauer [28] extended this result from a simple column to a branching tree.

McMahon [26] plotted $\log H$ against $\log D$ for trees selected for prodigious size, and found the expected slope of 2/3. This suggests that trees are indeed adapted to elastic rebound in the wind, rather than simply growing by a fixed amount in diameter and in height each year, a mechanism that would have produced a slope of 1 on his graph. So a study of allometry has produced an inference about adaptation. Furthermore, he found that large trees are about 1/4 the critical buckling height. However, of his 576 trees, 502 were from the American Forestry Association's "Social Register of Big Trees," for which inclusion is based on an index that sums height in feet plus circumference in inches [32]. So for McMahon's data, diameter is weighted over height by a factor of $12/\pi = 3.8$, close to the apparent safety factor.

Trees growing in a New Jersey forest (Figure 1) show dimensions that roughly parallel McMahon's critical buckling line, but with a much lower safety factor, averaging about 2, and as low as 1.2 for some spindly individuals.

The growth of an individual tree can be reconstructed by felling the tree and counting the annual rings in cross sections of the trunk at regular intervals. Figure 2 shows how the age at each height can be calculated by subtracting the age of the cross section at that height from the age at the base. The basal

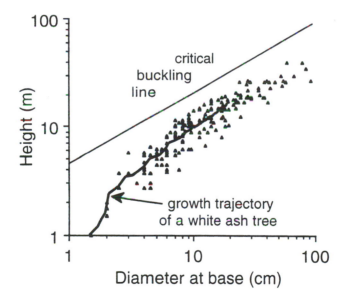

FIGURE 1 Allometry of height and diameter of forest trees, Princeton, NJ. The critical buckling line is the height above which a tree is likely to suffer elastic collapse ($H = 4.3\,D^{2/3}$ after McMahon [26]). The tiny triangles are 185 forest-grown trees of 15 species. On average they are about half the critical height. The trajectory of a white ash tree was reconstructed by Hoffman and Horn [16] by the technique illustrated in Figure 2. The trajectory is unsustainable early on, but becomes conservative as the tree grows.

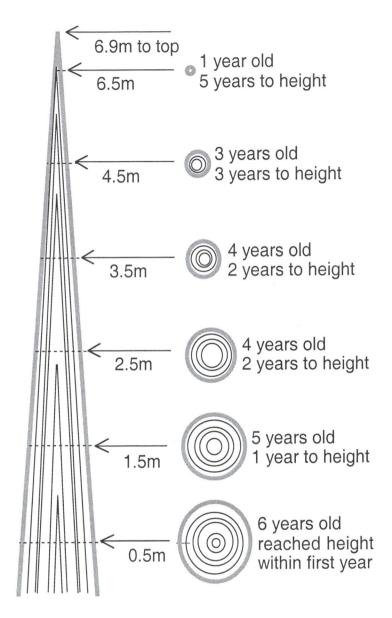

FIGURE 2   How to reconstruct the history of growth of the bole of a tree. This is a diagrammatic representation of the technique used to reconstruct the history of growth in height and diameter of the ash tree whose allometry is plotted in Figure 1. On the left is an idealized transverse section of the bole of a tree. On the right are cross sections taken at regular intervals. Their annual rings are counted. Subtraction of the age of each cross section from the age at the base gives the age at which the tree reached the height of that cross section. The diameter at each height and age can then be measured on the basal cross section [16].

diameter of the whole trunk at that age can be measured on the basal cross section. Hoffman and Horn [16] made these measurements on a 17-m tall white ash (*Fraxinus americana*), sectioned at 20-cm intervals. Their data are also plotted in Figure 1.

The early growth of the ash tree is clearly on a trajectory that, if continued, would intersect the critical buckling line. After a height of about 5 m, the tree grew more conservatively, toward becoming roughly parallel to the critical buckling line with a safety factor of about 2.

It is not my purpose to test McMahon's suggestion of elastic similarity critically against alternatives, though that is worth doing with similar data (see Niklas [30] and LaBarbera [22] for criticism, and LaBarbera [23] for technical suggestions). Rather, I simply assert that if trees are indeed designed for elastic stability, then they compromise safety when competing to remain in a canopy of increasing height. It is also not clear whether the difference in dimensions of my trees compared to McMahon's is due to a "strategic" physiological response to competitive setting, or to a direct physiological response of wind-induced thickening in his more open-grown trees [27, 29]. And of course wind-induced thickening may be viewed as a physiological response specifically evolved as a developmental mechanism to meet a strategic challenge (Bonner and Horn this volume). But none of these questions would have been raised without the allometric analysis.

2 LEONARDO'S RULE

In his notes for a treatise on painting, Leonardo da Vinci presented a number of insightful principles of ecology and developmental architecture for trees [35, 21]. Among these is the assertion that when a tree branches, the total cross-sectional area beyond the branch should equal the cross-sectional area before the branch. His reason for this principle was conservation of hydraulic function, which was made explicit by Shinozaki et al. [37, 38] in a model of a tree as a system of independent parallel pipes from root to leaf. The "pipe model" conjures images of a water distribution system of pipes of increasing multiplicity and decreasing diameter, but Shinozaki's model is more akin to a cable containing many bundled capillary tubes (Enquist et al. this volume). If trees were indeed such cables, then branching might be expected to be area-preserving. However, the hydraulics of many trees do not behave this simply [45].

Twigs may have varying amounts of pith central to the explicitly conductive xylem. In particular, at least in wet climates, trees with large or compound leaves often have thick twigs with extensive pith. As a pithy twig grows and branches, the additional xylem added by secondary thickening may initially be a relatively small proportion of the cross-sectional area. So branchings may continue almost the same cross-sectional area for each daughter twig as for the mother twig.

At the other extreme, the older trunk may have extensive nonconductive heartwood and little conductive sapwood. Furthermore, water conduction in trees with xylem vessels of large diameter (technically described as "ring-porous" wood) may depend on only the outermost annual ring, because the inner rings are vulnerable to cavitation and embolism [20, 46, 53]. For trees that use only the most recent annual ring, and that preserve vessel size and density (which many trees do not), hydraulic conservation at a branching node implies conservation of circumference rather than cross-sectional area.

To test these ideas I form an index that takes a particular value for each extreme alternative, and an intermediate value for area preserving. I select symmetrically bifurcating branches, measure distal diameters a and b, and proximal diameter c. Then I calculate the largest angle (C) in a triangle with sides a, b, and c. The index testing Leonardo's rule is, from the standard trigonometric formula, $\cos(C) = (a^2 + b^2 - c^2)/(2ab)$. Here is its interpretation. For very pithy twigs, I expect $a = b = c$; the triangle is equilateral, angle$C = 60°$, and $\cos(C) = 0.5$. If a branch obeys Leonardo's rule, $a^2 + b^2 = c^2$, the triangle is right, angle$C = 90°$, and $\cos(C) = 0.0$. For a ring-porous tree with xylem vessels of very large diameter, I expect that circumference and hence diameter is preserved, and $a + b = c$; the "triangle" is a straight line, angle$C = 180°$, and $\cos(C) = -1.0$.

In Figure 3 this index is plotted against proximal diameter for several species chosen to portray the range of behavior predicted above (after Tyree and Alexander [44]). Red cedar (*Juniperus virginiana*), with little pith and no explicit vessels, and sugar maple (*Acer saccharum*), with little pith and small vessels, were expected to obey Leonardo's rule throughout. Black walnut (*Juglans nigra*), with strong pith, strong heartwood, and medium vessels, was expected to have twigs of roughly common size for several levels of branching and to veer toward circumference preserving toward the trunk. Red oak (*Quercus rubra*), with large vessels, was expected to depart strongly from Leonardo's rule for large branches, and white oak (*Quercus alba*) to depart even more because it typically blocks past years' vessels with the growth of cellular tyloses.

Figure 3 is full of surprises. Among twigs, the pattern of departure from Leonardo's rule is in the expected direction, but stronger than expected, even in some species that have little recognizable pith. That is, the twigs and smaller branches are generally thicker than Leonardo's area-preserving rule predicts. Further interpretations must await more precise measurements, structured to disentangle the causes of variation at the small end of the scale. Toward the trunk, however, all species seem to obey Leonardo's rule, regardless of the different hydraulic permeabilities of their wood (see also Arastu [4] and Leigh [24]).

Apparently, cross-sectional area is roughly preserved in large branches of a wide variety of trees. An initial assumption made by West, Brown, and Enquist [48] (also this volume) is upheld, but the mechanism is more complicated than the first principles that they cite, specifically Shinozaki et al.'s [37, 38]

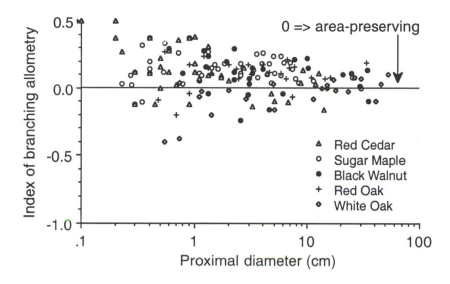

FIGURE 3 Index of branching allometry for branches of different sizes. The index (see text) varies from −1.0 if circumference is preserved at a branching, to 0.0 if area is preserved (Leonardo da Vinci's rule), to 0.5 if daughter branches are the same diameter as their mother. Large branches tend to obey Leonardo's area-preserving rule, but twigs and small branches tend to lie above the area-preserving line. Small daughter branches tend toward having a larger total cross-sectional area than their mother.

representation of a tree as a cable of bundled capillary vessels, all of them active. Specific conductivities in the hydraulically active wood are allometrically related to size of branch [54], but it is not yet clear that the allometry compensates in such a way as to rescue the hydrodynamic argument for area preservation. Leonardo's rule is also an explicit assumption in McMahon and Kronauer's [28] ax-head model for optimal support of a tree. And it is used below to estimate the volume of wood in the above-ground part of a tree. So an approximate confirmation of the area-preserving rule is empirically useful even though its conceptual basis is open for further study.

3 FRACTAL TREES

Trees grow by the proliferation and extension of buds that encapsulate the geometrical pattern of proliferation of future buds. Hence the growth of trees is inherently fractal, though a full account must include suppression and death of buds and branches as well as their origin [15, 47]. A very simple fractal model captures some of the biologically interesting features of the growth of real trees (see Horn [19], after Abelson and diSessa [2]). This model consists of fractal recursions of a trident, or turkey-track, on itself with parameters that imitate properties seen in real trees.

Figure 4 gives examples that justify these parameters. The first parameter is "growth deceleration," the ratio by which the turkey-track gets smaller with each recursion. The model pretends that a tree starts as a gigantic bud that produces smaller buds with each round of proliferation. This happens to be true for the elm twigs in Figure 4, but it is utterly false in the whole elm tree, and in all other trees as well. Nevertheless, it is surprising how realistic some of the fractal trees look given this initial falsehood. Real trees do not keep all the earlier branches. The "deceleration parameter" is due primarily to loss of earlier formed branches, and the resulting simplification of the main skeleton, rather than to a lesser extension of terminal twigs. A constant "deceleration parameter" might be produced by a "stable age distribution" in the population biology of terminal meristems, but this is less an explanation than an hypothesis for further exploration.

In practice, the "growth deceleration" is estimated by the ratio of two lengths—from the terminus of a twig to the first branching, and from that branching to the next branching at which another twig enters for its second branching. The elm in Figure 4 would give a growth deceleration of 1.0 between the longest terminal twig and the next branching, but a more realistic value of 2.1 to the following branching.

The second parameter is "branching order," the number of fractal recursions between trunk and terminal twig. It is intended to represent the order defined by Strahler [40] for the geomorphology of streams, where each terminal twig is defined as order one, and an $(n + 1)$th-order branch is defined wherever two nth-order branches come together. To determine the literal Strahler branching order for the trunk of a tree requires labeling every terminal twig and branch segment. However, a simple procedure does nearly as well. Follow a terminal twig, by definition order one, back to the trunk, and each time that what you are following joins something of its own size or larger, add one to the count. Do this for several twigs, take the largest number as the branching order for the tree, and you will seldom miss the Strahler order by more than one. The example of Figure 4 illustrates the common pattern that trees with compound leaves, like black walnut, have lower branching order (in this case 5), than trees of similar shape with simple leaves, like sugar maple (branching order in this case 8). The rachis of the compound leaf can be viewed as a disposable first-order "twig" that allows many leaves per "bud" without committing the tree to an elaborate scaffolding for further branching [12].

It is worth noting that the Strahler ordering system has some evil properties if branching is asymmetrical, as it is for trees that have "short shoots" (e.g., especially the Rosaceae, but spottily in many other families), twigs that leaf at the terminus year after year with little elongation and no branching [52]. Such twigs are of Strahler order 1, and variable numbers may join to form branches of order 2. In these cases, the Strahler order of the trunk gives little information about the number of functional units, i.e., first-order twigs, that

American Elm
Growth Deceleration

Black Walnut · Sugar Maple
Branching Order

Red Cedar
Branching Angle

Sassafras · White Ash
Terminal Dominance

Twisty Tuliptree

Normal Tuliptree

FIGURE 4 Real trees that illustrate the input parameters of a simple model of growth by fractal recursion of a modular twig. The parameters and details of their measurement are described in the text, along with comments about the particular trees in the figures.

the tree has. Appropriate labor-intensive techniques for analyzing such cases have recently been imported from geology to biology by Turcotte et al. [42].

The "branching angle" is obviously the typical angle of the first side twig from the terminal twig. In the example of Figure 4, a small difference in branching angle between two individual red cedar trees, compounded over and over with each branching, has produced shapes that range from that of an ornamental classic cypress (*Cupressus sempervirens*) to that of a hemlock (*Tsuga sp.*).

The "terminal dominance" is the degree to which the terminal twig exceeds or suppresses the growth of the next lateral twig. It is practically measured by the ratio of length of longer to shorter terminal twig. In the example of Figure 4, sassafras (*Sassafras albidum*) has low terminal dominance (ca. 1.1), and white ash has high (ca. 2.0). The high terminal dominance of ash is well adapted to compete with neighbors for upward growth in a crowded stand, whereas the low terminal dominance of sassafras is efficient at "exploring" and filling whatever open space is available, as befits its ecological role filling edges and small gaps (cf. Aarssen [1]). The twisty record of past explorations is seen in the larger branches of sassafras, and contrasts with the straight branches engendered by the high terminal dominance of ash. This contrast is repeated in the pictures of tuliptree (*Liriodendron tulipifera*)—an individual with an uncharacteristically low terminal dominance for its species, contrasted with a normal tuliptree.

When the model of a fractally recursive turkey-track is applied to measurements made on individual twigs, about one third of the computer-generated caricatures portray features that are at least vaguely reminiscent of appropriate species of tree [19]. For another third, there is something obviously wrong with the caricature, something that suggests an ad hoc addition to the model to suit that species. The remaining third of the caricatures hardly look like real trees at all, let alone like the species whose twigs were measured. Figure 5 presents some of the more successful caricatures. They really are caricatures. The computer explicitly draws a figure in two dimensions and fakes the third dimension by closing the branching angle in the twigs to represent an average of the measured angle if it were viewed from uniformly varied perspectives. Diameters of branches vary linearly with order for simplicity of programming. And the memory of reality against which the computer figures are judged is a two-dimensional projection of a three-dimensional shape.

The fake red maple (*Acer rubrum*) and sugar maple look little like the real trees, but their overall shapes differ in the same direction as real open-grown specimens of the two species—the red maple having a long trunk and a high branching structure, and the sugar maple spreading out with low branches nearly touching the ground. The resemblance of the overall outline of the fake maples to a maple leaf is coincidental. The fake American beech (*Fagus grandifolia*) has the species' characteristic peripheral concentration of twigs and leaves. The fake white ash and sassafras need only to have their branches turned upward to look quite like the real trees. The fake red cedar and white

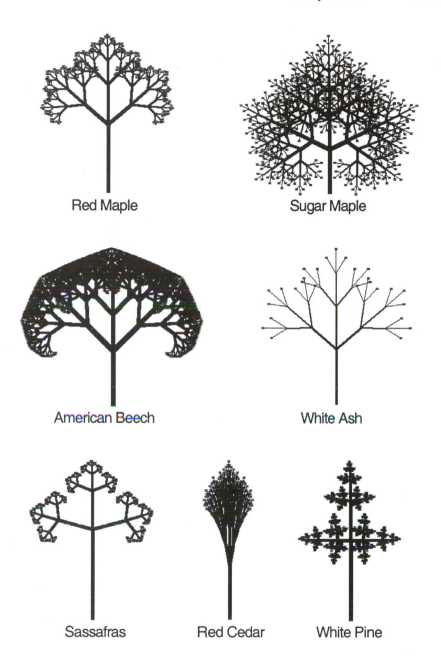

FIGURE 5 Fake trees from measurements made on real twigs, using the fractal model of Horn [19]. The basic turkey-track module is obvious in all but the red cedar. The turkey track is reduced and multiplied in an explicitly two-dimensional fractal, rather than a projection of a three-dimensional fractal. These examples were arbitrarily chosen from numerous candidates as having comparative features that were at least vaguely reminiscent of real trees of their species.

pine (*Pinus strobus*) not only look more like real conifers than like broad-leaved trees, but also exhibit a typical difference in shape between the two species.

The fractal turkey-track model and the trees of Figure 5 are very primitive when compared to other models and computer-generated trees (see Room et al. [36] for a review). Even Honda's [17] early computer-generated trees are more realistic, being explicitly three dimensional and including more realistic responses to gravity. Aono and Kunii [3] draw much more realistic trees, albeit with a model that has far more parameters. Prusinkiewicz and Lindenmayer [33] represent many species of trees and other plants with startling realism by fractal recursion of species-specific modules. Reffye et al. [34] portray an even greater variety of trees with equal realism. Bell [6] shows fractal reiteration of the qualitatively different developmental models of Hallé et al. [14] (also Tomlinson [41]). And whole natural landscapes are now routinely faked using fancy fractal insights in programs like Bryce [9].

Nevertheless, the pseudo-realism of some of the trees of Figure 5 is surprising, especially given the number of lies built into the simple program. Accordingly, it is worthwhile to explore some ecologically important properties of the fractal trees that are independent of the details of the computer model, and even independent of its general structure. One of these is the effect of terminal dominance on the dispersion of leaves. If terminal dominance is low (e.g., the beech tree in Figure 5), then the initial buds and their leaves are about the same distance from the first branching node, and successive recursions produce an advancing front of leaves in a more or less continuous shell at the periphery of the crown. Conversely, if terminal dominance is high (e.g., the sugar maple or white pine of Figure 5), then the initial leaves are at different distances from the first branching node, and successive recursions compound these differences to scatter leaves throughout the volume of the crown. Thus low and high terminal dominance, respectively, produce the shade-adapted "monolayer" and the sun-adapted "multilayer" of Horn [18]. A rough indicator of the dispersion of leaves in my computer-drawn trees is the coefficient of variation of their distances from the first branching node. This is tabulated for the trees of Figure 5 in Table 1, along with their other parameters. The numbers reinforce the visual impression that the dispersion of leaves in a tree is affected primarily by terminal dominance, interacting secondarily with branching angle and with some contribution from growth deceleration and branching order.

This correlation between the patterns of growth of individual twigs and the overall dispersion of leaves within the tree may help to explain why any particular species has a limited range of adaptability on a gradient of sun to shade. So there is in principle some conflict between optimizing a tree for growth in sun and optimizing it to grow in shade. Moreover, there is potentially an additional conflict among various criteria of optimality for the branching structure of a tree. Reasonable criteria could include minimal standing wood to support and supply maximally productive leaf area, minimal total

TABLE 1   Measured and derived parameters of fractal trees. These are the measurements of real twigs on real trees that produced the fake trees in Figure 5, and parameters that were measured on the computer-generated images. Measurements and parameters are described in the text.

Species of Twig and Fractal Tree	Measurements from Real Twig and Real Tree				Derived Parameters of Fake Tree		
	Branching Order	Growth Deceleration	Terminal Dominance	Branching Angle	Crown Diam./ Height	Crown Width/ Height	Leaf Disp. Index
Red Maple	5	1.7	1.3	50	.57	.86	.12
Sugar Maple	6	1.3	1.7	65	1.03	1.16	.26
American Beech	7	1.5	1.0	45	.70	1.17	.09
White Ash	3	1.3	1.3	50	.58	.97	.11
Sassafras	5	2.3	1.1	75	.52	.79	.14
Red Cedar	5	1.4	1.7	10	.31	.31	.13
White Pine	6	2.0	3.0	90	.66	.66	.47

developmental expenditure on both standing and shed wood to support such a leaf area at a given size of tree, or the latter criterion weighted by reproductive value and integrated over the lifetime of the tree, i.e., the fitness criterion recommended by Bonner and Horn (this volume) and by Kozłowski (this volume).

Even designing a minimal scaffold to hold a given geometrical dispersion of leaves is a recondite problem. As the number of leaves gets realistically large, the number of potential trees to be evaluated becomes intractably large [7]. Designing a protocol or a computer program to explore the possibilities efficiently, or even to narrow them down realistically, is itself a problem in the design of an efficient branching structure, i.e., an optimal tree. I have made some preliminary attempts with a protocol that is mainly a conceptual exercise in qualitative geometry. I assume that bundling twigs into larger and larger branches is necessary for support, and may convey additional economies of support and supply. Here is the algorithm. I attach the outermost leaves to nearest neighbors, their twigs to nearest neighbors, and so on until a central trunk is generated. Then I attach inner leaves in the same way, but I include all extant branches as potential nearest neighbors. The algorithm is loosely analogous to the "diffusion limited aggregation" model of Turcotte et al. [42]. It is also a primitive, inverse version of the elegant "constrained constructive optimization" of Schreiner et al. (this volume), which, as they have shown by heroic simulations, comes very close to generating an optimal final tree.

My tentative result is that only for some shade-tolerant monolayers, with their leaves concentrated at the edge of a bulky, spreading crown, is it possible to design an quasi-optimal branching structure that is an exact fractal recursion of a fixed module of decreasing size (a property which fractal folk variously define as "stationary" or "symmetrical"). This tree may provide a complete and conservative framework for a future tree that is also quasi-optimal. For all other trees—specifically trees with tall, thin crowns, and light-requiring multilayers—the quasi-optimal structure is not a fixed fractal recursion. The structure may be exactly fractal from leaves to twigs to second-order branches, but it soon simplifies, as branches of a given order join branches of a much higher order because they are closer than other branches of their own order. The simplified tree may be viewed as a fractal with side branches, but the complete analysis of Turcotte et al. [42] would be necessary to discover whether the parameters of self-similarity were consistent between twigs and trunk. The biological mechanism of this simplification usually involves shedding of shaded, interior branches. This loss of a previously constructed framework ensures that the developmental trajectory toward an optimal branching structure at a given size is itself less than fully efficient. Leigh [24] reviews this conflict between economy of growth and efficiency of form. This means that in framing a model of branching structure for lifetime fitness there are necessarily compromises in which side branches and other departures from fractal uniformity are important.

Further insights come from patterns of branching order, which can be measured on real trees independent of what my computer trees look like. When I measure branching order on trees near Princeton, New Jersey, the number is typically between four and six, with a maximum of ten for an individual open-grown sugar maple. In the drier climates of Lost Maples State Natural Area, Vanderpool, Texas, and near Palo Alto, California, the numbers average around nine—and twelve is not unusual. On the desert islands of southern Baja California Sur, numbers as high as thirteen and fifteen are found. Figure 6 plots casually gathered preliminary data. If the pattern is not a statistical fluke, branching order runs higher where climate is drier. At first this does not makes sense. Do trees branch more profusely in challenging dry climates than they do in salubrious wet climates? My speculative answer is just the opposite. I suspect that the trees in New Jersey grow so riotously that new growth shades the old, and extensive pruning of older twigs and small branches simplifies the standing structure. Conversely trees of dry climates tend to grow and branch so conservatively that they can keep more older interior branches as a framework for the new ones.

There is doubtless more to the interpretation, and the differences between multilayer and monolayer discussed above would run counter to the apparent pattern. However the fractal model suggests an explicit test, which I have not yet carried out. Using strict Strahler [40] ordering, but discounting "short-shoot" spurs [52], count the number of first-, second-, and third-order branch segments in a tree. Plot the logarithm of the number of branches of a given order against their order. If the tree develops by stationary recursion of a fixed module, the result will be a straight line. If the interior has self-pruned extensively and asymmetrically, then there will be a higher ratio between counts of first- and second-order branches than between counts of second and

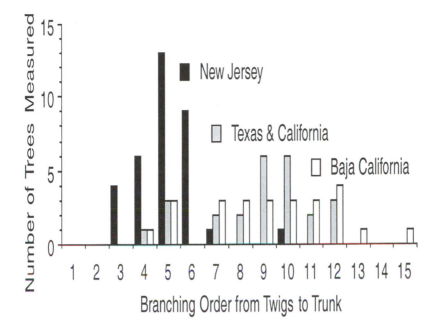

FIGURE 6 Branching order of trees on a climatic gradient. Branching order is higher where climate is drier. The trees with low branching order tend to have a large and distinct trunk, a few large branches, and a proliferation of branches toward the top or edges of the crown. Trees with high branching order tend to look shrubby, with branches dividing and reducing evenly through the whole volume even if twigs are concentrated at the periphery of the crown.

third. In some cases it may be necessary to enumerate deeper orders to detect self-pruning. Of course, this analysis concentrates on the branching structure per se, and may have little to say about the distribution of the crucial primary units of productivity, which may variously be leaf blades, compound leaves, or short shoots [8].

A semi-logarithmic graph of number of branch segments against their order has classically been used to estimate a species-specific and setting-specific "branching ratio" [5, 25, 31]. Steingraeber and Waller [39] pointed out that the branching ratio is not fixed, but they did not interpret this result. Whitney [50] explicitly compared branching ratios among species that differ in tolerance of shade. He found that what I would call multilayered trees had higher branching ratios, and that within a single tree the ratio was somewhat higher for low orders than for high. This result is consistent with my argument of more self-pruning in trees that have their leaves scattered throughout the crown than in trees with their leaves concentrated at the periphery.

The variation of real trees from a simple fractal model can be taken both as a caution against taking such models too seriously and as an opportunity to study the variation itself. In particular, the relations among terminal dominance of a twig, its variation within a tree, physiological performance of a tree, and ecological roles among trees are all worthy of more study. For additional inspiration, see especially Leonardo da Vinci [21, 35], Büsgen and Münsch [10], Wilson [52], Hallé et al. [14], Tomlinson [41], and Aarssen [1]. Other properties worthy of further study are branching order and the pattern of branching ratio within trees, both of which show intriguing preliminary patterns that may have important consequences.

4   DISCUSSION: CARBON BUDGET

The stoichiometry of photosynthesis and respiration requires that for a community to have a net influx of carbon dioxide and a net efflux of oxygen, there must be a net accumulation of carbon in the community. As simple as this notion is, it requires restatement and example for the general public, and even for naïve students of ecology, who have been brought up on the notion that we breathe in oxygen and exhale carbon dioxide, and that plants take in carbon dioxide and crank out oxygen. Accordingly, I usually extend the lessons from appropriate field exercises in my course to include the implications of our measurements for the carbon budget of the community. This has the added advantage of opening a discussion of the scaling of local human activities to their potential global consequences. In particular, students have their own data to begin to explore the dynamics of carbon being released to the atmosphere by human exploitation, carbon being extracted and stored by a community undergoing natural succession or managed restoration, and carbon at a steady state in long-undisturbed communities that do not have a specific mechanism to sequester it from oxygen as an incipient fossil fuel.

Figure 6 is interpretable as preliminary evidence that trees in dry climates tend to hold their interior branches for longer, and trees in wet climates tend to shed them earlier. If this is so, then a larger proportion of the primary productivity in trees of dry climates enters into long-term, above-ground storage of carbon. Trees of wet climates have a higher primary productivity, but they recycle part of it through shed branches, and so they store a smaller proportion. Of course there are other factors that will affect the relative storage of carbon. Differences in allocation to root versus shoot should augment the suggested climatic difference, though in a sufficiently wet and cold environment, the shed branches may be slow to decay, decreasing the suggested difference.

Whatever the tree's branching pattern details, if a tree obeys Leonardo's rule throughout, and if there is a regular allometry of branch length and diameter, then for a rough approximation of its volume, the tree may be collapsed into a cylinder with its measured height and basal diameter. These assumptions are implicit in an engaging introductory ecology exercise designed by

Weihe [49]. Weihe and his class measure the height and diameter of a tree, calculate the volume of its wood as a cylinder, and convert that volume to a weight of carbon using the published species-specific density of wood and the chemical formula for cellulose. They estimate the tree's age either from land-use history or from an increment boring. Then they compare the tree's average annual above-ground carbon storage with the annual carbon combustion of Weihe's automobile. Formal reports and discussions address a wide range of biological and social issues and their interactions, and include constructive criticism of assumptions, biases, potential inaccuracies, and missing information.

My classes do an exercise very like Weihe's, with the additions of an explicit test of Leonardo's rule and a species-specific consideration of lignin as well as cellulose. Our comparison is different. We compare the carbon stored above ground in the tree with the total carbon in the atmosphere above its crown. Our test of Leonardo's rule allows us immediately to assess whether the assumed cylinder overestimates, or more usually underestimates, the volume of the tree. For the mathematically inclined, departures from Leonardo's rule can be combined with simple measurements of branching order to increase the precision of the estimate by appropriately tapering the cylinder.

We typically estimate that a large, open-grown tree holds about two to seven times as much carbon as the atmosphere above it. This figure is of the same order of magnitude as the range of values for above-ground storage of carbon in trees in regional forests (i.e., several kg/m^2 [43]). Every year at least one group of students gets a value of about twenty by measuring a huge and particularly obese tuliptree on one of the central lawns of the Princeton University campus. Most students are routinely surprised at how large the numbers are, and they are equally surprised at how closely we can bracket the true figure between realistic upper and lower bounds. They can explain anomalies like the giant tuliptree on their own. The discussions invariably lead to substantive questions about global carbon balance and global climate change, and the students' previous investment in gathering and analyzing the data seems to motivate them to search for answers, even though this is not part of the formal assignment.

5   QUESTIONS AND ANSWERS

As I promised, these three examples are works in progress rather than definitive results. Nevertheless, all of them make simple measurements at a small scale, that of plant parts, and use allometry or fractal models to derive properties, insights, or significant questions at larger scales ranging from the whole plant to the community. In some cases, it is the fit to a particular model that is of interest. These cases justify a straightforward application of uniform allometry or fractal geometry to scale measurements at one level into properties at another. Some properties of the community may even be independent of the constituent species. In other cases, relative departures from a particular model are of more interest, especially for predicted differences from one competitive setting to another. In these cases appropriate scaling will also require explicit account of the dynamic relations between particular species with different developmental allometries.

Heights and diameters of trees growing in competition with one another are proportional as $H^3 \alpha D^2$, conforming to McMahon and Kronauer's [28] relationship for trees that have just enough diameter for their height to have a given safety against elastic failure under deformation. However, the safety factor is lower for these trees than for most of the trees reported by McMahon [26] that were growing alone. Dissection allows reconstruction of the developmental trajectory of a tree. A white ash in a dense stand was found to have an unsustainable trajectory early on, but to adopt a constant safety factor with increasing age, size, and dominance. This suggests that some trees compromise physical safety to maintain competitive standing. It also suggests a systematic program of research, gathering the same information factorially for several species in several competitive settings.

Total cross-sectional area is approximately preserved over branching nodes despite large interspecific differences in hydraulic architecture. This finding supports the generality of predictions of uniform allometry and a host of scaling laws from individual plants to whole communities [11] (also Enquist et al. this volume). Even where the approximation is poor, simple allometric data can be used to calculate upper and lower bounds on the above-ground volume of wood in a tree, and thus to bound estimates of above-ground carbon storage.

The representation of trees with a grossly over-simplified fractal model is of aesthetic interest for its modicum of reality, but of practical interest as a standard against which to measure departures and to interpret their ecological significance (this parallels a point made for vascular networks in the heart by Zamir, this volume). In particular, designing a tree for maximal performance with minimal structure often entails inherent conflicts in different ecological or developmental settings. Examples include: sun versus shade, varied competitive settings, early versus late in life, and over lifespan versus at any given stage of life. The most unrealistic aspect of my fractal model is representation of the tree as an ever-expanding population of individually smaller buds, rather than as a process of location-dependent births of buds and deaths of branches [6, 15]. But even this unreality is useful. The degree of departure from a uniformly symmetrical fractal pattern in standing branch orders is proposed as a simple measure of the pattern of deaths of branches. Or more detailed measures could be developed using the models reviewed by Turcotte et al. [42]. The pattern of mortality of branches affects the residence time for carbon stored in branches, and the relative contribution of gross community productivity to long-term carbon storage during succession.

Branching order appears to reach higher numbers in dry climates than in wet. A speculative interpretation is that trees in wet climates branch profusely but lose branches readily, while trees in dry climates branch conservatively and keep most branches. If true, this interpretation has the consequences for carbon storage during succession mentioned above. The speculation is worth testing directly by plotting the logarithm of the number of branches of a given order against that order, and using the departure from linearity as an estimate of branch deaths in dry versus wet climates. Simple and easy measurements on individual trees have uncovered a potentially important pattern at a regional scale, whose interpretation can be tested by simple, albeit very tedious, measurements.

Taken together, these exercises show that simple measurements of twigs and branches can test allometries that convert local measurements into parameters that have important effects on a regional, and ultimately a global, scale.

ACKNOWLEDGMENTS

These ideas have developed mainly in the field and classroom as I have taught Introductory Biology, Introduction to Environmental Science, Field Biology, Forest Ecology, and Terrestrial Plant Ecology, and advised the Junior and Senior Independent Work of many students. I am thankful to all of the students in these classes, and especially thankful to particular advisees: Robert Dahl, Arun Gosain, Una Smith, Peter Wisnovsky, Laura Lopez Hoffman, Indroneil Mukerjie, Rizwan Arastu, and Katherin McArthur. Field work was done at Princeton University's Stony Ford Center for Ecological Studies, established by Margaret and Millard Meiss, with continuing support from their daughter, Elinor Siner. Elizabeth Horn took dictation when I was more than 5 meters up in a tree. Thinking about this chapter brought back fond memories of Martin H. Zimmermann, who planted and nourished my intellectual interest in trees. I also thank the editors, some of whose phrases I have shamelessly appropriated in my revisions.

REFERENCES

[1] Aarssen, L. W. "Hypotheses for the Evolution of Apical Dominance in Plants: Implications for the Interpretation of Overcompensation." *Oikos* **74** (1995): 149–156.

[2] Abelson, H., and A. diSessa. *Turtle Geometry.* Cambridge, MA: MIT Press, 1980.

[3] Aono, M., and T. Kunii. "Botanical Tree Image Generation." *IEEE Comp. Graph. & Appl.* **4(5)** (1984): 10–34.

[4] Arastu, R. "Leonardo Was Wise: Trees Conserve Cross-Sectional Area Despite Vessel Structure." *(The National) J. Young Investigators* **1(1)** (1998): jyi.org/papers/Jenny/Arastu/.

[5] Barker, S. B., G. Cumming, and K. Horsefield. "Quantitative Morphometry of Branching Structure of Trees." *J. Theor. Biol.* **40** (1973): 33–43.

[6] Bell, A. D. *Plant Form: An Illustrated Guide to Flowering Plant Morphology.* Oxford: Oxford University Press, 1991.

[7] Bern, M. W., and R. L. Graham. "The Shortest Network Problem." *Sci. Am.* **260(1)** (1989): 84–89.

[8] Brown, J. H. In personal comments to me, has emphasized the importance of these differences, and in particular has cautioned that their effects may differ between my favorite mesic environments and his favorite xeric environments.

[9] Bryce 4 (Commercial computer program). Carpinteria, CA: Metacreations, 1999.

[10] Büsgen, M., and M. Münch. *The Structure and Life of Forest Trees*, transl. by T. Thompson. New York: Wiley, 1929.

[11] Enquist, B. J., J. H., Brown, and G. B. West. "Allometric Scaling of Plant Energetics and Population Density." *Nature* **395** (1998): 163–165.

[12] Givnish, T. J. "On the Adaptive Significance of Compound Leaves, with Particular Reference to Tropical Trees." In *Tropical Trees as Living Systems*, edited by P. B. Tomlinson and M. H. Zimmermann, 351–380. Cambridge: Cambridge University Press, 1978.

[13] Greenhill, G. "Determination of the Greatest Height Consistent with Stability that a Vertical Pole or Mast Can Be Made, and the Greatest Height to Which a Tree of Given Proportions Can Grow." *Proc. Cambridge Phil. Soc.* **4** (1881): 65–73.

[14] Hallé, F., R. A. A. Oldeman, and P. B. Tomlinson. *Tropical Trees and Forests: An Architectural Analysis.* Berlin: Springer-Verlag, 1978.

[15] Harper, J. L., and A. D. Bell. "The Population Dynamics of Growth Form in Organisms with Modular Construction." In *Population Dynamics*, edited by R. M. Anderson, B. D. Turner, and L. R. Taylor, 29–52. 20th Symp. Br. Ecol. Soc. London: Royal Society, 1978.

[16] Hoffman, L. L., and H. S. Horn. "Scaling Succession: From Interactions Between Individual Trees to Patterns Across Wood Lots." (1999): in prep. Based on Hoffman, L. L. "The Role of Initial Floristic Composition and Differential Growth Rates in Old-Field Succession." A.B. Thesis, Department of Ecology and Evolutionary Biology, Princeton University Princeton, NJ, 1996.

[17] Honda, H. "Description of the Form of Trees by Parameters of the Tree-like Body: Effects of the Branching Angle and Branch Length on the Shape of the Tree-like Body." *J. Theor. Biol.* **31** (1971): 331–338.

[18] Horn, H. S. *The Adaptive Geometry of Trees.* Princeton, NJ: Princeton University Press, 1971.

[19] Horn, H. S. *REALTREE: Simulation of Tree Growth by Recursion of a Modular Twig.* Madison, WI: WiscWare, 1989. Now out of print. RealTree.92 (Apple II, Macintosh, or DOS version) May be available from the

author (E-mail: hshorn@princeton.edu) if he does not get swamped by requests, or if he can figure out how to put it on the WWW.

[20] Huber, B. "Die Physiologische Bedeutung der Ring und Zerstreut-porigkeit." *Ber. Deutsch. Bot. Ges.* **53** (1935): 711–719.

[21] Kemp, M., ed. and transl., and M. Walker, transl. *Leonardo on Painting.* New Haven, CT: Yale University Press, 1989.

[22] LaBarbera, M. "The Evolution and Ecology of Body Size." In *Patterns and Processes in the History of Life*, edited by D. M. Raup and D. Jablonski, 69–98. Dahlem Konferenzen. Berlin: Springer-Verlag, 1986.

[23] LaBarbera, M. "Analyzing Body Size as a Factor in Ecology and Evolution." *Ann. Rev. Ecol. Syst.* **20** (1989): 97–117.

[24] Leigh, E. G. Jr. *Tropical Forest Ecology: A View From Barro Colorado Island.* Oxford, NY: Oxford University Press, 1999.

[25] Leopold, L. B. "Trees and Streams: The Efficiency of Branching Patterns." *J. Theor. Biol.* **31** (1971): 339–354.

[26] McMahon, T. A. "Size and Shape in Biology." *Science* **179** (1973): 1201–1204.

[27] McMahon, T. A. "The Mechanical Design of Trees." *Sci. Am.* **233** (1975): 93–102.

[28] McMahon, T. A., and R. E. Kronauer. "Tree Structures: Deducing the Principle of Mechanical Design." *J. Theor. Biol.* **59** (1976): 443–466.

[29] Niklas, K. J. *Plant Biomechanics: An Engineering Approach to Plant Form and Function.* Chicago, IL: University of Chicago Press, 1992.

[30] Niklas, K. J. *Plant Allometry: The Scaling of Form and Process.* Chicago, IL: University of Chicago Press, 1994.

[31] Oohata, S., and T. Shidei. "Studies on the Branching Structure of Trees. I. Bifurcation Ratio of Trees in Horton's Law." *Jap. J. Ecol.* **21** (1971): 7–14.

[32] Pomeroy, K. B., and D. Dixon. "AFA's Social Register of Big Trees: These Are the Champs." *Am. Forests* **72(5)** (1966): 14–35.

[33] Prusinkiewicz, P., A. Lindenmayer, J. S. Hanan, F. D. Fracchia, D. R. Fowler, M. J. M. deBoer, and L. Mercer. *The Algorithmic Beauty of Plants.* New York: Springer-Verlag, 1990.

[34] de Reffye, P., C. Edelin, J. Françon, M. Jaeger, and C. Puech. "Plant Models Faithful to Botanical Structure and Development." *Computer Graphics* **22(4)** (1988): 151–158.

[35] Richter, J. P., ed. *The Notebooks of Leonardo da Vinci*, vol. I. New York: Dover Publ., 1883. Reprinted 1970.

[36] Room, P., J. Hanan, and P. Prusinkiewicz. "Virtual Plants: New Perspectives for Ecologists, Pathologists, and Agricultural Scientists." *Trends in Plant Science* **1(1)** (1996): 33–38.

[37] Shinozaki, K., K. Yoda, K. Hozumi, and T. Kira. "A Quantitative Analysis of Plant Form—The Pipe Model Theory I. Basic Analyses." *Jap. J. Ecol.* **14(3)** (1964): 97–105.

[38] Shinozaki, K., K. Yoda, K. Hozumi, and T. Kira. "A Quantitative Analysis of Plant Form—The Pipe Model Theory II. Further Evidence of the Theory and Its Application to Forest Ecology." *Jap. J. Ecol.* **14(4)** (1964): 133–139.

[39] Steingraeber, D. A., and D. M. Waller. "Non-Stationarity of Tree Branching Patterns and Bifurcation Ratios." *Proc. R. Soc. Lond. B* **228** (1986): 187–194.

[40] Strahler, A. N. "Quantitative Analysis of Watershed Geomorphology." *Trans. Am. Geophys. Un.* **38** (1957): 913–920.

[41] Tomlinson, P. B. "Tree Architecture." *Am. Sci.* **71** (1983): 141–149.

[42] Turcotte, D. L., H. D. Pelletier, and W. I. Newman. "Networks with Side Branching in Biology." *J. Theor. Biol.* **193** (1998): 577–592.

[43] Turner, D. P., G. J. Koerper, M. E. Harmon, and J. J. Lee. "A Carbon Budget for Forests of the Conterminous United States." *Ecol. Appl.* **5(2)** (1995): 421–436.

[44] Tyree, M. L., and J. D. Alexander. "Hydraulic Conductivity of Branch Junctions in Three Temperate Tree Species." *Trees* **7** (1993): 156–159.

[45] Tyree, M. L., and F. W. Ewers. "The Hydraulic Architecture of Woody Plants." *New Phytol.* **119** (1991): 345–360.

[46] Tyree, M. T., and J. S. Sperry. "Vulnerability of Xylem to Cavitation and Embolism." *Ann. Rev. Plant Phys. Mol. Biol.* **40** (1989): 1–38.

[47] Waller, D. M., and D. A. Steingraeber. "Branching and Modular Growth: Theoretical Models and Empirical Patterns." In *Population Biology and Evolution of Clonal Organisms*, edited by J. B. C. Jackson, L. W. Buss, and R. E. Cook, 225–257. New Haven, CT: Yale University Press, 1985.

[48] West, G. B., J. H. Brown, and B. J. Enquist. "A General Model for the Origin of Allometric Scaling Laws in Biology." *Science* **276** (1997): 122–126.

[49] Weihe, P. "Tree Measurement and Carbon Cycling: A Laboratory Exercise." *Bull. Ecol. Soc. Am.* **178** (1997): 142–143.

[50] Whitney, G. G. "The Bifurcation Ratio as an Indicator of Adaptive Strategy in Woody Plant Species." *Bull. Torrey Bot. Club* **103(2)** (1976): 67–72.

[51] Whittaker, R. H., and G. M. Woodwell. "Structure, Production, and Diversity of the Oak-Pine Forest at Brookhaven, New York." *J. Ecol.* **57** (1969): 155–174.

[52] Wilson, B. F. *The Growing Tree.* Rev. ed. Amherst, MA: University of Massachusetts Press, 1984.

[53] Zimmermann, M. H. "How Sap Moves in Trees." *Sci. Am.* **208(3)** (1963): 132–142.

[54] Zimmermann, M. H. "Hydraulic Architecture of Some Diffuse-Porous Trees." *Can. J. Bot.* **56** (1978): 2286–2295.

Cell Size, Shape, and Fitness in Evolving Populations of Bacteria

Richard E. Lenski
Judith A. Mongold

1 INTRODUCTION

There is a long and substantial history of studying allometric scaling relationships in animals and plants, which is well represented by other chapters in this volume. These studies have relevance for many fields, from cardiovascular physiology to community ecology. From the perspective of evolutionary biology, scaling relationships are important because they provide an empirical focus for investigating the tension between structural constraints, on the one hand, and natural selection, on the other, as they vie to shape—quite literally—organisms and life histories.

Despite the breadth of research on allometric scaling, those organisms at the microscopic end of the scale, especially bacteria, have been largely ignored. The reasons for this include their lack of conspicuous morphology and relatively simple life histories, as well as the intellectual separation between the molecularly oriented field of microbiology and the organismal foci of zoology and botany. Thus, bacteriologists have typically pursued reductionist approaches to characterize one gene or pathway at a time, whereas quantitative traits of interest to organismal biologists, such as body size and growth rate, presumably depend on the interaction of many genes and pathways.

Scaling in Biology, edited by J. H. Brown and G. B. West.
Oxford University Press, 2000.   **221**

Yet, in spite of this historical schism, bacteria and other microorganisms offer important advantages for addressing basic questions about the morphology and performance of organisms, especially those questions that seek an evolutionary understanding. The structural simplicity of bacteria (relative to their macroscopic cousins) suggests that a truly integrative approach and understanding—one that spans the molecular, biochemical, physiological, and morphological dimensions—may be feasible [2, 7]. And from an evolutionary perspective, microorganisms offer unparalleled opportunities to observe the dynamics of evolution in action, as a consequence of their rapid generations and large populations, as well as the ability to measure directly the relative fitness of ancestral and derived types [3, 6].

The present chapter grows out of a study that was not originally intended to address issues of allometric scaling. Rather, this study was designed to investigate the dynamics of evolutionary adaptation and divergence in populations of bacteria during thousands of generations in a defined environment [8, 9]. Yet, as the experiment progressed, it became clear that the evolving bacteria were undergoing substantial changes in their morphology, while we as investigators were becoming increasingly interested in understanding the phenotypic basis of their evolutionary adaptation. In the following pages, we provide a progress report of what we have learned about the relationship between cell size, shape, and fitness in these evolving populations of bacteria, and we identify important questions that so far remain unanswered.

2   OVERVIEW OF THE EVOLUTION EXPERIMENT

Twelve experimental populations were founded with single cells from a clone of *Escherichia coli* B, and then serially propagated in a glucose-limited minimal medium at 37°C. (Glucose is the limiting resource, in the sense that its concentration can be doubled to give twice the cell yield.) Every day, 0.1 mL of each population was transferred into 9.9 mL of fresh medium; this 100-fold dilution and the resulting regrowth allowed ~ 6.6 generations of binary fission per day. The cells experienced lag, exponential growth, decelerating growth, and stationary phases every day, while the total size of each population fluctuated between $\sim 5 \times 10^6$ and $\sim 5 \times 10^8$ cells. There were some 10,000 generations during the first 1,500 days of this evolution experiment, which is still on-going. The ancestral strain was clonal (asexual), so that mutations provided all of the genetic variation in these populations. This mutational input was substantial, with each population having more than 10^9 spontaneous mutations during the experiment.

The ancestral strain is stored in suspended animation at $-80°C$, as are samples from each population obtained at 500-generation intervals. The capacity to freeze and then later revive the bacteria allows one to make contemporaneous measurements of the ancestral and derived cells. Various properties have been measured, including especially (for the purposes of this chapter) relative fitness and cell size. Ancestral and derived cells were always comparably acclimated to the experimental conditions, so that any significant differences between them have a genetic basis. Fitness was measured by placing the derived bacteria in competition with their ancestor, which carried a neutral genetic marker so that the two forms could be easily distinguished (by the color of their colonies on a suitable medium). The fitness of a derived population is expressed relative to the ancestor, and it is calculated as the ratio of their realized growth rates during competition. Fitness assays were performed in the same culture regime as used for the experimental evolution (except as otherwise noted). Cell size was determined using a Coulter counter, which measures the volume displaced by a particle as it passes through an aperture. The bacteria used for cell size determinations were from stationary-phase populations that had exhausted the available glucose and ceased growing. Measurements of cell shape were obtained specifically for this chapter using the methods described below.

3   CELL SIZE INCREASES WITH FITNESS...

During the evolution experiment, the fitness of the derived genotypes improved by approximately 50%, on average, relative to their common ancestor (Figure 1). Most of the gains occurred during the first 2,000 generations or so, with the rate of increase in fitness between generations 5,000 and 10,000 being an order of magnitude less than the rate in the first 1,000 generations [8].

The individual bacterial cells also became much larger during the experiment (Figure 1). At the end of 10,000 generations, the derived genotypes produced cells that were, on average, about twice the volume of the ancestral cells. The trajectory for cell size was similar to that for relative fitness over both coarse and fine time scales [4, 8].

4   ...WHILE NUMERICAL YIELD DECLINES

The increase in cell size was partially offset by a concomitant reduction in the numerical yield of the derived bacteria, defined as the number of cells produced at the end of the 24-h culture cycle (Figure 2). Note that yield reflects the number of cells produced by a population when grown in isolation, whereas fitness reflects the relative rates of increase in numbers when two populations compete for the same pool of resources. We can assume, to a very close approximation, that cell mass is directly proportional to cell volume. Given that a fixed quantity of glucose was provided and consumed, then the product of cell volume and numerical yield (i.e., total volumetric yield) provides a measure of the net energetic efficiency of the bacteria. Evidently, the derived bacteria were, on average, about 20% more efficient in converting glucose into cell mass than was their ancestor (Figure 2). Given that the same

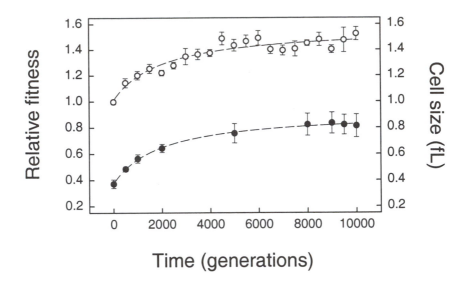

FIGURE 1   Changes in relative fitness and cell size during 10,000 generations of experimental evolution of *E. coli*. The left axis and open symbols indicate the mean fitness of the derived bacteria relative to their common ancestor. The right axis and filled symbols show the average cell size (volume: $1\,\text{fL} = 10^{-15}\,\text{L}$) in the same populations. For both traits, the dashed line gives the best fit of a hyperbolic model and the error bars are 95% confidence intervals. Data from Lenski and Travisano [8].

quantity of glucose was consumed over the same time interval by the derived and ancestral cells, whereas the derived cells fixed more of the glucose as biomass, then it also follows that the derived cells have a lower mass-specific metabolic rate.

5   ALLOMETRIC SCALING: ANAGENIC AND CLADOGENIC COEFFICIENTS

An important feature of the experimental approach is that one can directly measure the ancestral state of any organismal character, in contrast to making an indirect inference about the ancestral state using the comparative method (or the imperfect fossil record). Therefore, one can express the allometric scaling exponent for any character of interest by two different methods. The first method estimates the exponent, b, by regression analysis using character values from extant (or contemporaneous) populations. This method is the one typically used by comparative biologists, and we refer to the resulting values as cladogenic exponents, b_{cla}, because they are based on the variability among evolutionary clades. The second method calculates the scaling exponent from

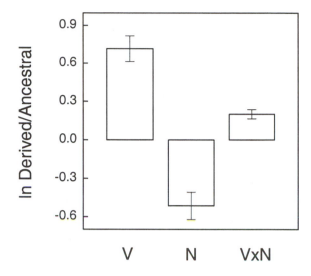

FIGURE 2   Relative values for the derived and ancestral bacteria in terms of cell volume (V), cell number (N), and their product ($V \times N$). Error bars are 95% confidence intervals.

the trend between ancestral and derived character states. The necessary data are available in the context of our evolution experiment, in which we observed anagenic change over time; hence, the scaling exponents obtained by this method are designated b_{ana}.

These two approaches can give very different scaling exponents, as is the case for fitness and cell size (Figure 3(a)). All twelve replicate populations had substantial gains in both fitness and cell size, but they diverged much more for the latter than the former [8]. Consequently, there was a very strong correlation between size and fitness within each population over time (anagenesis), but the relationship between size and fitness across replicate populations at any given time was much weaker (cladogenesis); the resulting scaling coefficients are significantly different. Thus, allometric relationships among traits that are manifest in extant populations may not correspond to those that prevailed during their evolutionary history.

The discrepancy between anagenic and cladogenic scaling coefficients is not somehow peculiar to fitness. The numerical yield of cells also shows a significant discrepancy (Figure 3(b)). It is interesting that the cladogenic scaling for numerical yield is almost exactly what one would expect under the view that energetic efficiency has evolved to some limiting constraint; i.e., cell number and size can each vary, but their product is essentially fixed. Thus, a discrepancy between anagenic and cladogenic coefficients might arise because the evolving populations were initially poorly adapted and, therefore, had

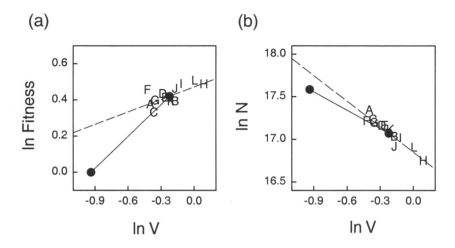

FIGURE 3   Allometric scaling coefficients calculated using two different approaches. Letters A–L denote 12 replicate populations after 10,000 generations; these values were used to calculate the cladogenic regression, b_{cla}, shown as the dashed line. Filled circles show the ancestral state and the mean of the 12 derived populations; the solid line indicates the anagenic trend, b_{ana}. (a) For fitness and cell volume (V), $b_{cla} = 0.229$ and $b_{ana} = 0.604$. The 95% confidence intervals for b_{cla} and b_{ana} are 0.057 to 0.402 and 0.536 to 0.672, respectively. Neither interval includes the estimate obtained using the other approach; hence, they are significantly different. (b) For numerical yield (N) and cell volume (V), $b_{cla} = -0.999$ and $b_{ana} = -0.707$. The confidence intervals for b_{cla} and b_{ana} are -1.246 to -0.751 and -0.773 to -0.641, respectively. Neither interval includes the alternative estimate, and so they are significantly different.

scope for improvement in both dimensions simultaneously. But as the populations became better adapted to the selective environment, they effectively exhausted the supply of new mutations that increased both size and number, thereby producing a constraint.

6   GLUCOSE TRANSPORT AS A TARGET OF SELECTION

It is surprising in two respects that cells became larger at the same time that they became more fit. On demographic grounds, a genotype's fitness in a serial transfer regime depends very strongly on its maximum population growth rate and, indeed, the derived bacteria have substantially higher maximum growth rates than did their ancestor [16]. Yet, comparative studies find a negative correlation between organismal size and maximum population growth rate [12]. Hence, one might expect that selection favoring higher maximum population growth rate would give rise to smaller individuals, whereas the opposite trend was seen. On physiological grounds, experiments strongly suggest that transport of glucose into the cells was an important target of selection during their evolution. In these experiments, the derived and ancestral bacteria competed with one another in various media that contained limiting substrates other than glucose [14, 15]. In general, the derived bacteria have higher and more uniform fitness values in those substrates that use the same mechanisms of transport as glucose than they did in substrates that depend on alternative transport mechanisms. Given selection for improved transport, one might expect that smaller cells would have an advantage by virtue of having a higher ratio of surface area to volume [17], but the opposite outcome was obtained.

Of course, the surface to volume ratio depends on the shape, as well as size, of the cells. Thus, we sought to determine whether the components of cell size had increased isometrically or allometrically.

7 METHODS USED TO MEASURE CELL SHAPE

Cell length and width were measured on a total of 420 individuals. Three clones were randomly chosen from each of the 12 experimental populations and stored (as described above). Each clone was then separately inoculated into the standard medium (except using a higher concentration of glucose to yield more cells and hence facilitate the microscopy), and each culture was incubated for one day, during which time it depleted the glucose and entered stationary phase. Three cultures of both reciprocally marked ancestral states were prepared in the same manner. Combining these six cultures with the 36 evolved clones gave a total of 42 separate cultures in each of two complete blocks of the experiment. Wet mounts of each culture were then prepared for phase-contrast microscopy using a thin film of 1.5% agarose on the slide surface [11]. Images were collected using laser-transmitted light on a Zeiss confocal microscope. From each culture preparation, five cells were chosen randomly from fields of view that were also chosen at random, and measurements were made on these cells (2 blocks × 42 cultures per block × 5 cells per culture). Because we had two cultures for each clone, we could partition the variation in cell size among the populations, among clones from the same population, and among repeated preparations of the same clone [13].

Before proceeding, it is necessary to confess a disconcerting aspect of the measurements. Assuming that cells are cylindrical, we calculated each cell's volume from its length and width; and we compared the distribution of inferred cell volumes with the distribution obtained with a Coulter counter, which measures cell volume directly. Unfortunately, the agreement was rather poor, with volumes calculated from the microscopic measurements being roughly twice those that were measured directly. Both techniques were calibrated using defined standards, so that does not seem to be the problem. However, several factors might produce a discrepancy, and they are not mutually exclusive:

(i) To the extent that cells are tapered, or their surfaces subtly convoluted, the approximation of a cylindrical shape using the microscopic data systematically overestimates cell volume.

(ii) Some artifact of preparation for microscopy may cause cells to appear larger; for example, they might be flattened by the cover slip.

(iii) The Coulter counter assumes that particles are spherical, and it might underestimate the volume of rod-shaped cells if they tend to pass through the aperture lengthwise.

Fortunately, in a relative sense, there is good concordance between the two sets of data. Both techniques indicated that the derived cells are, on average, roughly twice the ancestral volume. And across the twelve derived populations, there was a very strong correlation between cell volumes obtained from the same cultures by using the two different techniques ($r = 0.9702$, $p < 0.0001$). Thus, we are reasonably confident of these microscopic data in terms of what they indicate about changes in relative size and shape, even if not in terms of absolute size.

8   EVOLUTIONARY CHANGES IN CELL SHAPE...

All 12 derived populations produced cells that were, on average, longer and wider than those of their ancestor. However, they tended to increase in length more than in width, so that their ratio of length to width increased by about 15%, on average (Figure 4). As a consequence, the ratio of surface area to volume, calculated assuming a cylindrical shape, was significantly greater—about 2%, on average—than if the derived bacteria had increased in size *isometrically*.

Figure 5 shows representative micrographs of the ancestral bacteria (A), a typical derived population in which the cells became more elongated (B), and an atypical derived population in which the cells became more spherical (C). Thus, there was a tendency for the increase in cell volume to be offset by elongation, which improves the surface-to-volume ratio (see also Weiss et al. [17]). However, the elongation was not nearly sufficient to maintain the ancestral ratio of surface to volume. Instead, the surface to volume ratio in the derived cells decreased by 27%, on average, *relative to the ancestor*, as a consequence of the overall increase in size, and despite the tendency toward cell elongation.

9   ...AND SOURCES OF VARIATION

There was shape variation not only among the derived populations, but also sometimes among clones from the same population (Figure 6). Using variance components, we can calculate the heritabilities for various aspects of cell

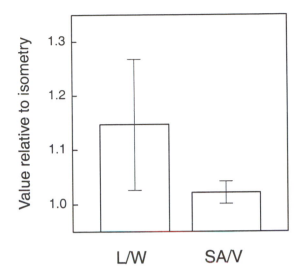

FIGURE 4   Changes in cell shape relative to isometry (geometric similarity) for the measured ratio of length to width (L/W) and the inferred ratio of surface area to volume (SA/V). Error bars are 95% confidence intervals.

shape. For each of four traits—length, width, ratio of length to width, and ratio of surface to volume (inferred assuming a cylindrical shape)—between 20% and 40% of the observed variation was heritable; the remaining 60% to 80% of the variation was due to nonheritable differences among individual cells (including variation between repeated cultures of the same clone). Differences among clones from the same population accounted for 50% to 70% of the heritable variation, with the remainder (between 5% and 20% of the total) indicating evolutionary divergence of the populations. For all four traits, the within-population variation among clones was highly significant ($p < 0.001$). The among-population variances were more equivocal due to fewer degrees of freedom, smaller signals, and the variation among clones (which becomes statistical noise when testing for variation among populations). Even so, cell width and the inferred ratio of surface area to volume varied significantly among populations ($p < 0.01$).

Individual cell measurements also allow us to express the average evolutionary change for these same four traits in terms of phenotypic standard deviations in the ancestral population. The most conservative trait by this measure was the ratio of length to width, which increased by only 0.6 standard deviations. The other traits changed by at least 1.8 standard deviations. Cell width was the most responsive trait, increasing by a factor of 3.2 phenotypic standard deviations during 10,000 generations.

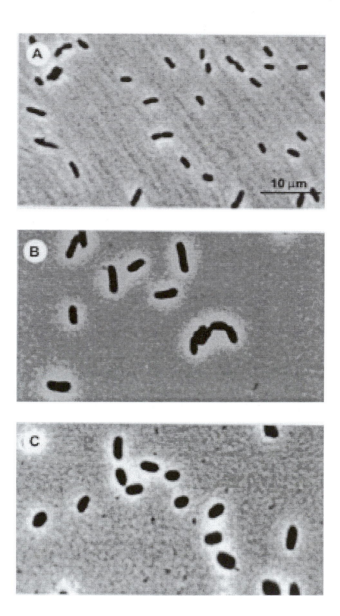

FIGURE 5   Micrographs showing evolved differences in size and shape between (A) the ancestral genotype [REL606], (B) a typical derived population in which the cells became more elongated as they also became larger [Ara–6], and (C) an atypical derived population in which the cells became more spherical as they became larger [Ara+5]. (A), (B), and (C) are depicted at the same scale.

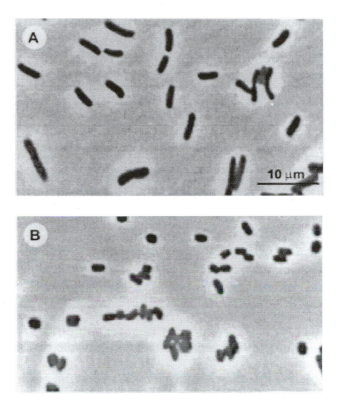

FIGURE 6   Micrographs showing heritable variation in cell size and shape between two clones from the same derived population [Ara+4]. Both panels are depicted at the same scale.

10   IS BIGGER BETTER?

It is unclear what advantage, if any, accrued to the larger cells in the evolutionary environment. Perhaps larger cells have greater metabolic reserves, so that they can respond more quickly to the sudden availability of nutrients at each daily transfer. Consistent with this hypothesis, the derived cells have a shorter lag prior to achieving exponential growth in fresh medium; on the other hand, this shorter lag phase contributes less to their improved fitness than does their faster exponential growth [16].

From another adaptive perspective, Brown et al. [1] developed a general model for the evolution of body size in which individual organisms are the nexus of a two-step process: energy acquisition from the environment, and subsequent conversion of the energy into offspring. Their model makes two basic assumptions. First, the rate of energy acquisition depends on *individual* metabolic rate, which scales positively with size. Second, the rate of energy conversion depends on the *mass-specific* metabolic rate, which scales negatively with size. As a consequence of this allometric tradeoff, selection favors an intermediate optimum size, the exact value of which will depend on the details of an organism's biology. Consistent with this model, the derived bacteria in our study—which are larger than their ancestor—are more effective at acquiring glucose on an individual basis and also have a lower mass-specific metabolic rate. Moreover, the initially rapid increase in cell size and the later plateau (Figure 1) could indicate directional selection when the bacteria are far from their optimal size, followed by stabilizing selection when they have reached the optimum. To evaluate this model further, it would be interesting to find another environment in which smaller bacteria were favored by selection, in which case the individual rate of energy acquisition should decline while the mass-specific metabolic rate should increase.

Alternatively, the larger cell size of the derived bacteria may not be adaptive *per se*. That is, there are other plausible explanations in which larger size evolves but is not an actual target of selection. For example, bacterial physiologists have shown a positive correlation between cell size and growth rate when bacteria are grown on different media. This effect is strictly phenotypic and does not involve genetic change. By extension, the larger cell volume of the derived bacteria in our evolution experiment could, in principle, simply indicate that they grow faster than did their ancestor. In other words, their larger size might be merely a correlated response to selection for faster growth (Figure 7(a)). In fact, we tested this hypothesis by forcing a derived genotype and its ancestor to grow at the same rate in separate chemostat vessels, and then measuring the resulting cell size, an experiment that was repeated over a broad range of growth rates [10]. We observed the previously documented relationship between cell size and growth rate for both the derived and ancestral types; that is, faster growing cells were also phenotypically larger. But the derived genotype produced significantly larger cells than did its ancestor even when the two strains grew at the same rate (Figure 7(b)). Thus, the purely phenotypic correlation between cell size and growth rate cannot fully explain the evolutionary increase in volume, and some other explanation (adaptive or otherwise) must be sought.

It is also worth considering that an adaptive change in the rate of a given metabolic process simultaneously changes the *relative* rate of that process and every other process within the cell, all else being equal. Thus, for example, faster glucose transport might cause an increase in cell size merely as a consequence of changing the relative rates of resource uptake and genome replication. This possibility calls to mind the logical difficulty of atomizing an organism into discrete traits [5]. We have not yet tested this hypothesis, but we may be able to do so in the future by first finding and then manipulating the genetic mutations responsible for higher fitness.

In conclusion, our findings have raised new questions, even as they have answered others. Do larger cells have an intrinsic advantage over smaller cells and, if so, why? Or have the bacteria evolved larger cell size merely as a

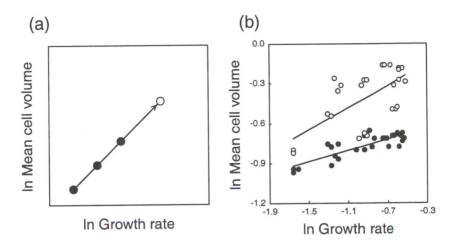

FIGURE 7   (a) Hypothetical explanation for an evolutionary change in cell size reflecting a strictly phenotypic correlation with growth rate. Filled circles: the ancestral strain's cell size increases with its growth rate. Open circle: a beneficial mutation that allows a derived genotype to grow faster produces an increase in cell size that extends the ancestor's phenotypic response. (b) This explanation is rejected by experiments in which ancestral and derived genotypes were forced to grow at the same rate in independent chemostats. Filled circles: mean cell size as a function of growth rate for the ancestral strain. Open circles: the same relationship for a derived genotype sampled after 2,000 generations. The derived genotype produced larger cells than the ancestor even when they were grown at the same rate. Data from Mongold and Lenski [10].

correlated consequence of some physiological change that, once identified, will be readily understood as enhancing fitness in the experimental environment?

11   ACKNOWLEDGMENTS

We thank Jim Brown and Geoffrey West for organizing the conference; Al Bennett, Jim Brown, and Mike Travisano for discussions; and Joanne Whallon for assistance with microscopy. This research has been supported by the National Science Foundation (DEB-9120006, DEB-9421237, and IBN-9507416).

REFERENCES

[1] Brown, J. H., P. A. Marquet, and M. L. Taper. "Evolution of Body Size: Consequences of an Energetic Definition of Fitness." *Amer. Natur.* **142** (1993): 573–584.

[2] Dykhuizen, D. E., and A. M. Dean. "Enzyme Activity and Fitness: Evolution in Solution." *Trends in Ecol. & Evol.* **5** (1990): 257–262.

[3] Dykhuizen, D. E., and D. L. Hartl. "Selection in Chemostats." *Microbiol. Rev.* **47** (1983): 150–168.

[4] Elena, S. F., V. S. Cooper, and R. E. Lenski. "Punctuated Evolution Caused by Selection of Rare Beneficial Mutations." *Science* **272** (1996): 1802–1804.

[5] Gould, S. J., and R. C. Lewontin. "The Spandrels of San Marco and the Panglossian Paradigm: A Critique of the Adaptationist Programme." *Proc. Roy. Soc. Lond. B* **205** (1979): 581–598.

[6] Lenski, R. E. "Experimental Evolution." In *Encyclopedia of Microbiology*, edited by J. Lederberg, vol. 2, 125–140. San Diego, CA: Academic Press, 1992.

[7] Lenski, R. E. "Molecules Are More Than Markers: New Directions in Molecular Microbial Ecology." *Molec. Ecol.* **4** (1995): 643–651.

[8] Lenski, R. E., and M. Travisano. "Dynamics of Adaptation and Diversification: A 10,000-Generation Experiment with Bacterial Populations." *Proc. Natl. Acad. Sci. USA* **91** (1994): 6808–6814.

[9] Lenski, R. E., M. R. Rose, S. C. Simpson, and S. C. Tadler. "Long-Term Experimental Evolution in *Escherichia coli*. I. Adaptation and Divergence During 2,000 Generations." *Amer. Natur.* **138** (1991): 1315–1341.

[10] Mongold, J. A., and R. E. Lenski. "Experimental Rejection of a Nonadaptive Explanation for Increased Cell Size in *Escherichia coli*." *J. Bacteriol.* **178** (1996): 5333–5334.

[11] Murray, R. G. E., R. N. Doetsch, and C. F. Robinow. "Determinative and Cytological Light Microscopy." In *Methods for General and Molecular Bacteriology*, edited by P. Gerhardt, R. G. E. Murray, W. A. Wood, and N. R. Krieg, 21–41. Washington, DC: American Society for Microbiology, 1994.

[12] Peters, R. H. *The Ecological Implications of Body Size.* Cambridge, UK: Cambridge University Press, 1983.

[13] Sokal, R. R., and F. J. Rohlf. *Biometry*, 3d ed. New York: Freeman, 1995.

[14] Travisano, M., and R. E. Lenski. "Long-Term Experimental Evolution in *Escherichia coli*. IV. Targets of Selection and the Specificity of Adaptation." *Genetics* **143** (1996): 15–26.

[15] Travisano, M., F. Vasi, and R. E. Lenski. "Long-Term Experimental Evolution in *Escherichia coli*. III. Variation Among Replicate Populations in Correlated Responses to Novel Environments." *Evolution* **49** (1995): 189–200.

[16] Vasi, F., M. Travisano, and R. E. Lenski. "Long-Term Experimental Evolution in *Escherichia coli*. II. Changes in Life-History Traits During Adaptation to a Seasonal Environment." *Amer. Natur.* **144** (1994): 432–456.

[17] Weiss, R. L., J. R. Kukora, and J. Adams. "The Relationship Between Enzyme Activity, Cell Geometry, and Fitness in *Saccharomyces cerevisiae*." *Proc. Natl. Acad. Sci. USA* **72** (1975): 794–798.

Does Body Size Optimization Alter the Allometries for Production and Life History Traits?

Jan Kozłowski

1 INTRODUCTION

Julian Huxley's [15] work on morphological allometries was later extended to life history traits [3, 6, 22, 28]. Calder [6] calls attention to three meanings of allometry: (i) as changes in proportions consequent to changes in body size, and related to design constraints; (ii) as size relationships observed along time scales of phylogeny; or (iii) as size relationships occurring in ontogeny (in growth and development). Natural selection can shape allometries in the third sense, and must obey allometries in the first sense. Phylogeny and this within-species evolution determine interspecific allometries, that is, allometries in the second sense.

Body size was usually considered an independent variable in allometric studies, that is, a characteristic which defines the values of other traits. It was implied that size itself shapes these allometric relationships through various design constraints or that a third variable correlates with both size and the feature under study. After it was realized that interspecific allometries can be phylogeny-dependent, in other words dependent on body plans or a common history, phylogenetically controlled tests were introduced to allometric studies (see Felsenstein [12], Harvey and Pagel [13], and Harvey this volume).

Scaling in Biology, edited by J. H. Brown and G. B. West.
Oxford University Press, 2000.   **237**

But adult size is not a given. It results from the development process. Under the same growth rate a larger size requires a longer development time, and this comes at a cost: delayed reproduction and an increased risk of dying without reproducing at all. On the other hand, larger individuals usually benefit from a better reproductive capacity (more energy channeled to reproduction), and better survival at least in terrestrial organisms. By this kind of reasoning, growth can be considered an investment in future reproductive success: it is worthwhile to allocate a spare calorie to growth instead of to immediate reproduction only if the size increase corresponding to this calorie increases the expected future reproductive allocation by more than one calorie [17]. Expected future reproduction must be weighted by the probability of surviving to a given age. It is obvious, therefore, that the decision whether to reproduce or to keep growing must be mediated by the mortality rate: under heavy mortality the investment in future reproduction through growth is likely to be lost. Larger size requires longer development time if the same growth rate prevails, but organisms that grow faster because of a larger production potential can reach the same size earlier. Optimal investment decisions should take into account both the mortality rate and the production rate.

With allocation models, such as illustrated in Figure 1, investment decisions can be studied. The logic behind such models is simple: organisms are limited by their available resources and must allocate them optimally to maximize fitness. Although in principle different resources can be considered, it seems reasonable to focus on energy first. Energy may be limited either by a resource scarcity or by the ability of an organism to process food, whichever comes first [30]. The energy stream coming into an animal may be used to meet maintenance costs, may be retained as tissue growth or storage, or may be channeled to reproduction. This complex problem of energy management is simplified in allocation models: (i) no distinction is made between external and internal limitations on energy input; (ii) it is usually assumed that maintenance costs are paid first; and (iii) it is assumed that surplus energy, equivalent under some assumptions to production rate $P(w)$, is a function of body size w as a result of assumptions (i) and (ii). Allocation models are built to find out what proportions of surplus energy should go to growth, reproduction, storage, repair, etc. Repair is often not considered explicitly. In fact, allocation to repair would be measured by the physiological ecologist mostly as heat production and part of the energy content of excreta, because repair means tissue replacement and maintenance of the immune system (but see Cichoń [10] for explicit considerations of allocation to repair).

The optimal proportions of surplus energy going to different sinks can change during life. Figure 1 shows a general scheme of an allocation model. Although the number of energy sinks considered is arbitrary, two of them cannot be removed: reproduction and growth. This is because reproductive allocation throughout life defines fitness, and growth changes body size, the important determinant of the production rate. Such a model, the simplest, is applied in this chapter to find the body size at which organisms should mature

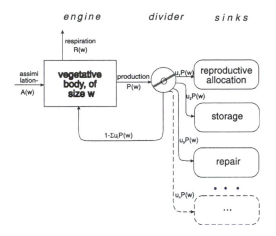

FIGURE 1   Basic scheme of an allocation model. An organism consists of an "engine," that is, soma providing resources, and a "divider" which allocates surplus resources (not used for maintenance) back to soma to allow for growth and to other "sinks" such as reproduction, storage, and repair. The amount of surplus energy $P(w)$ depends on body size w. The proportions u_i of energy going to different sinks are subject to optimization in order to maximize fitness. In this chapter, the simplest imaginable system is considered, with surplus energy allocated optimally between growth and reproduction.

to maximize fitness under a given mortality and a size-dependent production rate.

In the early 1990s Charnov started to link body size optimization with life history allometries [8], although he did not use explicit allocation models. External size-independent adult mortality is the driving force behind Charnov's model. Because different species have different adult mortalities, the optimal times of switching from growth to reproduction should also differ, and size at maturity (adult size in determinate growers like mammals) is a simple consequence of age at maturity. Heavy mortality makes early maturation— maturation at smaller size—optimal. This produces an allometric relationship between body size and mortality. Charnov shows that other allometries of life history features result from this mechanism as well.

The crucial assumption of Charnov's model is that animals share the same productivity; differences in productivity, if they exist, only scatter the points representing species around the predicted allometric lines without changing their slopes. Kozłowski and Weiner [19] show that this is not the case: varying the parameters defining productivity changes the slopes of the allometric lines for life history features. This is because optimal size depends not only on mortality but also on the function describing the dependence of the production rate on body size for a given species [17, 21]. Thus species with higher production have a larger optimal size even if there is no difference in mortality. This point will be explored here.

The difference between Charnov's and Kozłowski and Weiner's models has important consequences. Charnov is heuristically optimistic: everything depends on the distribution of the mortality rate alone and therefore the interspecific allometries he describes have some explanatory meaning. Thus we can manipulate the exponents arithmetically to eliminate adult body size and can find invariants (derived dimensionless variables no longer dependent on body size) which "imply deeper symmetries." Kozłowski and Weiner are heuristically pessimistic: because interspecific allometries depend on the distributions of many parameters important for body size optimization, "it is unlikely...that interspecific allometries are also valid intraspecifically and that natural selection can explain them functionally on this level." This chapter is aimed at pinpointing the crucial difference between the two models. The model used here is less general than the one in Kozłowski and Weiner [19]; it is intended to be as similar as possible to Charnov's model without losing the main point.

2 THE MODEL

Let us imagine a group of species with determinate growth (no growth after maturation) living in an aseasonal environment. Each species has a size-dependent production rate that can be described as a power function,

$$P(w) = aw^b \tag{1}$$

with a and b species-specific, where w is body size in energy units. Each species has its constant mortality rate m. To maximize lifetime reproductive output measured in energy units,[1] it is optimal to stop growing and to start allocating all production into offspring when

$$\frac{dP(w)}{dw} = m \tag{2}$$

(see Kozłowski [17] and Perrin and Sibly [21]). Incorporating Eq. (1) into condition (2) makes it possible to find the optimal size which satisfies the condition

$$w = \left(\frac{m}{ab}\right)^{\frac{1}{b-1}} \tag{3}$$

---

[1] So long as we are not interested in offspring size, this is a proper measure of fitness if the population has a constant size and its number is regulated through some density-dependent mechanisms early in life [20]. This kind of density dependence may be very common in nature. If the assumptions about population dynamics and density dependence are not fully satisfied, the results will be only approximate. If a population undergoes regular or irregular cycles of colonization, growth, and extinction, as in bacteria, other fitness measures should be applied.

and the resulting age at maturity

$$t = \frac{1}{a(1-b)} \left(w^{1-b} - w_0^{1-b} \right).$$ (4)

The initial size w_0 is set at 1 for simplicity. The dependence of the optimal size w on a, b, and m is shown in Figure 2. It is clear that all three parameters are important to the optimal size.

Optimality condition (2) is necessary but not sufficient [18]. The body size w satisfying it is optimal if $P(w)/m(w)$ is concave downward at w. For the case of size-independent mortality considered here it imposes a limit on b: it must be smaller than 1. Mortality rate decreasing with size in the vicinity of w additionally decreases the allowable upper limit of b.

Let us now assume that each parameter of a, b, m for a group of species is a normally distributed random variable with a certain mean and coefficient of variation. In Charnov's model only m is a random variable, and a and b are the same for all species. We can draw the parameters of a "species" using a random number generator. Then we can calculate the optimal life history for such a species: its adult size according to Eq. (3), its age at maturity according to Eq. (4), its life expectancy (equal to the reciprocal of the mortality rate), and its adult production rate after substituting optimal size into Eq. (1). After

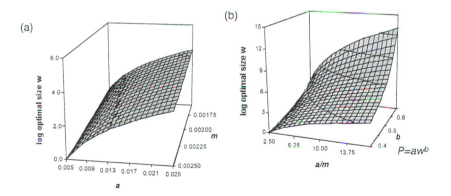

FIGURE 2 Optimal body size as a function of mortality rate m and constant a in the production equation $P = aw^b$ (a), or as a function of the ratio a/m and production exponent b (b). The value of b is constant and equal to 0.50 in (a). Note that optimal body size increases with an increase of the production constant and decreases with an increase in mortality (a). Optimal size increases with the ratio of the production constant to the mortality rate, and especially rapidly with an increase of the production exponent at large values of it (b). After Kozłowski and Weiner [19]. Reprinted with permission from The University of Chicago Press.

100 "species" are simulated we can plot those life history parameters against body size and calculate the slopes of the lines using regression analysis.[2]

The average values of the parameters, with one exception, do not change in this chapter, which is aimed at the effect of the parameters' variability. They are 0.015 for a, 0.50 for b, and 0.0002 for m. Although Charnov assumes $b = 0.75$ for all species, $b = 0.50$ is arbitrarily assumed here. This is to show that even at such a low within-species exponent it is still possible to get 0.75 interspecific scaling. The values of the averages do not affect the results qualitatively.

3 RESULTS

Let us first assume that a, b, and m have negligible variability (coefficients of variation CV = 0.1%). All the simulated species are virtually identical with respect to the optimal body size and the resulting adult production rate, which is represented by a dot in Figure 3(a). Let us now increase CV for the mortality rate to 20% (Figure 3(b)). The optimal body sizes differ now, with larger sizes for species with lower mortality. Differences in mortality are the only force driving optimal size, as in Charnov's model. Each species has its adult production rate according to the production equation common to all species. Each species is represented by a dot on the body size/production rate plane, and all points lie exactly on the straight line. This line represents the interspecific allometry for the production rate. It is not surprising that the production rate for the set of simulated species scales interspecifically and intraspecifically with the same slope.

Now we shall increase CV to 20% for a in the production equation, instead of for the mortality rate. This means that the differences in a are the only force driving optimal size. Although the points representing particular species also lie on a straight line (Figure 3(c)), the slope for the interspecific allometry of the production rate is now 1; that is, it is different from the common production exponent of 0.5. This is because species with higher a have larger optimal body size, and their adult production rates are high due to both large a and large body size. In other words, when a is increased the points move the same distance upward as to the right. For the case represented in Figure 3(b) the points moved more to the right than upward.

The coefficient of variation is further increased to 40% for the mortality rate (Figure 3(d)) or for parameter a (Figure 3(e)). This does not change the slopes (compare Figures 3(b) to 3(d) and 3(c) to 3(e)). The increase in CV has, however, a great effect on the size distribution: note that the lines in Figures 3(d) and 3(e) are much longer than in Figure 3(b) and 3(c), which means that the simulated species cover much broader size ranges if the parameters vary more.

---

[2]A Windows 95/Windows NT application for such simulations is supplied on request.

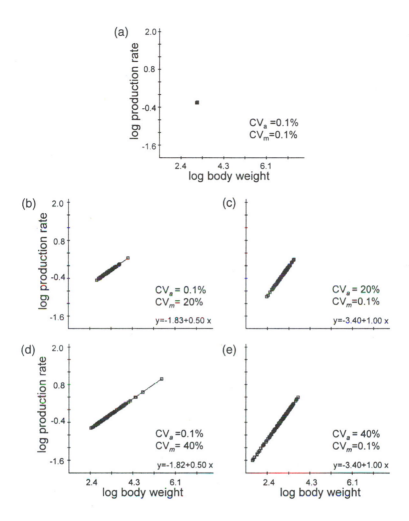

FIGURE 3   Interspecific allometries for production resulting from the model. For each simulated species the parameters of the production equation $P = aw^b$ and size-independent mortality m were generated from normal distributions, with expected values of 0.015 for a, 0.50 for b, and 0.0002 for m. Optimal body size was first calculated for a set of random parameters (horizontal coordinate of plots), and then the production rate (vertical coordinate) for an animal of this size with this set of parameters was calculated. Each dot in each graph represents one such "species." Parameter b is virtually the same for all species (coefficient of variation $CV_b = 0.1\%$), and the coefficients of variation for a (CV_a) and for m (CV_m) are given in each plot. Note that the interspecific slope for production differs from the common intraspecific slope (production exponent b) if the production coefficient varies between species (c) and (e).

Now we shall allow the mortality rate and a to vary simultaneously, with their CVs equal to 20% (Figure 4). There are now two forces driving optimal body size: differences in a and differences in the mortality rate. There are still allometries for the production rate (Figure 4(a)), as well as for age at maturity (Figure 4(b)) and adult life span (Figure 4(c)). In fact these allometries look more realistic because the points for individual species do not lie exactly on the straight line but are scattered around it. The interspecific slope for the production rate b_s is intermediate between the common production exponent (common intraspecific slope) b and unity.[3] Interestingly, the slopes for age at maturity and adult life span are $1 - b_s$, exactly as predicted by Charnov. However, b_s is not a common slope for all species as in Charnov's model, but rather a slope which depends on the ratio of the coefficients of variation for the mortality rate and production parameter a.

Because only two parameters were allowed to vary for the case illustrated by Figure 4, the points representing the relationship between adult life span and age at maturity lie exactly on a straight line with a slope almost equal to 1 (Figure 4(d)). Similarly, the residuals of the production rate (after the effect of body size is removed) plotted against the residuals of life span also lie exactly on a straight line (Figure 4(e)), as do the residuals of age at maturity plotted against the residuals of adult life span (Figure 4(f)).

Figure 5 represents the case in which a and m vary as in the previous examples, but the exponent b in production Eq. (1) is also allowed to vary, albeit with a lower CV, 10%.[4] There is now more scattering of individual points around the allometric lines, but there is still a strong interspecific relationship between the production rate and body size ($r^2 = 0.97$, Figure 5(a)), between age at maturity and body size ($r^2 = 0.63$, Figure 5(b)), and between adult life span and age at maturity ($r^2 = 0.61$, Figure 5(d)). Only adult life span is weakly correlated with body size (Figure 5(c)). Now the residuals of the production rate plotted against the residuals of life span do not lie exactly on the straight line (Figure 5(e)); nor do the residuals of age at maturity plotted against the residuals of life span (Figure 5(f)).

The projections of the points on the body size axis in Figures (3)–(5) shed some light on the size distributions resulting from body size optimization under random parameters. These distributions are depicted in detail in Figure 6. The left-hand column of histograms (a), (c), and (e) represent cases with 20% CV for the mortality rate and production parameter a, with negligible variation of production exponent b, taking values 0.5, 0.6, and 0.7. The

---

[3]Kozłowski and Weiner [19] show analytically that on average this slope should be exactly 0.75, that is, $(1 + b)/2$. Because we are playing with random numbers, the slope 0.73 differs from the predicted 0.75 by chance. If CV is greater for a than for m, the interspecific slope is closer to 1, and if CV for m is greater than CV for a, the interspecific slope is closer to the common intraspecific slope b.

[4]Under CV for b equal to 20% the range of optimal simulated sizes is too broad for any taxon. Figure 4(b) shows clearly that for b close to 1, which must appear if CV for b is high, the optimal size is indeed very large, and under b greater than 1 there is no optimal size.

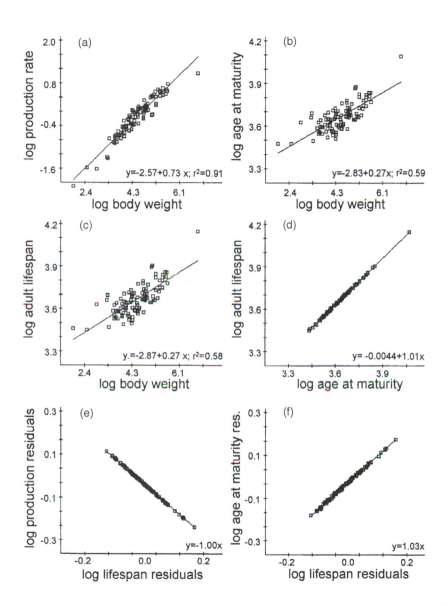

FIGURE 4   Interspecific allometries resulting from the simulation described in Figure 3 and in the text. The coefficients of variation are 20% for production constant *a* and the mortality rate, whereas all the species share the same production exponent *b*. The residuals in (e) and (f) were calculated after the effect of body size was removed. Mean values of the parameters as in Figure 3.

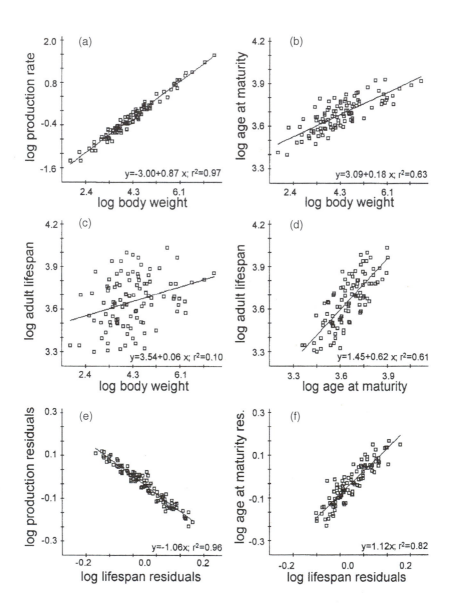

FIGURE 5 Interspecific allometries resulting from the simulation described in Figure 3 and in the text. The coefficients of variation are 20% for production constant a and the mortality rate, and 10% for production exponent b. The residuals in plots (e) and (f) were calculated after the effect of body size was removed. Mean values of the parameters as in Figure 3.

size distributions are symmetrical on the logarithmic scale. The right-hand column of histograms (b), (d), and (f) represent cases with 20% CV for m and a, and 10% or 8% CV for b. For average b equal to 0.6 and 0.7 the size distributions are strongly right-skewed (Figures 6(d) and (f)) even on the logarithmic scale.

4 DISCUSSION

Charnov's model [8] is a specific case of the model presented here, with mortality varying among the species and the production coefficients constant. The model presented here is a specific case of a more general model given by Kozłowski and Weiner [19]. The simplifications consist in assuming that adult mortality is size independent and that the size dependence of the production rate can be described by a power function (1). What all these models have in common is that they explicitly assume that body size is an outcome of optimizing life history through selection. Technically this means not only that mortality in Charnov's model or mortality and production in Kozłowski and Weiner's model depend on size, but also that adult size depends on them. This reciprocal dependence seems to have value in explaining interspecific life-history allometries.

Can Charnov be right in his predictions despite the fact that the assumptions in his model seem unjustified? This can be judged only on an empirical basis: if mortality is much more variable between species than production (by an order of magnitude at least), ignoring the differences between production rates would not introduce a substantial error. This does not seem to be the case. The fact that the same optimal size can result from high production and high mortality or from low production and low mortality explains the existence of animals with "fast" and "slow" lifestyles (high production, early maturation, and short life or low production, late maturation, and long life at a given size). For example, primate species living in tropical forests, having lower reproductive rates than species living in other habitats, represent a slow lifestyle [27]. For 48 species of placental mammals, those with high mortality rates for their size have short gestation lengths, early ages at weaning, early maturation, and large litters: these are the characteristics expected for a fast lifestyle [23, 24]. The fast-to-slow continuum is well illustrated in Figures 5(e) and 5(f). After the effect of body size is removed, a negative correlation between production rate and life span appears (Figure 5(e)), as well as a positive correlation between age at maturity and life span (Figure 5(f)). The left ends of the regression lines represent the "slow" edge of the continuum, and the right ends the "fast" edge. The labels "r-" or "K-selected" have been applied to species with these lifestyles, suggesting incorrectly that they represent adaptations to either increasing or stationary populations (see Kozłowski [16], Roff [26], and Stearns [29] for critiques of the classical r- and K-selection concept, and Caswell [7] for a critical examination of the con-

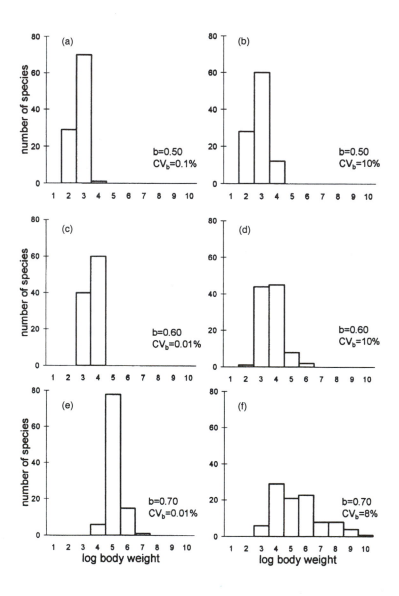

FIGURE 6 Size distributions resulting from the simulations described in Figure 3 and in the text, with 20% coefficients of variation for production constant a and the mortality rate, and the average values of these parameters as in Figure 3. The average values of production exponent b and its coefficient of variation CV_b are given in each histogram. Note that the distributions are skewed to the right on the logarithmic scale when average b is either 0.60 or 0.70, and that this exponent varies considerably between species.

nection between this concept and the population equilibrium problem). The results presented here demonstrate that both styles can evolve in stationary populations. But the necessary condition is that the mortality rate as well as the production parameters must vary between species, an assumption not satisfied in Charnov's model.

If species differ not only in mortality but also in production parameters, the average within-species exponent for the production rate should be lower than the interspecific one. Is there any evidence for this? Purvis and Harvey [25] show that maximum growth rate (a measure of the production rate) always scales with body size less steeply within than between species. Heusner [14] gives similar evidence for respiration: the average within-species exponent for respiration equals 0.67 ± 0.03 for 11 species of mammals, and the interspecific exponent is 0.776 ± 0.004. As shown by Kozłowski and Weiner [19], a steeper interspecific line should also occur for respiration. Interestingly, Heusner [14] argues that the 0.75 respiration exponent is a statistical artifact resulting from the variance of the respiration constant between species. Feldman and McMahon [11] acknowledge the difference between the intraspecific and interspecific exponents, but they are right that the 0.75 exponent is not a statistical artifact and that it tells us something about nature. The variance of the production constant is not enough to mechanically produce a steeper interspecific exponent. Larger species must have on average a higher constant of the production equation. Although this is the case for the species studied by Heusner, he does not provide any mechanism leading to such a pattern. Optimization of body size provides such a mechanism, as shown in this chapter for production and by Kozłowski and Weiner [19] also for respiration.

The model presented here generates distributions of optimal body sizes for species differing in their mortality rates and production parameters (Figure 6). If the mortality rate, production constant, and production exponent vary, the distributions are skewed to the right on the logarithmic scale provided that the production exponent is not very small. This result accords with natural patterns [1, 2, 4, 5, 9]. It is not surprising: if we look at Figure 2(b) we see that a large variation in the production equation exponent is likely to produce a broad range of values for it. The effect of this parameter is asymmetrical, in contrast to the effects of other parameters (compare Figure 2(a) and 2(b)). If the variability of the exponents in the production equation is decreased substantially, the size distribution becomes symmetrical on the logarithmic scale. If the model presented here is correct, the skewness of distributions in real taxa indicates that the exponents vary, which contradicts one of Charnov's assumptions.

In the simulation exercises presented here the mortality rate and production parameters were treated as noncorrelated random variables with normal distributions. A normal distribution results from the additive effects of many small causes. Related species will differ in the details of their morphology, physiology, and behavior that determine production or mortality at a given size. A lack of correlation between parameters was assumed for simplicity.

High production can be expected to bring about high mortality; for example, an animal chooses rich predator-filled places or does not spend much time keeping vigil. Studying the effect of such correlations can be an interesting next step, although quantitative rather than qualitative differences should result.

The status of production exponent b seems crucial. If it varies between species and can be treated as a random variable, as suggested here, there is no need to seek any explanation for its value on the interspecific level. Any functional interpretation is also removed for other interspecific allometries. The question of what determines the value of this exponent (or exponents of other physiological parameters) on the intraspecific level still remains. The paper by West et al. [31] or the chapter by Brown et al. in this volume are interesting attempts to explain it.

The value of parameter b must lie within some limits: the lower one below which a species would not be viable (the expected offspring number is too small to replace the parents), and the upper one (equal to 1 for size-independent mortality, or even less for mortality decreasing with size) above which there is no optimal body size [18]. Within these limits many small effects can define b, leading to normal distributions of this parameter between species. If the exponent proves relatively constant across different taxa, looking for functional explanations on the intraspecific level but common across the taxa is justified (see West et al. and Brown et al. in this volume). However, right-skewed size distributions will need an explanation other than the one suggested here. One thing seems certain: the interspecific and intraspecific exponents are not interchangeable in principle because, as shown in Figures (3)–(5), body size optimization moves the interspecific exponent for production up relative to the common or average intraspecific exponent even if all species share the same exponent. Thus an interspecific exponent cannot provide knowledge about a common (or average) within-species exponent, unless the species in the group share not just the exponent but the entire production equation.

ACKNOWLEDGMENTS

I thank J. Brown, M. Czarnoleski, P. Olejniczak, R. Świergosz, and J. Weiner for comments on an earlier version of the chapter, and M. Jacobs for helping to edit it. The work was supported by the Polish State Committee for Scientific Research, Grant No. 200/P04/96/11.

REFERENCES

[1] Blackburn, T. M., and K. J. Gaston. "The Distribution of Body Sizes of the World's Bird Species." *Oikos* **70** (1994): 127–130.

[2] Blackburn, T. M., and K. J. Gaston. "Animal Body Size Distributions: Patterns, Mechanisms, and Implications." *Trends in Ecol. & Evol.* **9** (1994): 471–474.

[3] Blueweiss, L., H. Fox, V. Kudzma, D. Nakashima, R. Peters, and S. Sams. "Relationships between Body Size and Some Life History Parameters." *Oecologia* **37** (1978): 257–272.

[4] Brown, J. H. *Macroecology.* Chicago, IL: Chicago University Press, 1995.

[5] Brown, J. H., and P. F. Nicoletto. "Spatial Scaling of Species Composition: Body Masses of North American Land Mammals." *Amer. Natur.* **138** (1991): 1478–1512.

[6] Calder, W. A., III. *Size, Function, and Life History.* Cambridge, MA: Harvard University Press, 1984.

[7] Caswell, H. "Life History Theory and the Equilibrium Status of Populations." *Amer. Natur.* **120** (1982): 317–339.

[8] Charnov, E. L. *Life History Invariants. Some Explorations of Symmetry in Evolutionary Ecology.* Oxford, NY: Oxford University Press, 1993.

[9] Caughley, G. "The Distribution of Eutherian Body Weights." *Oecologia* **74** (1987): 319–320.

[10] Cichoń, M. "Evolution of Longevity Through Optimal Resource Allocation." *Proc. Roy. Soc. Lond. B* **264** (1997): 1383–1388.

[11] Feldman, H. A., and T. A. McMahon. "The 3/4 Exponent for Energy Metabolism Is Not a Statistical Artifact." *Respir. Physiol.* **52** (1983): 149–163.

[12] Felsenstein, J. "Phylogenies and the Comparative Method." *Amer. Natur.* **125** (1985): 1–15.

[13] Harvey, P. H., and M. D. Pagel. *The Comparative Method in Evolutionary Biology.* Oxford: Oxford University Press, 1991.

[14] Heusner, A. A. "Energy Metabolism and Body Size. I. Is the 0.75 Mass Exponent of Kleiber's Equation a Statistical Artifact?" *Respir. Physiol.* **48** (1982): 1–12.

[15] Huxley, J. S. *Problems of Relative Growth.* London: Methuen, 1932.

[16] Kozłowski, J. "Density Dependence, the Logistic Equation, and r- and K-selection: A Critique and an Alternative Approach." *Evol. Theory* **5** (1980): 89–101.

[17] Kozłowski, J. "Optimal Allocation of Resources to Growth and Reproduction: Implications for Age and Size at Maturity." *Trends in Ecol. & Evol.* **7** (1992): 15–19.

[18] Kozłowski, J. "Optimal Initial Size and Adult Size of Animals: Consequences for Macroevolution and Community Structure." *Amer. Natur.* **147** (1996): 101–114.

[19] Kozłowski, J., and J. Weiner. "Interspecific Allometries Are By-Products of Body Size Optimization." *Amer. Natur.* **149** (1997): 352–380.

[20] Mylius, J., and O. Diekmann. "On Evolutionarily Stable Life Histories, Optimization, and the Need to Be Specific About Density Dependence." *Oikos* **74** (1995): 218–224.

[21] Perrin, N., and R. M. Sibly. "Dynamic Models of Energy Allocation and Investment." *Ann. Rev. Ecol. & System.* **24** (1993): 379–410.

[22] Peters, R. H. *The Ecological Implications of Body Size.* Cambridge, MA: Cambridge University Press, 1983.

[23] Promislow, D. E. L., and P. H. Harvey. "Living Fast and Dying Young: A Comparative Test of Charnov's Model." *J. Zoology* **237** (1990): 259–283.

[24] Promislow, D. E. L., and P. H. Harvey. "Mortality Rates and the Evolution of Mammal Life Histories." *Acta Oecologica* **12** (1991): 1–19.

[25] Purvis, A., and P. H. Harvey. "The Right Size for a Mammal." *Nature* **386** (1997): 332–333.

[26] Roff, D. A. *The Evolution of Life Histories. Theory and Analysis.* New York, London: Chapman & Hall, 1992.

[27] Ross, C. R. "The Intrinsic Rate of Natural Increase and Reproductive Effort in Primates." *J. Zoology* **214** (1987): 199–220.

[28] Schmidt-Nielsen, K. *Scaling. Why Is Animal Size So Important?* Cambridge, MA: Cambridge University Press, 1984.

[29] Stearns, S. C. *The Evolution of Life Histories.* Oxford: Oxford University Press, 1992.

[30] Weiner, J. "Physiological Limits to Sustainable Energy Budgets in Birds and Mammals: Ecological Implications." *Trends in Ecol. & Evol.* **7** (1992): 384–388.

[31] West, G. B., J. H. Brown, and B. J. Enquist. "A General Model for the Origin of Allometric Scaling Laws in Biology." *Science* **276** (1997): 122–126.

Why and How Phylogenetic Relationships Should Be Incorporated into Studies of Scaling

Paul H. Harvey

1 INTRODUCTION

Under what circumstances should we use phylogenetic information when examining scaling? I shall argue that we should use phylogenetic information whenever we can get it. That position is almost universally accepted by evolutionary biologists, and is widely accepted by comparative morphologists, physiologists, and even ethologists. Reticence remains among some ecologists, but force of example is showing its value even there. For example, we shall see that new patterns in community structure, which are generated in ecological rather than evolutionary time, are revealed when data are analyzed in a phylogenetic context. When those patterns are subjected to further scrutiny, they make good ecological and evolutionary sense.

How should we use phylogenetic information in studies of scaling? I shall argue that the answer depends on the type of phylogenetic information that is to hand, on our perception of evolutionary and coevolutionary models we wish to pitch the data against, and whether we wish to incorporate information on specified third variables which we think may influence scaling.

It is important to emphasize at the outset that, although phylogenetically controlled analyses may be preferable to cross-species comparisons, it does not follow that inferences drawn from cross-species comparisons are necessarily

Scaling in Biology, edited by J. H. Brown and G. B. West
Oxford University Press, 2000.   253

incorrect. For example, Kleiber's law which relates basal metabolic rate to the 0.75 power of body weight was derived from cross-species analyses, and has since been shown to hold under phylogenetically correct independent contrast analysis [16]. As a general rule, when data from many distantly related species spanning several orders of magnitude in body mass are incorporated into cross-species analyses that result in scaling relationships with high correlation coefficients (0.95 or above), phylogenetically corrected analyses are unlikely to radically alter conclusions. An approximate result is better than no result at all, and researchers might simply mention alternative interpretations of their results. Nevertheless, in practice, cross-species statistical analyses are based on the implicit assumption that a star phylogeny relates all the species in the sample, so that all pairs of species have the same most recent common ancestor. Incorporating any phylogenetic information, however approximate, into an analysis is likely to be an improvement on the assumption of a star phylogeny.

I do not argue in any way for primacy of phylogenetic *causes* of relationship over, say, environmental or other causes. A sample of ten pinniped species would be large bodied and have flippers, but ten rodents species would be small bodied and have feet. This does not, of course, mean the same forces that select for large size necessarily select for flippers over feet. A different independent origin of large size, say, among artiodactyls, demonstrates that this is not the case. The key is to define independent evolutionary events and determine whether the origin of one character, say, large size, has been accompanied by the origin of another character more often than we might expect by chance. If we are dealing with continuously varying characters like body size or gestation length, judicious use of phylogenetic information will identify many evolutionary changes of both characters and reveal whether they have evolved together or independently. Viewed in this way, phylogeny per se only causes a relationship in a statistical and not a biological sense. Phylogenetic information simply provides a means for determining whether two characters have changed together over evolutionary time.

This chapter is divided into two main parts. The first explains why species values do not constitute independent points for the statistical description of scaling, and then considers how scaling can be approached statistically using species values in a phylogenetic framework. I describe appropriate basic methods that have been suggested to alleviate the problem of nonindependence, and also respond to recent criticism of the most widely used such method. I conclude the section with the concept of qualitative shifts in scaling laws, and explain how such shifts may be associated with phylogenetic association, with extrinsic third variables, or a combination of the two. The second section describes an example from the ecological literature which shows that, when phylogenetic relationships are incorporated into analyses of the scaling of population density on body size, a pattern is revealed that results from processes occurring in ecological time. If phylogenetic relationships are important in the analysis of scaling laws in community structure, which they are, then it is difficult to envisage an area of scaling in biology where analyses should not take full cognisance of phylogenetic relationships.

2 SPECIES VALUES ARE NOT INDEPENDENT

Closely related species are more similar to each other through common descent in both body size and other morphological, anatomical, and life history or ecological characteristics [16]. It follows that species values do not constitute independent data points for the statistical analysis of scaling relationships: if we know the body size of an organism and the value of the variable that we are scaling on body size, then these provide good estimates for those of a close relative, such as a congener, but relatively poor estimates for a distant relative such as a species from a different order.

This realization has at least three consequences. First, scaling relationships estimated across species from logarithmically transformed data have unreliable standard errors [26]. Second, artifactual taxonomic or phylogenetically associated scaling relationships may be not recognized as such, and investigated as though real phenomena. Finally, the statistical significance of third variables which appear to be associated with residuals from scaling relationships may be overestimated, with the consequence that incorrect causation is inferred. I shall discuss each in turn, accompanied by examples.

2.1 CROSS-SPECIES SCALING RELATIONSHIPS CAN BE MISLEADING

Scaling relationships estimated from species values may be misleading: for the most part, although estimators are unbiased, the number of species in a taxon is an overestimate of the number of degrees of freedom in the sample so that standard errors cannot be trusted. The consequence here is that the probability of rejecting a correct null hypothesis in the data, such as a scaling exponent, may be increased markedly. In a pioneering simulation study which produced results that have subsequently turned out to be typical, Martins and Garland [26] took a phylogeny of 15 species and repeatedly evolved two characters (X and Y) independently by a Brownian motion process from the root of the tree to the tips. They then measured the cross-species correlation between X and Y and on 16% of occasions they rejected the null hypothesis of a correlation coefficient of zero at a probability level of 0.05. In other words, the Type I error rate was 16% instead of 5%.

The independent contrasts method of Felsenstein [10] was developed to deal with precisely the case simulated by Martins and Garland. Felsenstein pointed out, quite correctly, that a taxonomy containing orders, families, and so on is merely an approximation to a phylogeny. And in a phylogeny, differences between descendent daughter taxa from each node evolved independently of differences between daughter taxa from any other node. Those differences, or phylogenetically independent contrasts, form the data derived by Felsenstein's so-called independent contrasts method. Figure 1 gives a simple worked example of independent contrasts in which all branches of the phylogenetic tree are the same length. When branch lengths differ, contrasts must be standardized appropriately. Under a Brownian motion model if the variance accumulated in character X after one unit of time is σ_x^2, then after ν units of time the accumulated variance is $\nu\sigma_x^2$. Each X score and each Y score can therefore be scaled to have a mean of zero and standard deviation of 1 and contrasts estimated from those scaled scores. When Martins and Garland applied Felsenstein's independent contrasts method to their simulated data, their Type I error rate was reduced to 5%.

Of course, character evolution does not proceed by Brownian motion, and we can expect Felsenstein's independent contrast method to give approximate answers at best. If we have a good phylogeny, but an alternative model of character evolution is envisaged, then it can be incorporated into the analysis. Price [34] describes a model of character evolution for which cross-species and independent contrast analysis give almost the same results. If we lack confidence in our estimate of ancestral character states, it is always possible to carry out a sister taxon analysis in which pairs of sister taxa are compared, with every species being used in only one such comparison at most, and the potential information gained by estimating ancestral character states is discarded.

Two recent papers, while acknowledging the value of independent contrast techniques, claim to provide "mixed progress reports" (see Price [34] and Ricklefs and Starck [37]). In both cases, a variety of data sets are subjected to both cross-species (TIPS) and independent contrasts (PICS) analysis. As expected, slopes estimated under the two procedures are similar—they are, after all, estimates of the same thing—and, at least in Ricklefs' and Starck's, case correlation coefficients tend to be (absolutely) higher for TIP analyses than PIC analyses. This is exactly what we should have expected from first principles, and from Martins and Garland's simulation studies, if character evolution did proceed by Brownian motion.

Ricklefs and Starck go on to argue that the results of PIC analysis are "relatively insensitive to actual topology or branch lengths of a phylogenetic hypothesis," whereas Price points out in an aside that simulations along phylogenies in which lengths of deeper branches tend to be longer causes correlations between the results of TIP and PIC analyses to drop markedly. In fact, I believe that Ricklefs and Starck are mistaken, and that Price has inadvertently stumbled across an important feature of real phylogenies that makes PIC analyses even more fully justified. Phylogenies are produced by a speciation-extinction process. The simplest model for the generation of an extant clade is a constant-rates speciation-extinction process, in which the speciation rate has been higher than the extinction rate. If the extinction rate had been zero, younger and older branches (excluding branches containing tips) would have the same expected distribution of lengths. However, as

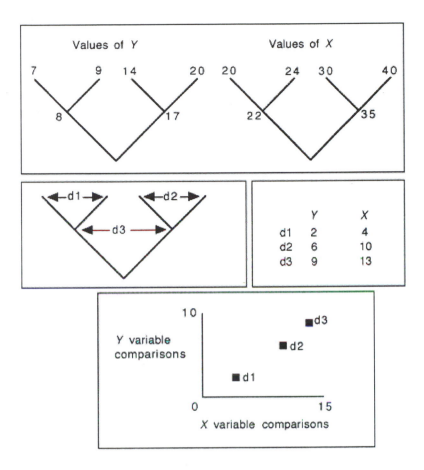

FIGURE 1   The comparative method of independent contrasts. Comparisons d1, d2, and d3 are independent of each other under a Brownian motion model of character evolution.

extinction rates increase, older branches become disproportionately longer because more recent branches will have had less time to go extinct than older branches, thereby leaving fewer undetected nodes. In fact, if a phylogeny was generated under a constant-rates process and then reconstructed from extant species, the reconstructed phylogeny would appear to accumulate lineages at a rate equal to the speciation minus the extinction rate in the distant past, but at the speciation rate in the more recent past [22]. It follows that real reconstructed phylogenies are likely to have deeper branches that are relatively longer.

TABLE 1   TIPS and PICS Type I error rates for simulations of uncorrelated characters evolving along 15-tip phylogenetic trees simulated with different extinction rates. One thousand trees were generated for each extinction rate.

Speciation Rate	Extinction Rate	TIPS Type I Error	PICS Type I Error
0.2	0.0	0.22	0.05
0.2	0.1	0.25	0.05
0.2	0.15	0.30	0.06
0.2	0.19	0.32	0.06

To examine whether phylogenies generated under a speciation-extinction process produce TIP analyses with higher Type I error rates, Harvey and Rambaut [17] simulated phylogenetic trees under constant-rates speciation-extinction processes with the extinction rate being at 0%, 50%, 75%, and 90% of the speciation rate for 1,000 runs each. Uncorrelated characters were evolved by Brownian motion along each of the trees, all of which were scaled to be 1000 time units long. As can be seen in Table 1, as extinction rates increase, the TIP analysis Type I error rates increased from 22% to 32% whereas the PIC analysis Type I error rates remained stable at about 5%. The extinction rate for the Plethodontid salamander mentioned above is about 87% of the speciation rate [31].

2.2   CHANGE IN SCALING EXPONENT WITH TAXONOMIC LEVEL OF ANALYSIS

A second consequence of treating species values as independent points for analysis results from the fact that, because closely related species are much more similar to each other, the ratio of true variance to error variance (arising from measurement error and intraspecific variation) increases as more distantly related species are compared (the real difference between two distantly related species tends to be larger than between two closely related species). Allometric exponents, estimated as standard least-squares regression slopes from logarithmically scaled data, may artifactually appear to increase as more distantly related species are compared. The reason for this is that Model I regression assumes no error variance in the X or independent variable [15]. When there is error variance, the value of the slope is underestimated. As the error variance decreases in proportion to the true variance, the slope becomes steeper. We might reasonably expect the error variance of species estimates to remain the same whatever the taxonomic level of analysis so that slopes would appear to increase as more distantly related species are compared.

The most notorious example is the scaling of brain size on body size among mammals where exponents relating species within genera were estimated between 0.2 and 0.4, whereas the relationship among species from different orders was either 0.67 or 0.75 (depending on author). Population genetic models have been constructed to explain this change in exponent with taxonomic level of comparison [24]. When an appropriate statistical model for slope estimation is used, which incorporates estimates of both true and error variance in both brain and body size, there is no evidence for a generalized increased slope with taxonomic level of comparison using logarithmically scaled data [32, 33]. The artifactual change in exponent with taxonomic level provides a clear example of the value of paying careful attention to the validity of the assumptions that underlie statistical models: research scientists spent several decades attempting to explain a phenomenon that did not exist!

2.3   THE THIRD VARIABLE PROBLEM

Correlational studies are notorious for relationships between two variables resulting from variation in an intervening third variable. Scaling relationships are no exception, and third variables can become involved in a number of ways.

Identified or unidentified third variables can be cited as the cause of a scaling relationship. For example, Western [44, 45] and others pointed out that life history variables in mammals and birds seem to show fairly regular scaling relationships with body size, and they focused on body size as the unifying cause of those allometric relationships. In that perspective, variation in gestation length, litter size, age at weaning, age at maturity, interbirth interval, life span, and so on were in some sense a consequence of variation in body size. Others argued that, in fact, body size was acting as a surrogate for a variable that correlated closely with it. Sacher [39] and Sacher and Staffeldt [40], for example, noticed that primates had unusually long gestation lengths and lifespans for their body size and suggested that, because primates are unusually encephalized, life history variables among mammals scale on brain size rather than body size. Alternatively, McNab [27, 28, 29] suggested that metabolic rate was the "key" variable and that species with high metabolic rates for their size lived fast and died young. In fact, when phylogenetically independent contrasts are calculated for life history variables, body size, metabolic rate, and brain size, partial correlation analysis shows that neither metabolic rate nor brain size contribute to life history variation independently of body size among either birds or mammals [1, 14, 19, 20, 21, 36, 43].

The focus on body size as a correlate of life history variables led to demographic reality being overlooked: on average, birth rates have to balance death rates in natural populations, independently of body size. This means that some components of fecundity must be related to components of mortality. And so it turned out [18]. In a series of papers using independent contrast analyses, my colleagues and I showed that partial correlations between many life history variables were significant when body size was held constant [20]. The partial correlation between two variables with body size held constant is, in fact, formally the same as the correlation between residuals of each variable regressed on body size. Some of the partial correlation coefficients were surprising when viewed in the light of then-current explanations for life history differences.

For example, taxa with long gestation lengths for their body size also had late ages of weaning, yet explanations for the precocial-altricial distinction had generally assumed that long gestation lengths resulted in the production of precocial offspring that became independent of their mother sooner after birth than did altricial offspring that were produced after a shorter gestation. A new theoretical structure for mammalian life history evolution was required to account for such findings. The signs of the various partial correlations relating life history variables to each other independently of body size were used by Charnov to formulate and partially test the first comprehensive model of mammalian life history evolution [3, 4]. When Charnov's model was further tested against both data and logic, it was found wanting in some respects [35], and a more satisfactory alternative has now been developed by Kozlowski and Weiner (see [23] and Kozlowski, this volume).

Some third variables might be expected to pull species away from scaling relationships. For example, testes produce both testosterone and spermatozoa. Larger-bodied species might be expected to have evolved larger testes both to maintain similar threshold concentrations of testosterone in a larger volume of blood, and to produce more sperm to counter the diluting effect of the larger female reproductive tract. However, males of species in which females typically mate with more than one male during a single estrus might be expected to have evolved larger testes for their body size in order to produce more sperm, thereby producing more tickets in the lottery of sperm competition. The expectation is that testes size increases with body size in general, but those species among which there is sperm competition will lie above the general scaling line, having larger testes for their body size. When this prediction (made by Short [41]) was tested for primates, it was found to hold and, what is more, the transition between promiscuousness among females with large-testes-size males and monogamy among females with small-testes-size males had occurred on at least four independent occasions [12, 13].

When dealing with deviations from scaling relationships associated with categorical variables, such as promiscuity versus monogamy, it is important to identify several cases of convergent or parallel evolution; otherwise, the association may be coincidental. For example, Martin [25] pointed out that primates have large brains for their body size and that primates are also precocial. On a plot of brain size against body size for mammal species with primates marked out, the association looked impressive. However, when Bennett and Harvey [1] repeated Martin's analysis looking for evolutionary transitions, the pattern vanished: other orders such as the Artiodactyla which contain precocial species have small brains for their body sizes, while the Carnivora which are altricial have relatively large brains. Indeed, when Bennett and Harvey [1] looked among orders of birds, they found the opposite to Martin's assertion—species belonging to orders which are precocial have relatively small brains while altrical orders have large brains, and this is a pattern which seems to have evolved on many occasions among birds. Bennett and Harvey's analyses were procrustean, having been performed before an adequate phylogeny of birds was available [42] and before Felsenstein's [10] method of independent contrasts had been developed.

3 ECOLOGICAL VERSUS EVOLUTIONARY TIME

Because closely related species tend to be more similar to each other on numerous unmeasured variables, I should argue that it is wise to control for these whenever possible by performing phylogenetically controlled scaling analyses. Of course, such analyses do not provide a panacea because we can never be sure that correlation implies causation—they simply ameliorate the problem. It has been argued repeatedly by ecologists that, since patterns of community structure develop in ecological rather than evolutionary time, phylogenetic control is not necessary (see, for example, [46, 47]). On first encounter, this is an alluring argument, and it is best countered by example for it is always possible to produce a hypothetical case, however unrealistic, to argue against the generality of any claim.

Perhaps the most widely measured ecological variable is population density, and one of the most well-established scaling relationships is that between population density and body size. Viable populations of larger-bodied species tend to occur at lower population densities than do those of smaller-bodied species. When Nee et al. [30] analyzed one of the most comprehensive data sets of population density in a phylogenetic context, they found that in certain circumstances the relationship was actually reversed: larger-bodied species lived at higher population densities. The data set they used was total counts of different bird species in the British Isles. Overall, larger-bodied species did indeed tend to occur at lower population densities and the exponent relating population density to body size did not differ from Damuth's [8, 9] oft-quoted value of minus 0.75. However, when congeneric species were compared, the sign of the relationship was reversed more often than not. Nee et al. suggested that this pattern might result from direct competition: closely related species might compete for the same food resources, in which case the larger-bodied species will displace the smaller-bodied species by direct competition.

Nee et al. [30] and Cotgreave and Harvey [7] suggested phylogenetically based analyses, which might allow preliminary tests of the competitive advantage explanation. Tribes within which interspecific competition was likely to reveal the clearest patterns would be those in which (i) the species had very similar lifestyles to each other and (ii) the species did not compete closely with species from other tribes. The first condition was likely to be met by tribes in which the constituent species had more recent common ancestors, and the second would be met by tribes whose sister taxa diverged in the most distant past. These ideas have been tested both for local bird communities, which have been completely censused, and for national estimates of total population size. Gregory [11] provides a summary of the results, which are significant far more often than would be expected by chance alone.

Cotgreave [5] took a more direct approach for testing the competitive advantage explanation. A riparian bird community in Arizona was censused for three years [2]. When Cotgreave analyzed the census data he found that across all species there was a negative correlation between population size and body mass, but that interspecific comparisons within the more deeply rooted tribes were more likely to show a positive correlation between population density and body mass. Furthermore, those tribes in which the larger species were more abundant also tended to be cavity nesters. Perhaps, then, those tribes were ones within which there was strong competition for nest sites. If the competitive advantage explanation was correct, we should predict that the provision of additional nesting sites would result in a disproportionate increase in population size of those tribes in which the larger species were more numerous. Fortunately, Bock et al. [2] reported the results of just such an experiment: nest boxes were added to some plots and others were left as controls. Further censuses were carried out in both experimental and control plots for the next three years. As predicted from the competitive advantage explanation, Cotgreave found a significant positive relationship between (i) the increase in mean population size within tribes in the experimental plots and (ii) the correlation between body weight and population size before the experiment. In summary, those tribes which "had strong internal interspecific competition before the experiment were the ones in which the largest species were the most abundant" (Cotgreave [5], p. 149). In a subsequent analysis, Cotgreave [6] found that tribes of Australian birds in which species share fewest resources are those in which the relationship between population density and body mass is most strongly negative, which again accords with the competitive advantage hypothesis.

A well-designed, phylogenetically based, and ecologically informative complement to investigations of the relationship between population density and body mass is reported by Robinson and Terborgh [38] who mapped the territories of more than 330 bird species along a successional gradient by the side of an Amazonian river. Species pairs in more than 20 genera showed nonoverlapping but contiguous territories, while others showed partially overlapping or totally overlapping territories. Congeneric pairs of species were chosen for reciprocal heterospecific song playback experiments (an experiment of a type which Cotgreave [6] suggested is necessary for more detailed testing of the competitive advantage hypothesis). One species of a pair from the nonoverlapping territories typically approached the speaker aggressively, and when this happened it was always the heavier species. Robinson and Terborgh argue that the larger of congeneric species pairs thereby occupies the more productive end of habitat gradients, and that it is the marked successional gradients typical of Amazonia which thereby help to explain the increased congeneric species richness of Amazonian bird communities. Where they occur, the larger species will be at higher densities as a consequence of contest competition for resources, in this case territories in the higher productivity habitats.

4 SUMMARY

Species do not constitute independent points for comparative analysis with the consequence that statistical estimates of scaling exponents have incorrect standard errors. Simulated and real examples of incorrect inferences of scaling exponents being drawn from cross-species analyses are given. Independent contrasts can provide satisfactory estimates of scaling exponents when phylogenetic information is available.

REFERENCES

[1] Bennett, P. M., and P. H. Harvey. "Brain Size, Development, and Metabolism in Birds and Mammals." *J. Zool.* **207** (1985): 491–509.

[2] Bock, C. E., A. Cruz, M. C. Grant, C. S. Aid, and T. R. Strong. "Field Experimental Evidence for Diffuse Competition Among Southwestern Riparian Birds." *Amer. Natur.* **140** (1992): 815–828.

[3] Charnov, E. L. R. *Life History Invariants: Some Explorations of Symmetry in Evolutionary Ecology.* Oxford: Oxford University Press, 1993.

[4] Charnov, E. L. R. "Evolution of Life History Variation Among Female Mammals." *Proc. Natl. Acad. Sci. USA* **88** (1991): 1134–1137.

[5] Cotgreave, P. "The Relation Between Body Size and Abundance in a Bird Community: The Effects of Phylogeny and Competition." *Proc. Roy. Soc. Lond. B* **256** (1994): 147–149.

[6] Cotgreave, P. "Population Density, Body Mass, and Niche Overlap in Australian Birds." *Func. Ecol.* **9** (1995): 285–289.

[7] Cotgreave, P., and P. H. Harvey. "Bird Community Structure." *Nature* **353** (1991): 123.

[8] Damuth, J. "Population Density and Body Size in Mammals." *Nature* **290** (1981): 699–700.

[9] Damuth, J. "Interspecific Allometry of Population Density in Mammals and Other Animals." *Biol. J. Linn. Soc.* **31** (1987): 193–246.

[10] Felsenstein, J. "Phylogenies and the Comparative Method." *Amer. Natur.* **125** (1985): 1–15.

[11] Gregory, R. D. "Phylogeny and Relations Among Abundance, Geographical Range, and Body Size of British Breeding Birds." *Phil. Trans. Roy. Soc. Lond. B* **349** (1995): 345–351.

[12] Harcourt, A. H., P. H. Harvey, S. G. Larson, and R. V. Short. "Testis Weight, Body Weight, and Breeding System in Primates." *Nature* **293** (1981): 55–57.

[13] Harvey, P. H., and A. H. Harcourt. "Sperm Competition, Testes Size, and Breeding System in Primates." In *Sperm Competition and the Evolution of Animal Mating Systems*, edited by R. L. Smith, 589–600. New York: Academic Press, 1984.

[14] Harvey, P. H., and A. E. Keymer. "Comparing Life Histories Using Phylogenies." *Phil. Trans. Roy. Soc. Lond. B* **332** (1991): 31–39.

[15] Harvey, P. H., and G. M. Mace. "Comparisons Between Taxa and Adaptive Trends: Problems of Methodology." In *Current Problems in Sociobiology*, edited by K. S. C. S. Group, 343–361. Cambridge, MA: Cambridge University Press, 1982.

[16] Harvey, P. H., and M. D. Pagel. *The Comparative Method in Evolutionary Biology.* Oxford: Oxford University Press, 1991.

[17] Harvey, P. H., and Rambaut. "Phylogenetic Extinction Rates and Comparative Methodology." *Proc. Roy. Soc. Lond. B* **265** (1998): 1691–1696.

[18] Harvey, P. H., and R. M. Zammuto. "Patterns of Mortality and Age at First Reproduction in Natural Populations of Mammals." *Nature* **315** (1985): 319–320.

[19] Harvey, P. H., D. E. L. Promislow, and A. F. Read. "Causes and Correlates of Life History Differences Among Mammals." In *Comparative SocioEcology*, edited by R. Foley and V. Standen, 305–318. Oxford: Blackwell, 1989.

[20] Harvey, P. H., A. F. Read, and D. E. L. Promislow. "Life History Variation in Placental Mammals: Unifying the Data with the Theory." *Oxford Surveys in Evol. Biol.* **6** (1989): 13–31.

[21] Harvey, P. H., M. D. Pagel, and J. A. Rees. "Mammalian Metabolism and Life Histories." *Amer. Natur.* **137** (1991): 556–566.

[22] Harvey, P. H., R. M. May, and S. Nee. "Phylogenies Without Fossils." *Evolution* **48** (1994): 523–529.

[23] Kozłowski, J., and J. Weiner. "Interspecific Allometries are By-products of Body Size Optimization." *Amer. Natur.* **149** (1997): 352–380.

[24] Lande, R. "Quantitative Genetic Analysis of Multivariate Evolution Applied to Brain: Body Size Allometry." *Evolution* **33** (1979): 402–416.

[25] Martin, R. D. "Relative Brain Size and Basal Metabolic Rate in Terrestrial Vertebrates." *Nature* **293** (1981): 57–60.

[26] Martins, E. P., and T. H. Garland. "Phylogenies and the Evolution of Continuous Characters." *Evolution* **45** (1991): 534–557.

[27] McNab, B. K. "Food Habits, Energetics, and Population Biology of Mammals." *Amer. Natur.* **116** (1980): 106–124.

[28] McNab, B. K. "Energetics, Body Size, and the Limits to Endothermy." *J. Zool.* **199** (1983): 1–29.

[29] McNab, B. K. "Food Habits, Energetics and Reproduction of Marsupials." *J. Zool.* **208** (1986): 595–614.

[30] Nee, S., A. F. Read, J. J. D. Greenwood, and P. H. Harvey. "The Relationship Between Abundance and Body Size in British Birds." *Nature* **351** (1991): 312–313.

[31] Nee, S., E. C. Holmes, R. M. May, and P. H. Harvey. "Extinction Rates Can Be Estimated from Molecular Phylogenies." *Phil. Trans. Roy. Soc. Lond. B* **344** (1994): 77–82.

[32] Pagel, M. D., and P. H. Harvey. "The Taxon Level Problem in Mammalian Brain Size Evolution: Facts and Artifacts." *Amer. Natur.* **132** (1988): 344–359.

[33] Pagel, M. D., and P. H. Harvey. "Taxonomic Differences in the Scaling of Brain on Body Size Among Mammals." *Science* **244** (1989): 1589–1593.

[34] Price, T. "Correlated Evolution and Independent Contrasts." *Phil. Trans. R. Soc. Lond. B* **352** (1997): 519–529.

[35] Purvis, A., and P. H. Harvey. "Mammal Life History Evolution: A Comparative Test of Charnov's Model." *J. Zool.* **237** (1995): 259–283.

[36] Read, A. F., and P. H. Harvey. "Life History Differences Among the Eutherian Radiations." *J. Zool.* **219** (1989): 329–353.

[37] Ricklefs, R. E., and J. M. Starck. "Applications of Phylogenetically Independent Contrasts: A Mixed Progress Report." *Oikos* **77** (1996): 167–172.

[38] Robinson, S. K., and J. Terborgh. "Interspecific Aggression and Habitat Selection By Amazonian Birds." *J. Anim. Ecol.* **64** (1995): 1–11.

[39] Sacher, G. A. "Relationship of Lifespan to Brain Weight and Body Weight in Mammals." In *C. I. B. A. Foundation Symposium on the Lifespan of Animals*, edited by G. E. W. Wolstenholme and M. O'Connor, 115–133. Boston, MA: Little, Brown, and Co., 1959.

[40] Sacher, G. A., and E. F. Staffeldt. "Relationship of Gestation Timer to Brain Weight for Placental Mammals." *Amer. Natur.* **108** (1974): 593–616.

[41] Short, R. V. "Sexual Selection and Its Component Parts, Somatic and Genital Selection, as Illustrated by Man and the Great Apes." *Adv. St. Behav.* **9** (1979): 131–158.

[42] Sibley, C. G., J. E. Ahlquist, and B. L. Monroe. "A Classification of the Living Birds of the World Based on DNA-DNA Hybridization Studies." *Auk* **105** (1988): 409–423.

[43] Trevelyan, R., P. H. Harvey, and M. D. Pagel. "Metabolic Rates and Life Histories in Birds." *Funct. Ecol.* **4** (1990): 135–141.

[44] Western, D. "Size, Life History, and Ecology in Mammals." *Afr. J. Ecol.* **17** (1979): 185–204.

[45] Western, D., and J. Ssemakula. "Life History Patterns in Birds and Mammals and Their Evolutionary Interpretation." *Oecologia* **54** (1982): 281–290.

[46] Westoby, M., M. R. Leishman, and J. M. Lord. "Further Remarks on Phylogenetic Correction." *J. Ecol.* **83** (1995): 727–734.

[47] Westoby, M., M. R. Leishman, and J. M. Lord. "On Misinterpreting the 'Phylogenetic Correction.'" *J. Ecol.* **83** (1995): 531–534.

Individual Energy Use and the Allometry of Population Density

Hélène Cyr

1 INTRODUCTION

Physiologists have known for a long time that the rate at which organisms function, for example how fast they eat, grow, or respire, is related exponentially to their body mass (i.e., physiological rate $\propto$ (body mass)^b). The exponents of these relationships (b) are remarkably constant, around 0.75 [15, 103]. Therefore, physiological rates in an individual (e.g., rates of feeding or growth) increase less than proportionately with increases in its body mass. In other words, per unit biomass (e.g., per kg of organisms) small organisms have higher physiological rates than larger organisms. These so-called allometric relationships are so prevalent that physiologists routinely extract them from their data before pursuing any other analysis. Viewed from an ecological perspective, however, these allometric relationships represent physiological "constraints" on individual organisms which are expected to determine, at least in part, the structure and functioning of populations, communities, and ecosystems. Ecologists are interested in using these physiological allometric relationships to predict the structure, dynamics, and interactions of complex assemblages of organisms in nature.

One intriguing allometric relationship found in nature is the density-body size relationship. Over a large range of body sizes, the average population

Scaling in Biology, edited by J. H. Brown and G. B. West.
Oxford University Press, 2000.

density at which individual species are found (D) decreases exponentially with increasing species body size (M) as:

$$D = a \times M^b, \tag{1}$$

with exponents (b) ranging between -0.73 and -1.05 [24, 32, 34, 104]. Because the rate at which individual organisms use energy (i.e., respire, R_i) scales allometrically with an exponent of 0.75 (i.e., $R_i \propto M^{0.75}$ [103]), some authors have concluded that the rate of energy use by whole populations ($R_p = D \times R_i$) is independent of species body size (i.e., $\propto M^0$ [33, 34]). This conclusion has been used to suggest that individual energy requirements ultimately limit maximum population densities in nature [6, 33, 34, 125]. This energetic equivalence hypothesis has remained very controversial, mostly because serious methodological problems have prevented a clear test of it (e.g., Blackburn and Gaston [10], Currie [23], Griffiths [53], Marquet et al. [87], Medel et al. [94]).

There are several possible sources of bias in current estimates of b, the exponent of density-body size relationships (see Eq. (1)). First, the exponent and fit of density-body size relationships are greatly dependent on the range of body sizes covered by the analysis. Overall density-body size relationships which cover a large range of species body size (> 6 orders of magnitude) have been criticized because they force different groups of organisms into a single relationship and do not include small rare species that are missed by current sampling strategies [9, 80]. The opposite criticism has also been expressed, that splitting the data into too many categories reduces the number of observations, the range of body sizes observed, and, therefore, the power of each analysis [15, 23]. Comparisons of density-body size relationships should be limited to similar ranges of species body sizes. Second, the nonindependence of data points in interspecific relationships has been shown to bias these analyses [61]. Negligible biases, however, are found in relationships that cover large ranges of body sizes, such as in the overall density-body size relationships [60]. Third, large differences can be found in exponents (b) calculated using different statistical methods (e.g., model I regression, major axis [53]). This bias is important in weak relationships (i.e., low r^2 [117]) that cover a small range of species body size, but is small in overall density-body size relationships with $r^2 = 0.75 - 0.98$ [24, 32, 34, 104].

A fourth possible source of bias comes from unaccounted sources of variability in the population density data. Global density-body size relationships describe "average" world-wide communities, and may provide poor descriptions of any specific community. The importance of energy use in limiting population densities should be tested locally, at the scale where species interact, rather than in "global communities." Individual populations vary as much as four orders of magnitude in global density-body size relationships (e.g., Peters and Wassenberg [104], Currie and Fritz [24], Cyr et al. [32]). This variability includes spatial variability among sites, for example in the amount of energy available, temporal variability (i.e., population dynamics), and the skewed pattern of species dominance found in natural communities. These different sources of variability must be partitioned before interpreting the exponents of density-body size relationships.

In this study, I test whether the exponents of local density-body size relationships in lakes of different productivity are consistent with the theoretical exponents predicted by the energetic equivalence hypothesis. I also test whether the exponent of the global density-body size relationship is biased by taxonomic differences in temporal variability and in community composition. Ideally, the three sources of population variability (spatial, temporal, community structure) should be compared in a single data set, but very few communities have been sampled thoroughly over many consecutive years. The present analysis is based on three separate data sets compiled from a wide range of studies.

2 DATA AND ANALYSES

Three different data sets were gathered to develop density-body size relationships in different lakes, and to compare population variability among years and among species within communities. The data sets include density measurements of phytoplankton, zooplankton, and fish populations from lakes studied during the International Biological Programme (IBP, 1965–1975) and from other published studies. Data from IBP studies were compiled from final reports deposited in 1972–1973 at the Freshwater Biological Station in England, from subsequent publications, and directly from the researchers. In cases of discrepancies between values reported in the IBP reports and in later publications, the most recent version of the data was used, assuming that sample processing had been completed and/or data had been revised. Other measurements of population densities were gathered from systematic searches through limnological and fisheries journals.

The first data set was used to compare density-body size relationships among lakes. It includes population density data for dominant phytoplankton, zooplankton, and fish populations (i.e., representing at least 1% of the density and/or biomass) covering as broad a range of species body sizes as possible in each lake. The second data set was used to compare temporal variability in phytoplankton, zooplankton, and fish populations. I selected data sets where annual mean population density was measured reliably over at least four consecutive years. The third data set was used to compare the range of population densities found in local communities of phytoplankton, zooplankton, and fish. It was restricted to well-studied communities where the annual mean density of all (or most) species were measured. Data sources are listed in Cyr et al. [31] and Appendixes A and B, respectively.

Mean annual population densities were calculated using all available data. Density measurements made on each sampling date were averaged over all sampling stations and weighted by the number of days in a year which were represented by each sampling date. Because algae and small invertebrates have short generation times (days to weeks), reliable measures of annual mean densities require intensive sampling. The phytoplankton and zooplankton species included in this study were sampled at least once a month, and in many cases fortnightly or even weekly, and provide reliable measures of annual mean population densities.

Species body sizes were obtained from the original studies or, when unavailable, were estimated from average values in the literature. Missing phytoplankton sizes were replaced by the geometric mean of cell volumes measured on the same species in other lakes (see references in Cyr et al. [32]). Missing rotifer and crustacean mean-body masses were obtained from Nauwerck [97] and McCauley [91], respectively. Missing fish sizes were obtained from Carlander [16], Scott and Crossman [115], and Balon [5]. Mean-body sizes were standardized to μg fresh mass, the units most often reported, assuming specific gravity of algae is $1 \text{ g} \times \text{cm}^{-3}$ and fresh mass of zooplankton is $10 \times$ dry mass. Any errors introduced by these approximations and conversions should be small relative to the range of body sizes being compared (4 orders of magnitude within taxonomic groups, 16 orders of magnitude overall).

2.1   LOCAL AND GLOBAL DENSITY-BODY SIZE RELATIONSHIPS

Density-body size relationships were tested using simple linear regression on $\log_{10}$ transformed variables [55] in individual taxonomic groups (algae, invertebrates, fish) and over the full range of body sizes as:

$$\log D = \log(a) + b \times \log M , \qquad (2)$$

where D is population density in individuals m^{-2} or individuals m^{-3}, M is species body size in μg fresh mass, and $\log(a)$ and b are fitted coefficients. Local density-body size relationships were developed in each of the 18 lakes for which data were available. Global relationships were developed using all data. The upper boundary of the global relationship was also determined using the method proposed by Blackburn et al. [11]. Maximum population densities were determined in $\log_2$ body size classes (set at 2 $\log_2$ unit intervals, e.g., $-1 \leq (\log_2 M) < 1$, where M is species body mass in μg fresh mass). The relationship between maximum population density in each size class and average body size were then tested using simple linear regression analysis. Equations presented in their exponential form (as Eq. (1)) were detransformed according to Sprugel [118].

2.2   TEMPORAL VARIABILITY

Interannual population variability was measured as the standard deviation of the logarithmically (base 10) transformed mean annual population density (STD($\log D$) [127]). The analysis was restricted to species that had nonzero annual mean densities for at least four consecutive years, a criterion that effectively focuses the analysis on abundant species. A minimum time frame of four years was chosen as a compromise between including a sufficient number of years to obtain a reliable measure of long-term variability [25] and finding data on a sufficient number of populations. In phytoplankton especially, weekly to monthly counts are very time-consuming and are seldom done regularly over many consecutive years. Since $STD(\log D)$ measures the proportional variation in population densities among years, it is not affected by differences in measurement units, and several measures of population densities were included in the analysis (e.g., individuals $\times m^{-2}$, individuals $\times m^{-3}$, biovolume $\times m^{-3}$, biomass $\times m^{-3}$).

Relationships between the magnitude of interannual variability in population density ($STD(\log D)$), species body size (in μg fresh mass), lake size (surface area in km^2; mean depth in meters), lake productivity (dummy variable with 0 for oligotrophic lakes and 1 for eutrophic lakes) and lake disturbance (dummy variable with 0 for unmanipulated lakes and 1 for manipulated lakes) were tested using multiple regression analysis with forward and backward variable selection [55]. Since no consistent measures of lake productivity were available, lakes were classified based on water column concentrations of chlorophyll or total phosphorus [126] or on comments from the authors. Lake disturbance was defined as one or several heavy impacts (usually, but not necessarily anthropogenic) during the monitoring period or less than three years prior to the first population density measurements. Lakes where disturbances were not mentioned may nonetheless be impacted in various ways by the presence of humans, but for the present study, the variability we find in these systems is considered background variability. A model with all main factors and all possible interactions was tested within taxonomic groups (phytoplankton, zooplankton, fish) and across all taxonomic groups. In the present study, I focus on relationships between interannual population variability and species body size (main effect and interaction terms). The importance of environmental factors on population variability is taken into account, but will be discussed in detail elsewhere [26].

2.3 COMMUNITY COMPOSITION

The range in population densities covered by different species within a community was measured using species-abundance distributions (Figure 1). For each community, a species-abundance distribution was built using the residuals from a density-body size relationship. The density-body size relationship (e.g., Figure 1(a)) accounts for systematic differences in population densities between small and large species that would artificially widen the species-abundance distributions. The width of a species-abundance distribution was measured as the difference between maximum and modal residuals in a frequency distribution with $0.5 \log_{10}$ size classes (e.g., $10^0 - 10^{0.5}$ μg fresh mass; Figure 1(b)). Modal values were preferred over means since they are less sen-

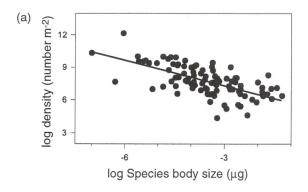

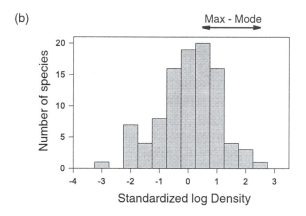

FIGURE 1 Method used to calculate the width of species-abundance relationships. This example shows the composition of the phytoplankton community in Lake Wingra (A. D. Hasler and J. F. Koonce unpublished, archived IBP data at University of Wisconsin). (a) Density-body size relationship (slope $= -0.78$, $n = 99$, $r^2 = 0.43$, $P < 0.001$). (b) Species-abundance distribution of standardized log Density (i.e., residuals from the density-body size relationship in (a)). The width of the species-abundance distribution (Max − Mode) is 2 orders of magnitude.

sitive to differences in sampling efforts (i.e., the presence of a few very rare species). This analysis focuses on variation among the most abundant species, in the right portion of the species-abundance distribution. The widths of species-abundance distributions were compared between phytoplankton, zooplankton, and fish communities using a median test [22].

All statistical analyses were performed on SAS for microcomputers, versions 6.08–6.10.

3 RESULTS

Data on 240 populations of phytoplankton, zooplankton, and fish from 18 well-studied lakes and reservoirs worldwide suggest a strong global relationship between annual mean population density and species body size (Figure 2). The overall relationship is best described by the following equation:

$$D = 10^{6.05\pm0.09} \times M^{-0.91\pm0.02}, \tag{3}$$

where D is annual mean population density of dominant species (numbers m^{-3}) and M is species mean-body size (μg fresh mass; statistics on log-linear relationship: $r^2 = 0.92, P < 0.0001$, residual mean square = 1.04). The upper boundary of the global density-body size relationship has a similar exponent of -0.93 ± 0.05. Mean annual densities of different populations varied two orders of magnitude on either side of the average global relationship (Figure 2).

The global density-body size relationship for aquatic organisms (Figure 2) represents a composite of relationships found in local communities (Figure 3). Local density-body size relationships were measured in each of the 18 lakes. These lakes vary more than 3 orders of magnitude in surface area (0.2–623 km^2) and more than an order of magnitude in mean depth (1.3–43 m). Water residence time ranges from two months to more than 11 years. Annual primary production covers as broad a range as is found across freshwater systems world-wide (4–1403 g C m^{-2} year^{-1}[28]). They also vary from pristine lakes

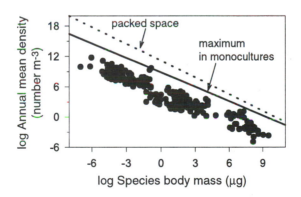

FIGURE 2 Density-body size relationship for lakes. This relationship includes phytoplankton, zooplankton, and fish species from 18 lakes worldwide for which volumetric densities (i.e., number m^{-3}) could be calculated (i.e., excludes zoobenthos; $\log D = 4.8 - 0.91 \log M, n = 240, r^2 = 0.92, P < 0.0001$). The solid line indicates maximum densities of organisms grown in monocultures ($\log D = 8.83 - 0.95 \log M$[36]). The dashed line indicates densities of organisms which would completely fill the available space ($\log D = 12 - 1 \log M$; assuming specific gravity of 1 g cm^{-3} for all organisms).

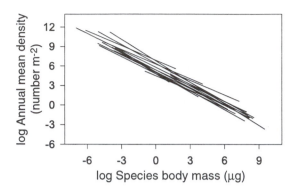

FIGURE 3   Comparison of density-body size relationships found in 18 local lake communities. Slopes range from -0.75 to -1.10, and intercepts from 4.5 to 6.9 ($r^2 = 0.78 - 0.98$).

located in arctic or alpine environments to heavily impacted lakes close to large cities. Lakes with such a broad range of environmental characteristics are expected to support communities that also range widely in size structure. Across the full range of body sizes, the exponents of local density-body size relationships ranged from -0.74 to -1.10 (median $= -0.93$), and were significantly related to primary productivity, anthropogenic impact, lake mean depth, and water residence time [31]. Together, these local density-body size relationships account for 3 orders of magnitude in population density in the global relationship (Figure 3).

The exponents of density-body relationships measured within taxonomic group differed according to the method of analysis used. Taxon-specific density-body size relationships measured using model I regressions were weak (i.e., low r^2) and had less negative exponents than overall density-body size relationships (Table 1). Model II regressions correct the bias in exponents estimated from weak relationships ($b_{II} = b_I/r$), and suggest more negative exponents that are consistent with those of overall relationships (Table 1). Global density-body size relationships calculated with model II regression had exponents of -0.95 ± 0.10 for algae, -0.71 ± 0.06 for invertebrates, and -1.24 ± 0.20 for fish. These values are generally consistent with the exponent of the overall global relationship (-0.93 ± 0.02). Similarly, in local communities the median exponents of taxon-specific density-body size relationships (-1.02 for algae, -0.65 for invertebrates, -1.44 for fish) are consistent with the median exponent of overall relationships (-0.95). When the taxon-specific analysis is extended to the full range of aquatic organisms using an analysis of covariance, a significant difference in exponents is found among taxonomic groups (analysis of covariance on algae, invertebrates and fish, $P = 0.02$ [32]). Algae and invertebrates share a common exponent of -0.54 ± 0.04, but fish

TABLE 1   Exponents (with standard errors) of density-body size relationships calculated over the full range of species (overall) and in individual taxonomic groups (algae, invertebrates, fish). These density-body size relationships are based on annual mean population density of dominant species (in numbers m^{-2}; note that these are different units than those used in Figure 2) and mean species body mass (in μg fresh mass). Local density-body size relationships were measured in 18 lakes, and the range and median (in parentheses) exponents are presented. Exponents calculated using model I and model II regressions are presented as b_I and b_{II}, respectively ($b_{II} = b_I/r$, where r^2 is the coefficient of determination).

	Global Relationships			Local Relationships		
	$b_I \pm SE$	r^2	$b_{II} \pm SE$	b_I	r^2	b_{II}
Overall	-0.89 ± 0.02	0.92	-0.93 ± 0.02	-1.10 to -0.74 (-0.93)	0.78 to 0.98	-1.20 to -0.77 (-0.95)
Algae	-0.64 ± 0.07	0.45	-0.95 ± 0.10	-1.35 to 0.21 (-0.79)	0.11 to 0.87	-1.64 to 0.53 (-1.02)
Invertebrates	-0.50 ± 0.04	0.51	-0.71 ± 0.06	-1.40 to 1.40 (-0.36)	0.002 to 0.97	-1.80 to 2.80 (-0.65)
Fish	-0.94 ± 0.15	0.58	-1.24 ± 0.20	-2.82 to -0.37 (-0.76)	0.11 to 0.96	-3.77 to -0.76 (-1.44)

have a much more negative exponent [32]. The exponents of taxon-specific density-body size relationships differ among taxa, and are either more negative or less negative than the expected -0.75.

Another portion of the variability in the global density-body size relationship can be attributed to year-to-year variation in population densities. The magnitude of interannual variability was compared in 249 populations of phytoplankton, zooplankton, and fish sampled for at least four consecutive years (Appendix A). These populations were gathered from a different set of lakes from the previous analyses. The lakes also ranged broadly in size, productivity and level of disturbance. The magnitude of interannual variability in population density ranged more than an order of magnitude within taxonomic groups (phytoplankton: 0.1–1.9, zooplankton: 0.04–1.2, fish: 0.03–1.3; Figure 4), but was not related to species body size ($P > 0.05$). Among taxonomic groups, $STD(\log D)$ decreased slightly with increasing species body size, from a geometric mean of 0.42 in phytoplankton, to 0.30 in zooplankton, and 0.26 in fish (0.15 in fish populations measured by mark-recapture techniques; Figure 4). This suggests that, on average, phytoplankton species vary almost 50 fold ($\pm 2 \times STD(\log D)$) in mean population density from year to year, zooplankton vary 15 fold and fish vary 4 to 10 fold. Despite this general trend, $STD(\log D)$ overlapped greatly among very different taxonomic groups (Figure 4). The magnitude of interannual population variability was significantly related to lake size, primary productivity and, in the case of small organisms, differed significantly between disturbed and nondisturbed lakes [26].

Uneven species representation in natural communities is the third principal source of variability in global density-body size relationships. Species-

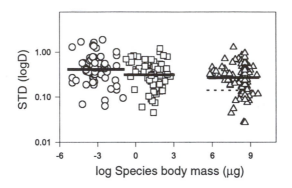

FIGURE 4   Interannual variability (STD(log D)) of 249 populations of phytoplankton (circles, $n = 63$), zooplankton (squares, $n = 91$), and fish (triangles, $n = 95$). STD(log D) does not vary significantly with species body size within taxonomic groups ($P > 0.05$). Solid lines indicate the geometric mean STD(log D) in phytoplankton (0.42), zooplankton (0.30), and fish (0.26). The dashed line is the geometric mean STD(log D) in fish populations sampled by mark-recapture method (0.15).

abundance distributions were compared in 28 well-studied communities of phytoplankton, zooplankton, and fish, where the annual mean density of 67 to 100% of the species had been measured (Appendix B). Phytoplankton communities contained between 58–205 species, zooplankton communities between 7–44 species, and fish communities between 6–40 species. These studies are from lakes and reservoirs around the world, which varied greatly in size and productivity, and are expected to support very different communities. The majority of communities were characterized by most species having relatively low densities, and few dominant or very rare species (e.g., Figure 1(b)). The width of species-abundance distributions varied several folds within taxonomic groups, but did not differ significantly between phytoplankton, zooplankton, and fish communities (median test, $P > 0.05$; Figure 5). On average, the density of dominant species was 1.5 to 2.5 orders of magnitude higher than the density of most species of similar body size, and this pattern was similar in phytoplankton, zooplankton, and fish (Figure 5). Therefore, the intercept of the global density-body size relationship drops when rare species are included, but its slope remains unchanged (Figure 6). Natural patterns of community composition can easily account for 4 orders of magnitude variation in global density-body size relationships (e.g., Figure 1(b)), but they do not bias its slope.

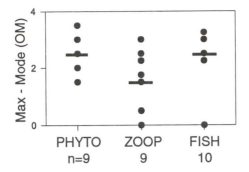

FIGURE 5   Comparison of the width of species-abundance distributions in phytoplankton (PHYTO), zooplankton (ZOOP), and fish communities. The width of each distribution (in orders of magnitude, OM) is measured as the difference between maximum and mode of standardized log Density (Max–Mode; see Figure 1). The number of communities analysed (n) is indicated for each taxonomic group (several points in the figure represent more than one community). Solid lines represent medians. Median widths of species-abundance distributions did not differ significantly among taxonomic groups (median test, $P > 0.2$).

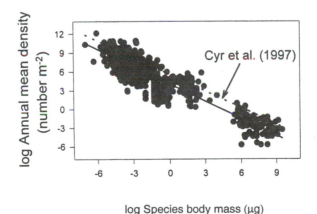

FIGURE 6   Global density-body size relationship, including rare species ($\log D = 4.1 - 0.92 \log M$; $n = 975$, $r^2 = 0.82$, $P < 0.001$). The global density-body size relationship for dominant aquatic species, which was developed from a largely independent data set, is shown for comparison ($\log D = 5.6 - 0.89 \log M$ [32]).

4 DISCUSSION

The energetic equivalence hypothesis predicts that the overall density-body size relationships in terrestrial and aquatic communities should have different exponents (b in Eq. (1)), due to fundamental differences in the structure of these communities. The hypothesis assumes that the same amount of energy is available to all species on a given trophic level, regardless of their body sizes. In terrestrial communities, density-body size relationships for carnivores, herbivores, and plants are expected to have similar exponents ($b = -0.75$), but different a coefficients (Eq. (1)) due to energy losses at each trophic transfer (e.g., Damuth [34]). Differences in a coefficients are also expected between endotherms and poikilotherms, which have very different metabolic requirements (e.g., Peters and Wassenberg [104] and Currie and Fritz [24]). The energetic equivalence hypothesis, therefore, suggests that the overall density-body size relationship in terrestrial communities should have an exponent (b) of -0.75, but a large scatter due to differences in a coefficients among groups of organisms. In contrast, body sizes in aquatic species are well correlated to their trophic levels. Large fish eat small fish, which eat invertebrates, which eat microscopic algae. Because energy is lost between each trophic level, the exponent of the overall density-body size relationship is expected to be more negative than -0.75. In very general terms, four trophic levels are spread over a 16-orders-of-magnitude range in species body size, from fish to algae (see Figure 2). Assuming that each trophic level spans on average 4 orders of magnitude in species body size and that energy is transferred among trophic levels with a 10% efficiency [110], the amount of available energy is expected to decrease exponentially with increasing body size, with an exponent of -0.25. In this example, we would expect a density-body size relationship with an exponent of -1 to result in energetic equivalence among populations ($D = R_p/R_i \propto M^{-0.25}/M^{0.75}$). The magnitude of the exponent is highly dependent on the efficiency of trophic transfers. In aquatic communities, the energetic equivalence hypothesis predicts an exponent of -0.75 within trophic level, but a more negative exponent across the full range of body sizes from algae to fish. Therefore, overall density-body size relationships are expected to have more negative exponents in aquatic than in terrestrial communities.

4.1 GLOBAL DENSITY-BODY SIZE RELATIONSHIPS

Overall density-body size relationships for aquatic organisms have exponents between -0.89 and -0.92, depending on which data set is used (areal or volumetric densities of dominant species, Table 1, Figure 2; all population densities including rare species, Figure 6), and have even steeper upper limits (-0.93 to -0.96). These overall exponents are indeed more negative than -0.75, but are similar to those reported in terrestrial communities (see Cyr et al. [32]). Within taxonomic groups, the exponents vary significantly among taxonomic groups, and are either more negative or less negative than -0.75 depending on the method of analysis used (Table 1). These exponents suggest that over the full range of species body sizes, large species use lower quantities of energy on average than small species (i.e., $b < -0.75$), but within taxonomic group large species use either more energy or less energy than small species [32]. These results are consistent with the conclusions of several studies of terrestrial animals (e.g., Maurer and Brown [88], Griffiths [53], Currie and Fritz [24], Silva and Downing [116]). Our best measures of the exponents of global density-body size relationships, both in aquatic and in terrestrial communities, do not support the energetic equivalence hypothesis.

The energy equivalence hypothesis assumes that an equal amount of energy is available to species of all sizes within a trophic level. This assumption is unlikely to hold in nature. Even within local communities, food availability varies greatly among species. For example, large species consume a larger range of food sizes than small species (e.g., Vézina [122], Cyr and Curtis [27]). Moreover, communities are generally composed of many more small species than large species (e.g., Griffiths [52], Brown and Nicoletto [14], Blackburn and Gaston [8]) so that small species may have access to a smaller portion of the available resources compared to large species. Energy is not equally available to all species in natural communities.

Organisms grown in monocultures, where food is supplemented, predators and diseases are controlled, and resources are shared with few other species (i.e., monocultures or artificially simplified communities), can presumably reach maximum densities. Under such ideal conditions, individual energy use is most likely to limit population densities. The densities of algae, aquatic invertebrates, and fish grown in monocultures are strongly correlated with species body sizes, with an exponent of -0.95 (Figure 2, [36]). This exponent is closer to -1, which is expected for perfectly packed organisms (dashed line in Figure 2, assuming equal tissue density from algae to fish; $1 - 1.2$ g cm^{-3} in phytoplankton [109]; $1.02 - 1.08$ g cm^{-3} in fish [83]) than to the theoretical slope of -0.75, suggesting that space is more likely limiting the maximum density of aquatic organisms than individual energy use. Even under food saturation, allometric differences in individual metabolic rates cannot account for the exponent of the relationship between population densities and species body sizes.

Natural population densities are, on average, 4 orders of magnitude lower than in monocultures, and even the most abundant species in nature are 10 times less dense than cultured populations (Figure 2). The only exceptions in this data set are populations of blue-green algae, which in some lakes reach densities similar to those grown in monoculture (e.g., two points on the line in Figure 2). Low densities of natural populations cannot simply be attributed to the presence of other species sharing the available resources. When the contribution of all species is added, the total biomass of natural communities is still lower than biomasses grown in cultures, and it decreases slightly with increasing body size (Figure 7, [30]). Phytoplankton communities, especially when dominated by cyanobacteria, reach biomasses close to the predicted maximum,

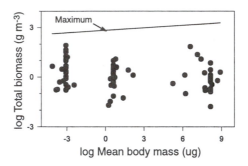

FIGURE 7   Total biomass (B) of phytoplankton, zooplankton, and fish communities measured in lakes world wide varies only weakly with species mean-body mass ($\log B = 0.14 - 0.04 \log M; n = 84, r^2 = 0.05, P = 0.04$; data sources are listed in Cyr and Peters [30]). The maximum biomass grown in monocultures is shown for comparison (recalculated from Duarte et al. [36]).

but zooplankton and fish communities attain at most 10 times lower biomasses than organisms of comparable sizes grown in culture (Figure 7). This pattern is different from the slope of +0.25 predicted by the energetic equivalence hypothesis (biomass = density × body size $\propto M^{-0.75} \times M$). The discrepancy between the observed density-body size relationship and that predicted by the energetic equivalence hypothesis cannot be attributed to differences in the number of small and large species sharing resources in natural aquatic communities.

Global density-body size relationships leave much unexplained variation among natural populations (Figure 2, [32]) which may bias its exponent. This variation compounds spatial and temporal population variability, as well as the uneven representation of species in communities.

4.2 SPATIAL VARIABILITY

The magnitude of among-site variability in population densities is not trivial. Differences in the slopes and intercepts of local density-body size relationships studied in 18 lakes world wide, account for 3 orders of magnitude variation in the global relationships (Figure 3). The size structure of individual communities were well described by log-linear density-body size relationships (Figure 3, [31]), and suggest, in contrast to the conclusions of Blackburn and Gaston [9], that log-linear relationships are not simply artifacts of the large continental or global scale at which they are developed. Log-linear (or exponential) density-body size relationships are found in local communities.

The exponents of local density-body size relationships ranged between −0.75 and −1.10, and were significantly related to lake characteristics [31]. The exponents (b) and coefficients a of density-body size relationships (Eq. (1))

varied most strongly with differences in lake primary productivity [31]. Not surprisingly, population densities were higher in more productive lakes. Increases in population densities, however, were larger in phytoplankton than in zooplankton or fish, a pattern that resulted in more negative exponents. These results are consistent with the view that in aquatic systems, transfers of energy to higher trophic levels (i.e., from small to larger organisms) become less efficient with increasing productivity [47, 112]. Alternatively, these results are also consistent with the observation that the ratio of microbial to algal biomass increases with decreasing primary productivity [43, 46], and that microbial communities provide an important energy subsidy to higher trophic levels in unproductive aquatic systems. The energetic equivalence hypothesis assumes that an equal amount of energy is available to species of all sizes, and that large individuals simply require more energy than small individuals [34]. The results presented here, however, suggest that differences in energy availability to the community (primary and microbial productivity) and in the efficiency of trophic transfers are important factors limiting the density of populations in lakes.

4.3 INTERANNUAL VARIABILITY

The magnitude of interannual variability $(STD(\log D))$ in aquatic populations varied between 0.03 and 1.9 (Figure 4), a range similar to that measured in terrestrial populations [21, 114]. Within taxonomic groups, interannual variability in population density differed more than an order of magnitude among species, and was not related to species body size ($P > 0.05$; Figure 4). Mean $STD(\log D)$ decreased slightly between phytoplankton, zooplankton, and fish populations, but despite large differences in reproductive cycle and life span, many fish populations were just as variable from year to year as zooplankton or phytoplankton populations (Figure 4). Interannual variability in natural populations introduces much variability in global density-body size relationships, but is unlikely to bias their exponents and a coefficients (Eq. (1)).

4.4 COMMUNITY COMPOSITION

It has long been recognized that species are not represented evenly in natural communities [107, 121]. Most communities are composed of one or a few dominant species with the majority of species being several orders of magnitude less abundant. Early comparisons of species-abundance distributions across a wide range of taxonomic groups (benthic diatoms, phytoplankton, moths, birds) found very similar variances of the log-normal models fitted to these distributions [69, 107], but these similarities were later shown to be mathematical necessities of the canonical log-normal models [89]. The quasi-universal use of log-normal models to describe species-abundance distributions has been criticized because these models fit poorly in many communities (e.g., Hughes [68] and Lambshead and Platt [79]). Moreover, most of the early and of the more recent [85, 110, 119] comparisons of variances were based on data from a few types of communities (mostly birds) sampled at different spatial and temporal scales, with techniques likely to have biased the relative representation of different species (e.g., diatoms on artificial substrates, moths attracted to light trap). Perhaps not so surprisingly, few generalities have emerged as to how species-abundance distributions vary in different types of communities.

There are two reasons to expect the width (or variance) of species-abundance distributions to vary systematically among taxonomic groups of different body sizes. First, most factors which are thought to affect average and minimum viable population densities (e.g., dispersal rate, home range, energy use, life history parameters) are highly correlated with species body size [15, 103]. Communities of small species, therefore, could be structured quite differently from communities of large species. Second, even if we assume that all communities follow perfectly the canonical log-normal model of species-abundance distributions, the variance in population densities is expected to increase with increasing species richness in communities with fewer than $\sim$ 200 species [89, 119]. Since small taxa are generally more speciose than large taxa [90], we would expect communities of smaller organisms to have wider species-abundance distributions.

In the present data set, the degree of species dominance, measured as the width of a species-abundance distribution, varied 2 to 4 fold among similar types of communities (vertical distribution of points in Figure 5), but did not differ systematically between phytoplankton, zooplankton, and fish communities. These results suggest that, on average, biomass is partitioned similarly among species in phytoplankton, zooplankton, and fish communities regardless of major differences in body size, physiological rates, life cycles, and species richness (phytoplankton communities had 70–205 species, zooplankton communities had 7–44 species, fish communities had 6–40 species). Natural patterns of community composition, therefore, are expected to contribute the same amount of variation to global density-body size relationships along the full range of species body size. The best data available suggest that the under-representation of rare species in published global density-body size relationships has overestimated their a coefficients (intercept in Figure 6) but has not biased their exponents (slope in Figure 6).

5 CONCLUSION

The exponents of global density-body size relationships in both aquatic (-0.89 to -0.92) and terrestrial systems (-0.92 to -0.98) [24, 34, 104] are steeper than predicted by the energetic equivalence hypothesis (-0.75) [33, 34]. A large portion of the variability around these global relationships can be attributed to interannual population variability and to the uneven representation of species in communities. However, because both sources of variability are equally important in different taxonomic groups, they are unlikely to bias the exponent of the global density-body size relationship. Current global density-body size relationships do not support the energetic equivalence hypothesis.

Similar results were found in local communities. In contrast with predictions from the energetic equivalence hypothesis, the exponents of local density-body size relationships were not constant, but varied with lake characteristics. Differences in energy availability and/or in trophic transfer efficiencies are important in limiting population densities.

Natural population densities, therefore, are not simply determined by physiological requirement of the individuals but, not too surprisingly, are also affected by interactions among organisms and between organisms and their environment. The body sizes of organisms found in a community is well related to some community and ecosystem functions (e.g., productivity of algal communities [73]; respiration of planktonic communities [1]) but not others (e.g., grazing of zooplankton communities [29]). In order to link individual physiological "constraints" to overall patterns of population densities in nature, we will need to consider the relative importance of different ecological interactions and environmental constraints on organisms of different body sizes. Many of these relationships are still poorly known.

ACKNOWLEDGMENTS

I would like to acknowledge the early inputs of R. Peters and J. Downing with whom this adventure into large-scale patterns of community structure started. I thank J. Brown and G. West for inviting me to this conference and giving me the opportunity to pull together different pieces of research. I especially thank J. Brown for useful feedback and suggestions on an earlier version of this chapter.

APPENDIX A

Data gathered to measure interannual variability in population density. n_{sp} is the number of species for which data were available in each lake, n_{yr} is the number of consecutive years for which nonzero population density were measured in each species.

TABLE 1   Phytoplankton

Lake	n_{sp}	n_{yr}	Source
Grane Langsø, Denmark	1	4	Nygaard [100]
Leven, UK	5	4-9	Bailey-Watts [4]
Loch Neagh, UK	4	9-11	Gibson and Fitzsimons [45]
Windermere South, UK	1	45[a]	Maberly et al. [84]
Tjeukemeer, Neth.	8	6-7	Moed and Hoogveld [96]
Muggelsee, Germany	1	12	Nixdorf and Hoeg [99]
Neusiedlersee, Austria	11	4	Dokulil [35]
Kinneret, Israel	1	4	Pollingher and Serruya [106]
Sagami, Japan	8	5	Saito [113]
4 lakes, New Zealand	5-6	4	Duthie and Stout [37]

TABLE 2   Zooplankton

Lake	n_{sp}	n_{yr}	Source
Aleknagik, AK, US	4	10	IBP reports (unpublished)
Loch Neagh, UK	13	5-10	Fitzsimons and Andrew [42]
Rutland, UK	2	5	Harper and Ferguson [58]
4 lakes, BC, Canada	2	10[b]	Walters et al. [123]
Lake	n_{sp}	n_{yr}	Source
Constance, Switzerland	7	14-25	Einsle [40]
Bowland, ON, Canada	19	6	Keller et al. [76]
Neusiedlersee, Austria	4	4	Herzig [66]
Ontario, Canada/US[c]	6	4-6[d]	Johannsson and O'Gorman [70]
Kasumigaura, Japan	6	4	Hanazato and Aizaki [57]
Kinneret, Israel	4	5-17	Gophen [48], Gophen and Landau [49],Gophen et al. [50]

[a]Two years missing.
[b]May–November data.
[c]Several sites considered separately.
[d]July–October data.

TABLE 3   Fish

Lake	n_{sp}	n_{yr}	Source
Leven, UK	1	5	Thorpe [120]
Aleknagik, AK, US	2	14	IBP reports (unpublished)
St-George, ON, Canada	3	7	McQueen et al. [92]
Lake 226, 2 basins, ON, Can.	1	11	Mills and Chalanchuk [95]
Chequamegon Bay, Can./US	1	16	Bronte et al. [13]
22 lakes, AR, US	1	4-7	Bailey [3]
Pyhäjärvi, Finland	1	8	Helminen et al. [63]
Wilson, MN, US	2	5-7	Johnson [71]
Oneida, NY, US	1	14	McQueen et al. [93]
Huron, ON, Canada	1	33	Henderson et al. [65]
Ontario, Canada/US[a]	1	8	O'Gorman et al. [101]
Kingston Bassin, Can./US	7	8-20	Christie et al. [19]
2 lakes, NWT, Canada	1	7-8	Johnson [72]
Little Rock, 2 basins, WI, US	3	6	Eaton et al. [38]
Michigan, US	5	12	Eck and Wells [39]
Red Lakes, MN, US	2	51	Pereira et al. [102]
Windermere, 2 basins, UK	1	37	Bagenal [2]
Sharpe, SD, US	4	7-11	Elrod [41], Nelson and Walburg [98]
2 lakes, ON, Canada	1-3	6-9	Mills et al. [95]
Great Central, BC, Can.	1	8	LeBrasseur et al. [81]
Klicava, Bohemia	8	5-6	Holcik [67]
Escanaba, WI, US	10	14-30	Kempinger and Carline [78]
Rainy, Can./US	1	22	Chevalier [18]
2 lakes, SD, US	3	7-17	Nelson and Walburg [98]

[a]Several sites considered separately.

APPENDIX B

Data gathered to measure the width of species-abundance distribution in freshwater communities. n_{sp} is the number of species for which density data were available, n_{tot} is the number of species known to exist in the lake.

TABLE 1   Phytoplankton

Lake	n_{sp}	n_{tot}	Source
Albufera, Spain	131	131	Romo and Miracle [111]
Banyoles, Spain	76	77	Planas [105]
Buda, Spain	58	58	Comín [20]
Encanizada, Spain	154	205	Comín [20]
Esrom, Denmark	92	131	Jónasson and Kristiansen [74]
Lanao, Philippines	70	70	Lewis [82]
Pääjarvi, Finland	82	107	Granberg [51]
Tancada, Spain	90	104	Comín [20]
Wingra, US	99	103	Hasler and Koonce [62]

TABLE 2   Zooplankton

Lake	n_{sp}	n_{tot}	Source
Balaton, Hungary	24	24	Zánkai and Ponyi [128, 129]
Mergozzo, Italy	44	44	de Bernardi and Soldavini [7]
Pääjarvi, Finland	31	31	Granberg [51]
Pietra, Italy	19	19	Bonacina and de Bernardi [12]
Port-Biehl, France	7	7	Rey and Capblancq [108]
Queen Elizabeth Res., Eng.	20	20	IBP final report
Queen Mary Res., England	19	20	IBP final report
Tjeukemeer, Netherlands	12	14	IBP final report
Wingra, US	22	22	Hartman and Teraguchi [59]

TABLE 3   Fish

Lake	n_{sp}	n_{tot}	Source
Annie, FL, US	13	19	Werner et al. [124]
George, Uganda	25	32	Gwahaba [54]
Kariba, Rhodesia	29	40	Balon [5]
Lawrence, MI, US	14	19	Hall and Werner [56]
Long Pond, ON, Canada	22	22	Mahon and Balon [86]
Lower Poole, ON, Canada	12	12	Keast and Fox [75]
Red Deer, ON, Canada	6	6	Chadwick [17]
Sirena, FL, US	11	14	Werner et al. [124]
Texoma Res., OK/TX, US	24	24	Gelwick and Matthews [44]
Turkey, ON, Canada	6	9	Kelso [77]

REFERENCES

[1] Ahrens, M. A., and R. H. Peters. "Plankton Community Respiration: Relationships with Size Distribution and Lake Trophy." *Hydrobiologia* **224** (1991): 77–87.

[2] Bagenal, F. B. "Experimental Manipulations of the Fish Populations in Windermere." *Hydrobiologia* **86** (1982): 201–205. [Appendix A, Table 3]

[3] Bailey, W. M. "A Comparison of Fish Populations Before and After Extensive Grass Carp Stocking." *Trans. Am. Fish. Soc.* **107** (1978): 181–206. [Appendix A, Table 3]

[4] Bailey-Watts, A. E. "A Nine-Year Study of the Phytoplankton of the Eutrophic and Non-Stratifying Loch Leven (Kinross, Scotland)." *J. Ecol.* **66** (1978): 741–771. [Appendix A, Table 1]

[5] Balon, E. K. *Fishes of Lake Kariba, Africa: Length-Weight Relationship, a Pictoral Guide.* Hong Kong: T. F. H. Publications, 1974. [Text; Appendix B, Table 3]

[6] Begon, M., L. Firbank, and R. Wall. "Is There a Self-Thinning Rule for Animal Populations?" *Oikos* **46** (1986): 122–124.

[7] de Bernardi, R., and E. Soldavini. "Long-Term Fluctuations of Zooplankton in Lake Mergozzo, Northern Italy." *Mem. Ist. Ital. Idrobiol.* **33** (1976): 345–375. [Appendix B, Table 2]

[8] Blackburn, T. M., and K. J. Gaston. "The Distribution of Body Sizes of the World's Bird Species." *Oikos* **70**: 127–130, 1994.

[9] Blackburn, T. M., and K. J. Gaston. "A Critical Assessment of the Form of the Interspecific Relationship Between Abundance and Body Size in Animals." *J. Anim. Ecol.* **66** (1997): 233–249.

[10] Blackburn, T. M., and K. J. Gaston. "Some Methodolocial Issues in Macroecology." *Amer. Natur.* **151** (1998): 68–83.

[11] Blackburn, T. M., J. H. Lawton, and J. N. Perry. "A Method of Estimating the Slope of Upper Bounds of Plots of Body Size and Abundance in Natural Animal Assemblages." *Oikos* **65** (1992): 107–112.

[12] Bonacina, C., and R. de Bernardi. "Struttura di Comunita' e Parametri Demografici del Popolamento Zooplanctonico." In *Il Lago di Pietra del Pertusillo: Definizione delle sue Caratteristiche Limno-Ecologichepp*, 99–132. Pallanza: Istituto Italiano di Idrobiologia, 1978. [Appendix B, Table 2]

[13] Bronte, C. R., J. H. Selgeby, and D. V. Swedberg. "Dynamics of a Yellow Perch Population in Western Lake Superior." *N. Am. J. Fish. Manag.* **13** (1993): 511–523. [Appendix A, Table 3]

[14] Brown, J. H., and P. F. Nicoletto. "Spatial Scaling of Species Composition: Body Masses of North American Land Mammals." *Amer. Natur.* **138** (1991): 1478–1512.

[15] Calder, W. A. *Size, Function, and Life History.* Cambridge, MA: Harvard University Press, 1984.

[16] Carlander, K. D. *Handbook of Freshwater Fishery Biology*. Ames: Iowa State University Press, 1969.

[17] Chadwick, E. M. P. "Ecological Fish Production in a Small Precambrian Shield Lake." *Env. Biol. Fish.* **1** (1976): 13–60. [Appendix B, Table 3]

[18] Chevalier, J. R. "Changes in Walleye (*Stizostedion vitreum vitreum*) Population in Rainy Lake and Factors in Abundance, 1924–1975." *J. Fish. Res. Board Can.* **34** (1977): 1696–1702. [Appendix A, Table 3]

[19] Christie, W. J., K. A. Scott, P. G. Sly, and R. H. Stus. "Recent Changes in the Aquatic Food Web of Eastern Lake Ontario." *Can. J. Fish. Aquat. Sci.* **44** (Suppl. 2) (1987): 37–52. [Appendix A, Table 3]

[20] Comín, F. A. "Caracteristicas Fisicas y Quimicas y Fitoplancton de las Lagunas Costeras, Encanizada, Tancada y Buda (Delta del Ebro)." *Oecologia Aquatica* **7** (1984): 79–162. [Appendix B, Table 1]

[21] Connell, J. H., and W. P. Sousa. "On the Evidence Needed to Judge Ecological Stability or Persistence." *Amer. Natur.* **121** (1983): 789–824.

[22] Conover, W. J. *Practical Nonparametric Statistics*. New York: Wiley, 1980.

[23] Currie, D. J. "What Shape Is the Relationship Between Body Size and Population Density?" *Oikos* **66** (1993): 353–358.

[24] Currie, D. J., and J. T. Fritz. "Global Patterns of Animal Abundance and Species Energy Use." *Oikos* **67** (1993): 56–68.

[25] Cyr, H. "Does Inter-Annual Variability in Population Density Increase with Time?" *Oikos* **79** (1997): 549–558.

[26] Cyr, H. In preparation.

[27] Cyr, H., and J. M. Curtis. "Zooplankton Community Size Structure and Taxonomic Composition Affects Size-Selective Grazing in Natural Communities." *Oecologia* **118** (1999): 306–315.

[28] Cyr, H., and M. L. Pace. "Magnitude and Patterns of Herbivory in Aquatic and Terrestrial Ecosystems." *Nature* **361** (1993): 148–150.

[29] Cyr, H., and M. L. Pace. "Allometric Theory: Extrapolations from Individuals to Communities." *Ecology* **74** (1993): 1234–1245.

[30] Cyr, H., and R. H. Peters. "Biomass-Size Spectra and the Prediction of Fish Biomass in Lakes." *Can. J. Fish. Aquat. Sci.* **53** (1996): 994–1006.

[31] Cyr, H., J. A. Downing, and R. H. Peters. "Density-Body Size Relationships in Local Aquatic Communities." *Oikos* **79** (1997): 333–346.

[32] Cyr, H., R. H. Peters, and J. A. Downing. "Population Density and Community Size Structure: Comparison of Aquatic and Terrestrial Systems." *Oikos* **80** (1997): 139–149.

[33] Damuth, J. "Population Density and Body Size in Mammals." *Nature* **290** (1981): 699–700.

[34] Damuth, J. "Interspecific Allometry of Population Density in Mammals and Other Animals: The Independence of Body Mass and Population Energy-Use." *Biol. J. Linns. Soc.* **31** (1987): 193–246.

[35] Dokulil, M. "Seasonal Pattern of Phytoplankton." In *Neusiedlersee: The Limnology of a Shallow Lake in Central Europe*, edited by H. Loffler,

203–232. The Hague: Dr. W. Junk bv Publishers, 1979. [Appendix A, Table 1]

[36] Duarte, C. M., S. Agusti, and R. H. Peters. "An Upper Limit to the Abundance of Aquatic Organisms." *Oecologia* **74** (1987): 272–276.

[37] Duthie, H. C., and V. M. Stout. "Phytoplankton Periodicity of the Waitaki Lakes, New Zealand." *Hydrobiologia* **138** (1986): 221–236. [Appendix A Table 1]

[38] Eaton, J. G., W. A. Swenson, J. H. McCormick, T. D. Simonson, and K. M. Jensen. "A Field and Laboratory Investigation of Acid Effects on Largemouth Bass, Rock Bass, Black Crappie, and Yellow Perch." *Trans. Am. Fish. Soc.* **121** (1992): 644–658. [Appendix A, Table 3]

[39] Eck, G. W., and L. Wells. "Recent Changes in Lake Michigan's Fish Community and Their Probable Causes, with Emphasis on the Role of the Alewife (*Alosa pseudoharengus*)." *Can. J. Fish. Aquat. Sci.* **44** (Suppl. 2) (1987): 53–60. [Appendix A, Table 3]

[40] Einsle, U. "The Long-Term Dynamics of Crustacean Communities in Lake Constance (Obersee, 1962–1986)." *Schweiz. Z. Hydrol.* **50** (1988): 136–165. [Appendix A, Table 2]

[41] Elrod, J. H. "Abundance, Growth, Survival, and Maturation of Channel Catfish in Lake Sharpe, South Dakota." *Trans. Am. Fish. Soc.* **1** (1974): 53–58. [Appendix A, Table 3]

[42] Fitzsimons, A. G., and T. E. Andrew. "The Seasonal Succession of the Zooplankton of Lough Neagh, 1968–1978." *Monographiae Biologicae* **69** (1993): 281–326. [Appendix A, Table 2]

[43] Gasol, J. M., P. A. D. Giorgio, and C. M. Duarte. "Biomass Distribution in Marine Planktonic Communities." *Limnol. Oceanogr.* **4(2)** (1997): 1353–1363.

[44] Gelwick, F. P., and W. J. Matthews. "Temporal and Spatial Patterns in Littoral-Zone Fish Assemblages of a Reservoir (Lake Texoma, Oklahoma-Texas, U.S.A.)." *Env. Biol. Fish.* **27** (1990): 107–120. [Appendix B, Table 3]

[45] Gibson, C. E., and A. G. Fitzsimons. "Periodicity and Morphology of Planktonic Blue-Green Algae in an Unstratified Lake (Lough Neagh, Northern Ireland)." *Int. Revue Ges. Hydrobiologia* **67** (1982): 459–476. [Appendix A, Table 1]

[46] del Giorgio, P. A., and R. H. Peters. "Patterns in Planktonic P:R Ratios in Lakes: Influence of Lake Trophy and Dissolved Organic Carbon." *Limnol. Oceanogr.* **39** (1994): 772–787.

[47] Gliwicz, Z. M., and A. Hillbricht-Ilkowska. "Efficiency of the Utilization of Nannoplankton Primary Productivity by Communities of Filter Feeding Animals Measured in Situ." *Int. Ver. Theor. Angew. Limnol. Verh.* **18** (1972): 197–203.

[48] Gophen, M. "Temperature Dependence of Food Intake, Ammonia Excretion and Respiration in *Ceriodaphnia reticulata* (Jurine) (Lake Kinneret, Israel)." *Freshwater Biol.* **6** (1976): 451–455. [Appendix A Table 2]

[49] Gophen, M., and R. Landau. "Trophic Interactions Between Zooplankton and Sardine *Mirogrex terraesanctae* Populations in Lake Kinneret, Israel." *Oikos* **29** (1977): 166–174. [Appendix A, Table 2]

[50] Gophen, M., S. Serruya, and P. Spataru. "Zooplankton Community Changes in Lake Kinneret (Israel) During 1969–1985." *Hydrobiologia* **191** (1990): 39–46. [Appendix A, Table 2]

[51] Granberg, K. "Seasonal Fluctuations in Numbers and Biomass of the Plankton of Lake Pääjarvi, Southern Finland." *Ann. Zool. Fenn.* **7** (1970): 1–24. [Appendix B, Table 1 and Table 2]

[52] Griffiths, D. "Size-Abundance Relations in Communities." *Amer. Natur.* **127** (1986): 140–166.

[53] Griffiths, D. "Size, Abundance, and Energy Use in Communities." *J. Anim. Ecol.* **61** (1992): 307–315.

[54] Gwahaba, J. J. "The Distribution, Population Density and Biomass of Fish in an Equatorial Lake, Lake George, Uganda." *Proc. Roy. Soc. Lond. B* **190** (1975): 393–414. [Appendix B, Table 3]

[55] Gujarati, D. *Basic Econometrics.* New York: McGraw-Hill, 1978.

[56] Hall, D. J., and E. E. Werner. "Seasonal Distribution and Abundance of Fishes in the Littoral Zone of a Michigan Lake." *Trans. Am. Fish. Soc.* **106** (1977): 545–555. [Appendix B, Table 3]

[57] Hanazato, T., and M. Aizaki. "Changes in Species Composition of Cladoceran Community in Lake Kasumigaura During 1986–1989: Occurrence of *Daphnia galeata* and Its Effect on Algal Biomass." *Jpn. J. Limnol.* **52** (1991): 45–55. [Appendix A, Table 2]

[58] Harper, D. M., and A. J. D. Ferguson. "Zooplankton and Their Relationships with Water Quality and Fisheries." *Hydrobiologia* **88** (1982): 135–145. [Appendix A, Table 2]

[59] Hartman, J. M., and M. Teraguchi. "Pelagic Zooplankton 1968–1969." Unpublished archived data, Institute of Environmental Studies, Publications, Information and Outreach, University of Wisconsin–Madison, Madison, WI. [Appendix B, Table 2]

[60] Harvey, P. H. Personal communication.

[61] Harvey, P. H., and M. D. Pagel. *The Comparative Method in Evolutionary Biology.* Oxford: Oxford University Press, 1991.

[62] Hasler, A. D., and J. F. Koonce. "Number and Biomass of Phytoplankton Species in Lake Wingra (1970–1971)." Unpublished archived data, Institute of Environmental Studies, Publications, Information and Outreach, University of Wisconsin–Madison, Madison, WI. [Appendix B, Table 1]

[63] Helminen, H., H. Auvinen, A. Hivonen, J. Sarvala, and J. Toivonen. "Year-Class Fluctuations of Vendace (*Coregonus albula*) in Lake Pyhajarvi, Southwest Finland, During 1971–1990." *Can. J. Fish. Aquat. Sci.* **50** (1993): 925–931. [Appendix A, Table 3]

[64] Hemmingsen, A. M. "Energy Metabolism as Related to Body Size and Respiratory Surfaces, and Its Evolution." Reports of the Steno Memorial Hospital and Nordisk Insulin Laboratorium **9** (1960): 6–110.

[65] Henderson, B. A., J. J. Collins, and J. A. Reckahn. "Dynamics of an Exploited Population of Lake Whitefish (*Coregonus clupeaformis*) in Lake Huron." *Can. J. Fish. Aquat. Sci.* **40** (1983): 1556–1567. [Appendix A, Table 3]

[66] Herzig, A. "The Zooplankton of the Open Lake." In *Neusiedlersee: The Limnology of a Shallow Lake in Central Europe*, edited by H. Loffler, 281–336. The Hague: Dr. W. Junk bv Publishers, 1979. [Appendix A, Table 2]

[67] Holcik, J. "Changes in Fish Community of Klicava Reservoir with Particular Reference to Eurasian Perch (*Perca fluviatilis*), 1957–1972." *J. Fish. Res. Board Can.* **34** (1977): 1734–1745. [Appendix A, Table 3]

[68] Hughes, R. G. " A Hypothesis Concerning the Influence of Competition and Stress on the Structure of Marine Benthic Communities." In *Proceedings of the 19th European Marine Biology Symposium*, edited by P. E. Gibbs, 391–400. Cambridge, UK: Cambridge University Press, 1984.

[69] Hutchinson, G. E. *A Treatise on Limnology*. New York: Wiley, 1967.

[70] Johannsson, O. E., and R. O'Gorman. "Roles of Predation, Food, and Temperature in Structuring the Epilimnetic Zooplankton Populations in Lake Ontario, 1981–1986." *Trans. Am. Fish. Soc.* **120** (1991): 193–208. [Appendix A, Table 2]

[71] Johnson, F. H. "Responses of Walleye (*Stizostedion vitreum vitreum*) and Yellow Perch (*Perca flavescens*) Populations to Removal of White Sucker (*Catostomus commersoni*) from a Minnesota Lake, 1966." *J. Fish. Res. Board Can.* **34** (1977): 1633–1642. [Appendix A, Table 3]

[72] Johnson, L. "Homeostatic Characteristics of Single Species Fish Stocks in Arctic Lakes." *Can. J. Fish. Aquat. Sci.* **40** (1983): 987–1024. [Appendix A, Table 3]

[73] Joint, I. R., and A. J. Pomroy. "Allometric Estimation of the Productivity of Phytoplankton Assemblages." *Marine Ecology Progress Series* **47** (1988): 161–168.

[74] Jonasson, P. M., and J. Kristiansen. "Primary and Secondary Production in Lake Esrom. Growth of *Chironomus anthracinus* in Relation to Seasonal Cycles of Phytoplankton and Dissolved Oxygen." *Int. Rev. Ges. Hydrobiologia* **52** (1967): 163–217. [Appendix B, Table 1]

[75] Keast, A., and M. G. Fox. "Fish Community Structure, Spatial Distribution and Feeding Ecology in a Beaver Pond." *Env. Biol. Fish.* **27** (1990): 201–214. [Appendix B, Table 3]

[76] Keller, W., N. D. Yan, T. Howell, L. A. Molot, and W. D. Taylor. "Changes in Zooplankton During the Experimental Neutralization and Early Reacidification of Bowland Lake Near Sudbury, Ontario." *Can. J. Fish. Aquat. Sci.* **49** (Suppl. 1) (1992): 52–62. [Appendix A, Table 2]

[77] Kelso, J. R. M. "Fish Community Structure, Biomass, and Production in the Turkey Lakes Watershed, Ontario." *Can. J. Fish. Aquat. Sci.* **45** (Suppl. 1) (1988): 115–120. [Appendix B, Table 3]

[78] Kempinger, J. J., and R. F. Carline. "Dynamics of the Walleye (*Stizostedion vitreum vitreum*) Population in Escanaba Lake, Wisconsin, 1955–1972." *J. Fish. Res. Board Can.* **34** (1977): 1800–1811. [Appendix A, Table 3]

[79] Lambshead, P. J. D., and H. M. Platt. "Structural Patterns of Marine Benthic Assemblages and Their Relationship with Empirical Statistical Models." In *Proceedings of the 19th European Marine Biology Symposium*, edited by P. E. Gibbs, 371–380. Cambridge, UK: Cambridge University Press, 1984.

[80] Lawton, J. H. "What Is the Relationship Between Population Density and Body Size in Animals?" *Oikos* **55** (1989): 429–434.

[81] LeBrasseur, R. J., C. D. McAllister, W. E. Barraclough, O. D. Kennedy, J. Manzer, D. Robinson, and K. Stephens. "Enhancement of Sockeye Salmon (*Oncorhynchus nerka*) by Lake Fertilization in Great Central Lake: Summary Report." *J. Fish. Res. Board Can.* **35** (1978): 1580–1596. [Appendix A, Table 3]

[82] Lewis, W. M. J. "A Compositional, Phytogeographical and Elementary Structural Analysis of the Phytoplankton in a Tropical Lake: Lake Lanao, Philippines." *J. Ecol.* **66** (1978): 213–226. [Appendix B, Table 1]

[83] Loundes, A. G. "The Density of Some Living Organisms." *Proc. Linn. Soc. Lond. II* (1937): 62–73.

[84] Maberly, S. C., M. A. Hurley, C. Butterwick, J. E. Corry, S. I. Heaney, A. E. Irish, G. H. M. Jaworski, J. W. G. Lund, C. S. Reynolds, and J. V. Roscoe. "The Rise and Fall of *Asterionella formosa* in the South Basin of Windermere: Analysis of a 45-Year Series of Data." *Freshwater Biol.* **31** (1994): 19–34. [Appendix A, Table 1]

[85] Magurran, A. E. *Ecological Diversity and Its Measurement.* Princeton, NJ: Princeton University Press, 1988.

[86] Mahon, R., and E. K. Balon. "Ecological Fish Production in Long Pond, a Lakeshore Lagoon on Long Point, Lake Erie." *Env. Biol. Fish.* **2** (1977): 261–284. [Appendix B, Table 3]

[87] Marquet, P. A., S. A. Navarrete, and J. C. Castilla. "Body Size, Population Density, and the Energetic Equivalence Rule." *J. Anim. Ecol.* **64** (1995): 325–332.

[88] Maurer, B. A., and J. H. Brown. "Distribution of Energy Use and Biomass Among Species of North American Terrestrial Birds." *Ecology* **69** (1988): 1923–1932.

[89] May, R. M. "Patterns of Species Abundance and Diversity." In *Ecology and Evolution of Communities*, edited by M. L. Cody and J. M. Diamond, 81–120, 1975.

[90] May, R. M. "How Many Species Are There on Earth?" *Science* **241** (1988): 1441–1449.

[91] McCauley, E. "The Estimation of the Abundance and Biomass of Zooplankton in Samples." In *A Manual on Methods for the Assessment of Secondary Productivity in Fresh Waters*, edited by J. A. Downing and F. H. Rigler, 228–265. Oxford: Blackwell, 1984.

[92] McQueen, D. J., M. R. S. Johannes, J. R. Post, T. J. Stewart, and D. R. S. Lean. "Bottom-Up and Top-Down Impacts on Freshwater Pelagic Community Structure." *Ecol. Monogr.* **59** (1989): 289–309. [Appendix A, Table 3]

[93] McQueen, D. J., E. L. Mills, J. L. Forney, M. R. S. Johannes, and J. R. Post. "Trophic Level Relationships in Pelagic Food Webs: Comparisons Derived from Long-Term Data Sets for Oneida Lake, New York (USA), and Lake St. George, Ontario (Canada)." *Can. J. Fish. Aquat. Sci.* **49** (1992): 1588–1596. [Appendix A, Table 3]

[94] Medel, R. G., F. Bozinovic, and F. F. Novoa. "The Mass Exponent in Population Energy Use: The Fallacy of Averages Reconsidered." *Amer. Natur.* **145** (1995): 155–162.

[95] Mills, K. H., S. M. Chalanchuk, L. C. Mohr, and I. J. Davies. "Responses of Fish Populations in Lake 223 to 8 Years of Experimental Acidification." *Can. J. Fish. Aquat. Sci.* **44** (suppl. 1) (1987): 114–125. [Appendix A, Table 3]

[96] Moed, J. R., and H. L. Hoogveld. "The Algal Periodicity in Tjeukemeer During 1968–1978." *Hydrobiologia* **95** (1982): 223–234. [Appendix A, Table 1]

[97] Nauwerck, A. "Die Beziehunger zwichen Zooplankton und Phytoplankton im See Erken." *Symb. Bot. Ups.* **17** (1963): 1–163.

[98] Nelson, W. R., and C. H. Walburg. "Population Dynamics of Yellow Perch (*Perca flavescens*), Sauger (*Stizostedion canadense*), and Walleye (*S. vitreum vitreum*) in Four Main Stem Missouri River Reservoirs." *J. Fish. Res. Board Can.* **34** (1977): 1748–1763. [Appendix A, Table 3]

[99] Nixdorf, B., and S. Hoeg. "Phytoplankton—Community Structure, Succession and Chlorophyll Content in Lake Muggelsee from 1979 to 1990." *Int. Revue Ges. Hydrobiologia* **78** (1993): 359–377. [Appendix A, Table 1]

[100] Nygaard, G. "Seasonal Periodicity of Planktonic Desmids in Oligotrophic Lake Grane Langso, Denmark." *Hydrobiologia* **211** (1991): 195–226. [Appendix A, Table 1]

[101] O'Gorman, R., R. A. Bergstedt, and T. H. Eckert. "Prey Fish Dynamics and Salmonine Predator Growth in Lake Ontario, 1978–1984." *Can. J. Fish. Aquat. Sci.* **44** (Suppl. 2) (1987): 390–403. [Appendix A, Table 3]

[102] Pereira, D. L., Y. Cohen, and G. R. Spangler. "Dynamics and Species Interactions in the Commercial Fishery of the Red Lakes, Minnesota." *Can. J. Fish. Aquat. Sci.* **49** (1992): 293–302. [Appendix A, Table 3]

[103] Peters, R. H. *The Ecological Implications of Body Size.* Cambridge, UK: Cambridge University Press, 1983.

[104] Peters, R. H., and K. Wassenberg. "The Effect of Body Size on Animal Abundance." *Oecologia* **60** (1983): 89–96.

[105] Planas, M. D. "Composicion, Ciclo y Productividad del Fitoplancton del Lago de Banyoles." *Oecologia Aquatica* **1** (1973): 3–106. [Appendix B, Table 1]

[106] Pollingher, U., and C. Serruya. "Phased Division of *Peridinium cinctum* F. Westii (*Dinophyceae*) and Development of the Lake Kinneret (Israel) Bloom." *J. Phycol.* **12** (1976): 162–170. [Appendix A, Table 1]

[107] Preston, F. W. "The Commonness, and Rarity, of Species." *Ecology* **29** (1948): 254–283.

[108] Rey, J., and J. Capblancq. "Dynamique des Populations et Production du Zooplancton du Lac de Port-Bielh (Pyrenees Centrales)." *Ann. Limnol.* **11** (1975): 1–45. [Appendix B, Table 2]

[109] Reynolds, C. S. *The Ecology of Freshwater Phytoplankton.* Cambridge, UK: Cambridge University Press, 1984.

[110] Ricklefs, R. E. *Ecology*, 3rd ed. New York: W. H. Freeman, 1990.

[111] Romo, S., and M. R. Miracle. "Population Dynamics and Ecology of Subdominant Phytoplankton Species in a Shallow Hypertrophic Lake (Albufera of Valencia, Spain)." *Hydrobiologia* **273** (1994): 37–56. [Appendix B, Table 1]

[112] Sager, P. E., and S. Richman. "Functional Interaction of Phytoplankton and Zooplankton Along the Trophic Gradient in Green Bay, Lake Michigan." *Can. J. Fish. Aquat. Sci.* **48** (1991): 116–122.

[113] Saito, S. "Seasonal Succession of Phytoplankton in Sagami Reservoir from 1973 to 1977." *Jap. J. Limnol.* **39** (1978): 147–155. [Appendix A, Table 1]

[114] Schoener, T. W. "Patterns in Terrestrial Vertebrate Versus Arthropod Communities: Do Systematic Differences in Regularity Exist?" In *Community Ecology*, edited by J. Diamond and T. J. Case, 556–586. New York: Harper & Row, 1986.

[115] Scott, W. B., and E. J. Crossman. "Freshwater Fishes of Canada." Bulletin, Fisheries Research Board of Canada, 1973.

[116] Silva, M., and J. A. Downing. "The Allometric Scaling of Density and Body Mass: A Nonlinear Relationship for Terrestrial Mammals." *Amer. Natur.* **145** (1995): 704–727.

[117] Sokal, R. R., and F. J. Rohlf. *Biometry*, 2nd ed. San Francisco: W. H. Freeman, 1981.

[118] Sprugel, D. G. "Correcting for Bias in Log-Transformed Allometric Equations." *Ecology* **64** (1983): 209–210.

[119] Sugihara, G. "Minimal Community Structure: An Explanation of Species Abundance Patterns." *Amer. Natur.* **116** (1980): 770–787.

[120] Thorpe, J. E. "Trout and Perch Populations at Loch Leven, Kinross." *Proc. Roy. Soc. Edinb. Sect. B* **74** (1972): 295–314. [Appendix A, Table 3]

[121] Tokeshi, M. "Species Abundance Patterns and Community Structure." *Adv. Ecol. Res.* **24** (1993): 111–186.

[122] Vézina, A. F. "Empirical Relationships Between Predator and Prey Size Among Terrestrial Vertebrate Predators." *Oecologia* **67** (1985): 555–565.

[123] Walters, C. J., D. C. E. Robinson, and T. G. Northcote. "Comparative Population Dynamics of *Daphnia rosea* and *Holopedium gibberum* in Four Oligotrophic Lakes." *Can. J. Fish. Aquat. Sci.* **47** (1990): 401–409. [Appendix A, Table 2]

[124] Werner, E. E., D. J. Hall, and M. D. Werner. "Littoral Zone Fish Communities of Two Florida Lakes and a Comparison with Michigan Lakes." *Env. Biol. Fish.* **3** (1978): 163–172. [Appendix B, Table 3]

[125] West, G. B., J. H. Brown, and B. J. Enquist. "A General Model for the Origin of Allometric Scaling Laws in Biology." *Science* **276** (1997): 122–126.

[126] Wetzel, R. G. *Limnology*, 2nd ed. Philadelphia, PA: Saunders College Publishing, 1983.

[127] Williamson, M. *The Analysis of Biological Populations.* London: Edward Arnold, 1972.

[128] Zankai, N. P., and J. E. Ponyi. "The Quantitative Proportions of Rotifera Plankton in Lake Balaton, in 1967." *Annal. Biol. Tihany* **37** (1970): 291–308. [Appendix B, Table 2]

[129] Zankai, N. P., and J. E. Ponyi. "The Biomass of Rotatoria in Lake Balaton." *Annal. Biol. Tihany* **40** (1973): 285–292. [Appendix B, Table 2]

Diversity and Convergence: Scaling for Conservation

William A. Calder

1 INTRODUCTION

"...conservation planning for many endangered bird species will be seriously (maybe hopelessly) hampered by a lack of critical information..."

—Verner, 1992

"...the paucity of data available for most species, communities, and ecosystems often forces biologists and policy makers to make decisions without any quantitative information."

—Doak and Mills, 1994

"Age at cessation of breeding was based on data from the captive eld's deer population of Montgomery Zoo...because there were no data on life-span and breeding structure for the natural population."

—Song, 1996

The consensus seems to be pretty clear that conservation biology suffers from scarcity of information which would be basic for protecting biologi-

Scaling in Biology, edited by J. H. Brown and G. B. West.
Oxford University Press, 2000. **297**

cal diversity, particularly for management of rare, threatened, and endangered species. Rare species are not only the most vulnerable, but usually the least studied. Some enter our scientific awareness without any documents at all, such as the giant muntjac (*Megamuntiacus vuquangensis*), leaf deer (*Muntiacus sp.*), and saola (*Pseudoryx nghetinhensis*) in southeast Asia, the Panay cloudrunner (*Crateromys heaneyi*) in the Philippines, and seven species of monkeys recently discovered in Brazil, all unknown before the 1990s [33, 42, 53, 57, 73]. For other species, so few individuals survive, that the risks of capture and marking for demographic studies are a "taking" that would further jeopardize survival. Some species are declining so rapidly that there is not time for long-term studies that could normally yield the proper information.

An information deficit, in the face of immediate threats, forces conservation biology to be a "crisis discipline," "an inexact science" in which "uncertainty is inherent" and "probabilistic, rather than prescriptive answers to problems...[are] the norm" [46]. Science has a strong theoretical basis, and is expected to provide supporting numbers, in contrast with folklore, emotion, and demagoguery. A conservation biologist trying to prepare a protection plan without data on the species' biology is like an underwriter issuing an insurance policy in ignorance of the applicant's age, family status, and medical history! Political and economic support for protection and management are diminished by public perception that "junk science," not the "facts," are driving government conservation policies and regulations!

The status of a species or population is quantitative. Continued decline in abundance triggers petitions for listing as *threatened* or *endangered*. Once a species is listed, the Endangered Species Act [ESA; § 1533 (f) (1)] mandates development and implementation of a recovery plan, incorporating: "(i) a description of...management actions as may be necessary...for the conservation and survival of the species; (ii) objective, measurable criteria which, when met, would result in a determination...that the species be removed from the list; and (iii) estimates of time required...to achieve the plan's goal and to achieve intermediate steps toward that goal."

To meet such requirements and deal with the uncertainty of not having enough quantitative information on the species' life history and population dynamics, conservation biology has responded with *adaptive management*, "which consists of managing according to a plan by which decisions are made and modified as a function of what is known and learned about the system..." [52]. To prevent interventions which make things worse, adaptive management proceeds with caution, using "...information from other species and our general knowledge about population dynamics to build a range of conceptual models about how the system works..." [52].

Two types of biological information needed for recovery planning are often unavailable for species most at risk. First, within populations, we lack baseline data on abundance, genetic diversity, reproductive success, availability of suitable habitat and food, and local threats to survival. These are habitat- and population-based, so must come from field work. Second, however, are life history characteristics for which a first approximation or estimate can be made without confrontation or long-term harassment. Resource and area requirements and reproductive potential (e.g., time to sexual maturity, survivorship, reproductive life span, fecundity, and intrinsic rate of increase) can often be "borrowed" from better known species of similar body size and/or phylogeny, the focus of this chapter.

"All I really needed to know...I learned in kindergarten" [32]; all we know about functional biology we learned from biodiversity of "unusual animals" whose "unique characteristics which seem to make [a] particular animal singularly suited for a specific scientific purpose, as if the animal had been designed with this purpose in mind" [64]. The instrumental value of biodiversity for physiological and biomedical research was reviewed by Schmidt-Nielsen two decades before *biodiversity* entered our vocabulary [55]. The short generation time of fruit flies (*Drosophila*) expedited genetics. Guinea pigs (*Cavia*) revealed the role of vitamin C. Knowledge of kidney function emerged from studies of active secretion in aglomerular nephra of goosefish (*Lophius*), and urine concentration by long-looped nephra of desert kangaroo rats (*Dipodomys*). Triumphs in neurophysiology came from experimental advantages found in *Venus* clams (for acetyl-choline assay) and giant axons of the squid (*Loligo*) and tetrodotoxin from puffer fish (Tetraodontidae) for understanding the ionic basis of nerve function. The fat sand rat, *Psammomys obesus*, is a model for sugar diabetes research [63, 64].

Biodiversity's intellectual resources have also provided the body-size ranges represented in data for deriving the scaling relationships of allometry, one of the most widely used frameworks of comparative biology (lucid introduction: Schmidt-Nielsen [65]). Many of these relationships have been derived with data points from 100 to 200+ species over a wide range in size. If that is not diversity, what is? The present volume describes many of the basic insights and some of their applications to human welfare.

Thus we comparative biologists have a special debt to the very existence of diversity. We should contribute to the protection of the diversity from which our ideas have sprung. How can we contribute? Perhaps by adapting our framework to the filling of gaps in life history information, lack of which handicaps conservation planning.

Size profoundly influences the quantitative details of life histories [10, 14, 22, 54]. Body size is the most influential and correlatable quantitative trait in a class of organisms, its allometry often accounting for more than 75% of the interspecific variability in physiological, morphological, and ecological data (Figure 1). Allometry could be a useful "tool" for conservation biology, as a means of deriving predictive values to substitute for missing data. This has already been done to a limited extent, e.g.: "...estimates of [minimum viable population density] can be used to make provisional decisions about the size of conservation areas" [68], or: "Based on rough calculations of the amount of habitat needed to support *an animal of this size*, it is thought that

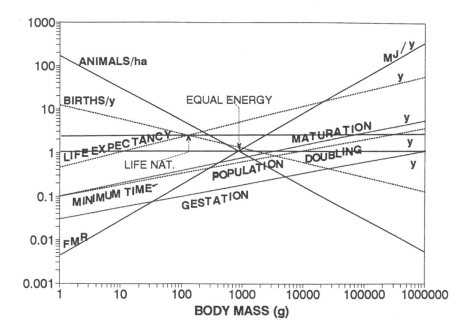

FIGURE 1   Sample empirical size regressions of primary consumer mammals. "Energy equivalence" is seen if the product of density per species (animals/ha) [28] times field metabolic rate (FMR) is size-independent, $M^{-3/4} \times M^{3/4} = M^0$. This essentially size-independent product is the upper bound of population metabolism at a level approximated by the crossing of FMR and density, FMR increase countering density increase, and vice versa). For replacement, annual fecundity times percent survival to reproduction times life expectancy should be 2 per monogamous pair (note that the regressions cross at ~ 2.5 "LIFE NAT").

probably only a few hundred saola survive [in the mountainous forests along the Vietnam-Laos border]" (italics added) [77].

Max Kleiber [40, 41] and Samuel Brody [5] stimulated six decades of pondering their empirical scaling of metabolic rates, which scaled as mass to the 3/4-power. Scaling relationships are derived with data from as broad a range in body size and as many species as possible. The number of species for which metabolic rate (or some other variable) has actually been measured at the time of the analysis is only a subset of the total species richness—at a crude fractal level. Hopefully it represents all extant species (cataloged, recently discovered, and undiscovered species).

Use of Kleiber's $M^{3/4}$ metabolic scaling has been questioned because it was based on few species, many domesticated and isolated from continuing natural selection. The data set increased in size as more species were studied, and the recent, larger data base gave scaling exponents slightly different from

$M^{3/4}$ [2, 36]. Still, $M^{3/4}$ remains a useful approximation that is consistent with scalings of metabolic support functions, as well as ecological scalings [14]. For applications proposed here, exact exponents are less crucial than they might be in theoretical applications (see below).

Biological scaling is not a substitute for field research, but it can generate realistic surrogate[1] values (*sensu* Congdon and Dunham [25]) as a practical hypothetical base for planning and an expedient start for adaptive management. Its predictions could be applied to the design of recovery plans for endangered species, determination of resource needs, anticipation of potential rates of population growth and recovery times, and reduction of guesswork about population viability and reserve sizes. Size effects can insert themselves as methodological bias, but also give insight into adequacy of sites and sampling and plausible recovery trajectories. Conservation-relevant scaling could also play a part in impact assessment and the credibility of conservation policy.

This approach is directed toward preliminary answers to four questions significant to protection of endangered/ threatened species:

1. What is minimum viable population size?
2. If protection and critical habitat designation are adequate, how much time will be required to achieve interim and final goals?
3. How might census accuracy be affected by animal size?
4. How can allometric predictions be refined for greater predictive precision?

2 MINIMAL VIABLE POPULATION SIZE AND DENSITY

There is good reason to strengthen the Endangered Species Act to focus on a higher goal than just survival of a minimum viable population, but until there is legislation that provides for full recovery, minimum viable population size (N_{MVP}) is the standard (but vague) reference level for conservation biology. Estimates of N_{MVP} are few and slow in coming. N_{MVP} is the product of minimum viable population density (D_{MVP}) times area of distribution in the metapopulation. Habitat area is a scaling matter for land management, landscape ecology, and restoration ecology. D_{MVP} is a topic more directly within the scope of the other scaling, to body size. How is D_{MVP} related to average population density (D_{AP} animals/km^2)?

2.1 EUTHERIAN MAMMALS

D_{AP} is size-dependent, for herbivorous mammals [28]:

$$D_{\mathrm{AP}} = 91.2\, M^{-0.73} = 14{,}125\, \mathrm{m}^{-0.73} \qquad (1)$$

---

[1]Surrogate is "a person or (usu.) thing that acts for, or takes the place of another; a substitute" (Oxford Universal Dictionary [51]).

in the familiar format of $Y = a M^b$, wherein Y is individuals per km^2, a is the scaling constant, or the antilog of the Y-intercept in a least-squares linear regression vs. body mass ($M = $ kg, m $=$ g). According to the trophic pyramid of numbers, density decreases at higher trophic levels [50, Pp. 70–81], so when Damuth pooled data for mammals from all trophic levels ($n = $ 467 spp.), the secondary consumer data diluted the density scaling to 60% of that for herbivores of equal size:

$$D_{AP} = 52.5 \, M^{-0.78} = 11,486 \, \mathrm{m}^{-0.78} . \tag{2}$$

Silva et al. [70] divided data from 364 mammal species into three size classes, which separately gave weaker correlations than when they pooled all mammals. This yielded an average density at 1 kg only 36% of that in Eq. (2):

$$D_{AP} = 17.9 \, M^{-0.69} = 2103 \, \mathrm{m}^{-0.69}. \tag{2a}$$

While these correlations are highly significant ($p = 0.0001$ for Eq. (2a)), the scaling of density is complex, size accounting for only 55 to 74% of the variability. A single linear regression of log values is therefore somewhat tenuous, so it is perhaps best to use it with caution for comparison of relative values. The scaling (or nonscaling?) of density with size has been a hot topic (review: Silva et al. [70]).

Silva and Downing [68] derived a closely parallel D_{MVP}/size relationship from 147 published mammalian population densities of endangered, vulnerable, and rare populations, minimum densities in cyclical populations, and minimum populations of otherwise common species as the data base:

$$D_{MVP} = 8.51 \, M^{-0.68} = 933 \, \mathrm{m}^{-0.68} . \tag{3}$$

Their -0.68 scaling was not significantly different from a -0.75 scaling exponent for D_{AP}. Silva and Downing concluded that "mammal populations that have fallen to 10% of their long-term average density can be considered to be near to their minimum viable level." Comparing their D_{MVP} regression with D_{AP} from Eq. (2), the normal "safety factor" (intercepts a_{AP}/a_{MVP}) is about 6-fold at 1 kg. However, given the marked difference in constants (52.5 vs. 18.8) of Eqs. (2) and (2a), it might be useful to make relative comparison between D_{MVP} and the D_{AP} relationship with almost identical scaling exponents, from the same authors [68, 70] with consistent methods and more data overlap. This cuts the a_{AP}/a_{MVP} "safety factor" to a modest 2.2. Taken together, until the difference in D_{AP} is explained, Eqs. (2) and (2a) can be used to bracket the possible range in natural D_{AP} population sizes and anticipate area requirements.

2.2 BIRDS

Avian D_{AP} [39] (Figure 2; $n = 564$, $r^2 = 0.18$, $p = 0.0001$) was:

$$D_{AP} = 3.09 \, M^{-0.49} = 91.2 \, \mathrm{m}^{-0.49} . \tag{4}$$

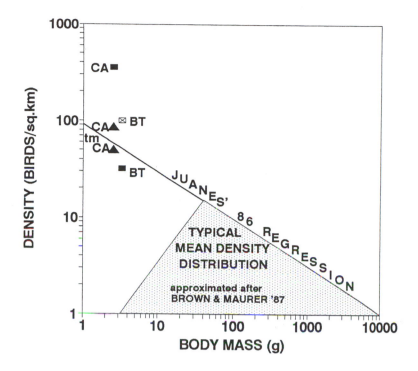

FIGURE 2 Presumed maximum densities increase with decreasing bird size, but when less abundant species of all sizes are included, size-abundant relationships are better represented by polygons than by regression lines [7]. CA = calliope hummingbird in post-logging vegetative recovery, no feeders [18], BT = broad-tailed hummingbirds with feeders for trapping [15].

Calder and Calder [17, 18] confirmed this at the small size extreme of the hummingbird *Stellula calliope*. The comparable regression by Silva et al. [70] ($n = 564$ spp. (also), $r^2 = 0.252$, $p = 0.0001$) was:

$$D_{AP} = 1.39\, M^{-0.60} = 90.2\, m^{-0.60}. \tag{4a}$$

Densities of birds in an old-growth Douglas Fir forest in the southern part of the Oregon Coastal Range (converted to birds/km², as above, from Carey et al. [19], using species masses from Dunning [30]) scaled with an order of magnitude less in abundance, which is apparently characteristic of mature conifer forests (Figure 3):

$$D_{AP} = 0.104\, M^{-0.68} = 11.4\, m^{-0.68}. \tag{4b}$$

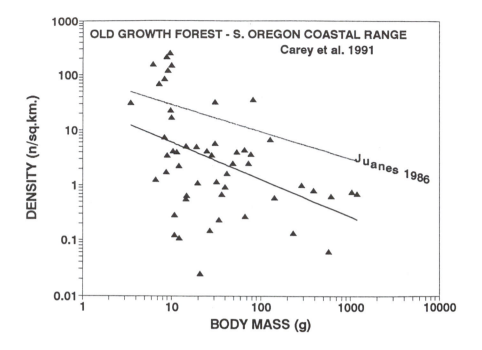

FIGURE 3 Data on densities of birds in an old-growth forest (from Carey et al. [19]) show the Brown-Maurer triangular size-abundance distribution. Even with "dilution" by rarer species, the regression is somewhat steeper than those of Juanes [39] and Cotgreave and Harvey [26], although statistically, these slopes do not differ significantly.

These relationships:

1. Have exponents approaching the inverse of metabolic scaling.
2. Scale with an exponent identical to the relationship between mass and estimated minimum viable population density in 143 populations of minimal, rare, and endangered mammal species.

Note also that Eq. (4b) is lower and steeper than Juanes' [39] regression, but closer to the slope of Eq. (4a), and the $M^{-0.60}$ for 434 species pooled from 49 communities [26]. (Note that the intercepts for Eqs. (4) and (4a) are almost identical expressed at the hypothetical 1-g vs. 1-kg level, diverging because of the exponent difference.)

These avian regressions are significant at $p < 0.01$, but size alone accounted for only 18% of the variability in density [39], 15% from Cotgreave and Harvey [26], 25% from Silva et al. [70], and 19% from Carey et al. [19].

These relationships all indicate that small mammals attain much higher densities than birds, at 1-kg size, higher by factors of 13 to 31. This is consistent with the high ratios of mammalian to avian biomass (kg/ha) in forests [4]. The difference may be, in part, an artifact of census methods (see below), but if it is approximately correct, this could be related to greater mobility with greater spread over the resources. It could also reflect greater specialization, which should appear as higher species richness, with more species to compete for resources in a particular size range (tiny vs. small insects or fruits).

Avian D_{MVP} does not appear to have been analyzed this way. Speculating, if avian D_{MVP} is, as for mammals, about one-sixth of D_{AP}, but similar in scaling exponent, we would have:

$$D_{\mathrm{MVP}} = 1.39/6\, M^{-0.604} = 0.23\, M^{-0.604}\,. \tag{5}$$

This is half that for bird densities in old-growth Douglas Fir (Eq. (4b)). However, a lower safety factor might be adequate for persistence, perhaps because avian life spans are longer, allowing more time for reproductive success. (Conversely, mammals of a given size have higher natality which compensates for greater mortality; see next section.)

3 RECOVERY: POPULATION DOUBLING TIME

The goal of an endangered species recovery plan is, of course, to increase abundance to the level of sustainable, genetically diverse population. This requires an adequate area of suitable habitat, protected and/or restored, with other causes of past decline, such as alien competition, overharvesting, pollution, and anthropogenic climate change controlled. To secure all of this, the public must be educated, convinced by reasonable projections, that public funds would be justified and well spent in the recovery. Success will be evidenced by population trends that confirm predicted recovery times. Data on population changes are expressed different ways and span various periods, but for comparative/predictive purposes, these can be standardized as population doubling times (t_{2n}). I will now apply "...information from other species and our general knowledge about population dynamics" [52], to compare t_{2n}, derived from published compilations of intrinsic growth rates with t_{2n} previous successful recoveries.

The intrinsic growth rate (r_m) is "the effective compounded growth rate at which the population is capable of growing" [44]. At a particular r_m, expressed as growth rate per annum, a population can double in $\ln 2/r = 0.69/r$ years. From this, we calculate r_m from census data, and t_{2n} from r_m (or vice versa), doubling time is:

$$t_{2n} = 0.69/r_m\,. \tag{6}$$

3.1 MAMMALS

Several authors have derived allometric relationships between r_m and body mass. Since the dimension of t_{2n} is *time*, and r_m is the *reciprocal of time*, the fractal/material-distribution explanation of quarter-power scaling ($1/4\,ps$) [76] would lead us to expect t_{2n}, to scale as $M^{1/4}$, and r_m to scale as $M^{-1/4}$. Using Cole's [24] equation, Hennemann [37] estimated r_m and derived this allometry (see Figure 4):

$$r_m, y^{-1} = 0.80\,M^{-0.26}. \tag{7a}$$

Adding 38 species of marine mammals, Schmitz and Lavigne [66] obtained a lower but parallel scaling:

$$r_m, y^{-1} = 0.69\,M^{-0.26}. \tag{7b}$$

However, Caughley and Krebs (using published lab and field values) [21] and Thompson (also estimating from Cole's 1954 equation) [72] found scalings closer to $M^{-1/3}$ than $M^{-1/4}$. Charnov [22] pointed out that the Caughley and Krebs regression was comprised of two clusters, which were results from ($n = 4$) small mammal species in the laboratory and ($n = 4$) large ungulates in the field. Thompson's data set included a more diverse assemblage of mammals ($n = 42$ eutherians) than Caughley and Krebs. Converting his reduced major axis regression to least squares gives a $M^{-1/3}$ slope like the Caughley and Krebs. Which scaling should we accept as our model, the $M^{-1/4}$ of Hennemann and Schmitz and Lavigne, which fits $1/4\,ps$, or the $M^{-1/3}$ of Caughley and Krebs and Thompson? For an interim settlement, combining all eutherian values tabulated by the above authors yields:

$$r_m = 0.99\,M^{-0.33} \tag{7c}$$

and the reciprocal function (Figure 4):

$$t_{2n} = 0.69\,M^{0.33}. \tag{8}$$

May and Rubenstein [44, Eq. (1.2)] provided an alternative way of estimating from life history variables for average lifetime production of female offspring (R_0 = annual fecundity × life expectancy) and age at first reproduction (t_{mat}, y):

$$r_m \simeq \ln R_0 / t_{\mathrm{mat}}. \tag{9}$$

Substituting the fecundity allometry from Allainé et al. [1] and t_{mat} from Calder [14]:

$$r_m = 3.3\,M^{-0.21}. \tag{7d}$$

This predicts r_m which is, at 1 kg, four times that found by Henneman [37], giving a more optimistic, rapid population doubling (t_{2n}).

As was the case for population density, quantitative differences between regressions perhaps bracket a range of predicted values for a particular body

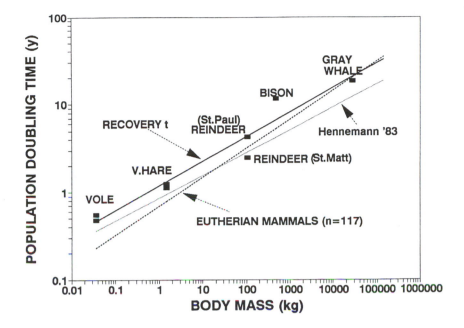

FIGURE 4   Population doubling times of mammals, calculated from Henne-mann [37] and a pooling of $n = 117\,R_m$ values from several authors, and from recovery time mammals (squares).

size. They also challenge further study that could refine and clarify the scaling of r_m. We should bear in mind that the algebra of complex scalings which incorporate 2 or more subcomponent $1/4\,ps$ scalings may result in something not recognizable as $1/4\,ps$, without violating the assumption of $1/4\,ps$. So until we see what goes into r_m scaling, we may come to realize that some scalings in a range from $M^{-0.33}$ to $M^{-0.21}$ are actually compatible with $1/4\,ps$!

3.2   BIRDS

Avian incubation takes about 40% as much time as mammalian gestation, according to scaling comparisons [3, 58]:

$$\text{birds' incubation period, da.} = 28.9\,M^{0.20}, \tag{10}$$

$$\text{mammals' gestation period, da.} = 66.3\,M^{0.26}. \tag{11}$$

Even with this initial time-saving via oviparity, birds do not match the repro-ductive output of mammalian viviparity and lactation. From Eq. (9), avian r_m scales as $\sim 1/3$ of mammal r_m:

$$\text{avian } r_m, y^{-1} = 1.12\,M^{-0.17}. \tag{12}$$

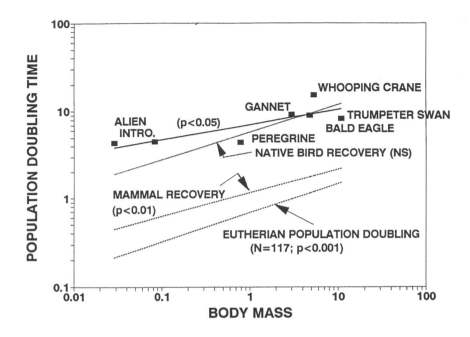

FIGURE 5   Population doubling times and recovery times of native birds, with and without population growth estimates for introduced house sparrows and starlings (to supply small bird data points), compared with mammalian recoveries and minimum population doubling times.

Consequently, population doubling times (t_{2n}) at r_m are about three times as long for birds as for mammals, compared at the 1 kg size, using the optimistic r_m in Eq. (7d) (Figure 5). This is likely because birds are handicapped by later onset of first breeding, shorter breeding seasons, and lower annual natality, but may also reflect differences in census techniques (see below).

4   RECOVERY TIMES FOR ENDANGERED AND THREATENED SPECIES

Mean t_{2n} provides a useful baseline expression for quantitative evaluation of recoveries by listed species and for recovery time estimates, as required by ESA § 1533 (f) (1)(iii). The problem is that the relatively few recoveries do not span a wide range in body sizes for deriving robust scaling relationships.

4.1 MAMMAL RECOVERIES

For a preliminary approximations, I have calculated t_{2n} for recovery of the bison (*Bison bison*) and gray whale (*Eschrichtius robustus*), and for the introduction of reindeer (*Rangifer tarandus*) to two Pribilof islands and two population cycling species, the 36 g meadow vole (*Microtus pennsylvanicus*) and the 1.5-kg varying hare (*Lepus americanus*) to derive a respectable $1/4\,ps$ scaling ($r^2 = 0.91$, $p < 0.01$; see Figure 4):

$$\text{recovery } t_{2n} = 1.17\,M^{0.28}\,. \tag{13}$$

Thus realized t_{2n} in some successful mammalian population expansions and recoveries has been only slightly longer than the $t_{2n} = 0.69\,M^{0.33}$ (Eq. (8)), the potential calculated from r_m.

4.2 BIRD RECOVERIES

As a first approximation (Figure 5), mean avian recovery t_{2n} scales in parallel with potential at r_m. For the trumpeter swan, whooping crane, bald eagle, peregrine falcon, gannet, and introductions to North America of the European starling and house sparrow:

$$\text{recovery } t_{2n} = 7.1\,M^{0.17} \tag{14}$$

($r^2 = 0.65$, $p < 0.05$). This takes ca. 11 times as long (has a higher intercept) for a hypothetical 1-kg bird [using the more optimistic May and Rubenstein t_{2n} [44, Eq. (1.2)].] Because the scaling exponents are so similar, this difference by an order of magnitude would exist between recovery and potential doubling across the size range of birds.

The following preliminary conclusions are offered from scaling analysis:

1. Avian t_{2n} values from notable recoveries and alien introductions (regressed together to include an adequate size range) took about 6 times as long as in mammalian recoveries, compared at 1-kg body size. Prolongation of recovery from a population decline would give birds more time for loss of genetic heterozygosity, resulting in greater danger of an "extinction vortex" entry.
2. Slopes differ as well, but much of the differences may be reduced by a growing data base.
3. Avian and mammalian slopes may differ because possible error in my assumption that introduced aliens can expand their populations more easily because of greater genetic diversity (if introduced stocks were mixed) or because they were isolated from adapted enemies, predators, and diseases in their homelands.

5   CENSUS AND SCALING: HOW MANY INDIVIDUALS SURVIVE?

Accurate monitoring is necessary for providing the "objective, measurable criteria" called for by the ESA. To know population size and individual survival, some form of census is necessary. Large animals in open habitats can be counted by aerial survey, but census of small and closed-forest species may entail detectability problems, which could complicate or bias a census. We must understand both conceptual and dimensional aspects of techniques and their weaknesses. Detectability of a bird's song is a function of its output power, song structure, distance from the census taker, and environmental attenuation. Sound intensity attenuates in proportion to the square of distance, with added attenuation due to obstruction and absorption by vegetation, atmospheric conditions, and frequencies of sound used [35, 48, 62]. Estimates of forest bird densities can be no more reliable than the reliability of distance estimates. Pyke and Rector [56], for example, recommend that only detection by sight is acceptable in point counts.

In general, sound power output (mW/cm^2) at 1-m distance (P) of bird songs and calls is proportional to distances across claimed areas (A), both of which correlate strongly with body mass (P: $r^2 = 0.862$, $p < 0.001$, $n = 32$; A: $r^2 = 0.629$, $p < 0.001$, $n = 86$):

$$P = 108\, M^{1.14}, \tag{15}$$

$$A = 1068\, M^{1.17}. \tag{16}$$

The statistically indistinguishable exponents ($p > 0.25$) of A and P gave an effectively size-independent ratio [13]:

$$P/A = 1.28\,\mathrm{m}^{-0.03}. \tag{17}$$

The center frequency of bird songs is inversely related to body mass ($M^{-0.24}$) [75]. Thus small birds' songs will be audible for considerably shorter ranges, so censuses relying on auditory detection are likely to be proportionately biased toward larger-bodied species from a wider area (or smaller birds will be under counted). A census taker normally does not hear birds from equal areas, but areas that vary according to the loudness of each species (Figure 6). Ravens are also visible at greater distances than are gnatcatchers. Carey et al. [19] corrected for variation in detectability by calculating effective radius of detection for each species, and calculated the area sampled around each point. From distributions of birds detected over distance, they calculated the third-quartile detection distance for each species, which approximated the shoulder in curve of detections vs. distance, indicating a marked decline in number of birds detected. Huff and Raley [38] truncated detection distances at 50 m to avoid biases. Their estimates of detection distances for the ten commonest species provide a comparison of field detection distances with what theory predicts. Figure 7 shows distances compared to regressions for distances at

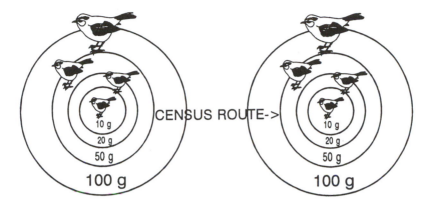

FIGURE 6 Larger birds produce louder vocalizations. Consequently detection distances increase with bird size, so at any stop on a linear census route, the effective area censussed is greater for larger birds.

which songs would be attenuated progressively to intensities of 50, 40, and 30 db ("traffic," "radio low," and "whisper," respectively). The voices of the two largest birds (varied and hermit thrushes) were apparently less powerful than predicted for size, or perhaps emit in frequencies or patterns attenuated significantly in excess of geometric spread. These aspects of size and sound output have implications for census techniques and apparent abundance.

6 PREDICTIONS: IMPROVING ACCURACY

The foregoing cases of ecological allometries provide a framework, with general consistency between exponents for related functions. However, to make the grand-scale consistency of allometry useful beyond the revelation of basic relationships, we need a means for fine-tuning the rather crude predictions for particular species that come from the interesting variability that "gets swept under the rug" (as a colleague put it) by allometric plots. Fortunately, the variability itself has patterns that reflect adaptive departures from the general regression plots, adaptations which occur as suites that may be characteristic of the genus, family, or some higher taxa. By "suites" I refer to the inevitability that life history traits must be compatible. Since species survival requires replacement, a short-lived mouse species must have a higher annual fecundity than a longer-lived wapiti. A species that evolved with one trait that is slow or fast for its size must have its other traits appropriately geared up or down. This reality is the basis for what I propose as the fine tuning of current or newly derived allometric relationships to provide surrogate values

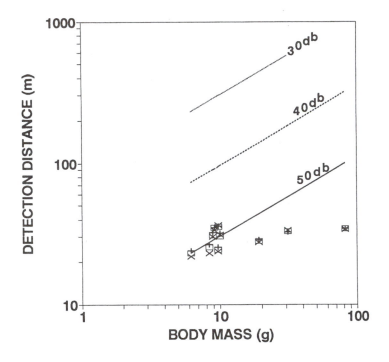

FIGURE 7   Auditory detection distances in old-growth forest [38] are all less than distances normally used in transect or point-count techniques. The threshold for hearing the smaller birds approximated attenuation to 50-db sound power intensity.

for the life history variables otherwise missing in conservation planning, in three steps: (1) *derivation* of broadest relationships from published allometry or data, (2) *prediction* from these relationships, using body size of the species of concern, and (3) *fine-tuning adjustment* of those predictions, using observed/predicted ratios to correct for phylogenetic and adaptive departures peculiar to taxon, habitat, or diet.

6.1   DERIVATION

For practical use in conservation applications, we need to derive more general scaling relationships and to expand existing ones, adding new data to compilation of values from the literature. This should span as wide a range in body sizes as possible, so that size can swamp other sources of variation. The math is simple, as straightforward as the geometry of spheres (see Brown et al. this volume), and the same as for species-area relationships (see Rosenzweig [61]). Rather than duplicate Brown's discussion of allometric methodology (Brown this volume), I note only some specifics used here. Least squares (LSR or

Model I regression) is the most appropriate correlation technique for allometry [22, 60], and the simplest (included in standard spreadsheet programs).

Practical application rests on these methods and their assumptions:

1. Body mass is the best expression of size and the most frequently and accurately recorded basis for correlations. Thanks to Dunning [30] and Silva and Downing [69], body masses have been gathered for 65% of the extant bird species and almost 60% of extant mammal species. For the other 35–40%, other measures of size [e.g., head and body length in mammals [67] or snout-vent length in reptiles] can be used, with caution, to estimate mass with sufficient accuracy for this purpose.
2. Empirical allometric relationships are best derived by using all available data (vs. limiting use to only one species or an average of species per genus or family), log-transformed, and regressed by the method of least squares [13, 14].
3. Radiation of body sizes within different taxonomic groups has led to a convergent evolution of sizes in independent lineages. This convergence has not been considered to the extent accorded radiative speciation in the phylogenetic method. Rates of such evolution have probably been faster in smaller-sized species [43], which tend to be more abundant in both numbers of species and sizes of populations [6]. Therefore arguments for using only one species or an average of species per genus or family might actually distort rather than correct for assured nonindependence of congeners, by giving all taxa, species rich or species poor, the same influence.
4. While many species entries are limited to small sample sizes or regional bias, errors from using them are negligible if the body size range in the regression data is several orders of magnitude, as is usually possible.
5. Life history variables are assumed to be strongly interrelated so that when one trait deviates from predicted allometric relationships, the direction and magnitude of deviation in other traits can be predicted [8, 11, 14, 22].
6. Phylogenetic effects are not distinguished by restricting inclusion to one species or an average of species per genus or family (points 2 and 3 above). All species will be included without identification by making a common correction to the allometric prediction.

6.2 PREDICTION

This easiest step in the process simply requires solving the appropriate allometric equation for the size of the animal for which information is missing and needed. One cautionary reminder is that general log-log allometric plots (Figure 1) make relationships look tighter than they really are, visually de-emphasizing variability. Thus simply plugging in a species' mass cannot predict accurately. Adjustment from allometric generalizations is necessary to reflect the phylogenetic and/or adaptive departures of the taxon in question to yield more realistic estimates.

6.3   ADJUSTMENT OF ALLOMETRIC PREDICTIONS

Different life history traits are clearly interdependently coadjusted to each other. Rather than evolving independently in functional isolation, the traits must be functionally matched to each other, as documented by direct correlations between traits, eliminating size effects (see Grosberg [34], Millar and Zammoto [47], Read and Harvey [60]). If one trait deviates from the class-wide central tendency, other traits must conform or compensate to re-match the inputs and outputs. For example, the minimum reproduction for persistence in natural selection is replacement of the two parents. Note that plots of life expectancy and annual birth rate in Figure 1 cross at about two offspring per life span. Charnov [22] provides a theoretical basis for this. Given that no species (besides our own) is taking over the world, the net reproduction rate (R_0) must be ~ 1 and other traits must be adjusted accordingly. Note also that plots of field metabolic rate (FMR) and population density cross at about 800 kilojoules hectare per year. Denser populations of the more common smaller mammals and less abundant larger ones approach energetic equivalence because each individual has a reciprocally lower total energy intake. Similar arguments can be made for growth rates and other life history traits which are expressed in parallel power laws.

Phylogenetic effects are treated without their actual identification by multiplying by the ratio of observed/predicted ratio (O/P) for whatever trait is known from the rare or endangered species, the closest documented relative, or mean departure of the taxon from the appropriate general allometric equation. Consequently, if we know how a species deviates from the general allometric equation for one variable, this observed value can be divided by the allometric-predicted value to give an observed-to-predicted ratio. This ratio can then be used as a factor for adjusting allometric predictions for other life history values. This assumes that if one trait has been stepped up or slowed down, other life history traits of the species of concern will have changed in similar proportions. The most useful O/P ratios are those which contain the dimension of time, because (a) life history plays out on a time scale, (b) times are reciprocals of rates and rates reflect capacity to allocate energy and materials between growth, survival, and reproduction—i.e., Charnov's [22] "production" $\propto M^{3/4}$, and (c) larger individuals can coexist (herd) on a common area with slower turnover [14].

7   TRIAL APPLICATIONS

As examples, we can consider two species of birds, "atypical" in how they diverge from the general allometries in related life history traits. Insight from patterns in the relatively slow "pace of living" of their life histories can then be extended to two species of whales, the blue whale (*Balaenoptera musculus*) and the gray whale (*Eschrichtius robustus*). Ironically, body sizes of these largest living mammals was not taken into account when evaluating its gestation period and maturation time, life history traits which reveal a stepped-up pace of living. This, in turn, has implications for the likely population growth rate and recovery from near extinction.

The outlier birds are kiwis and hummingbirds. Their eggs require longer incubation than predicted from egg size. Each day, eggs lose water by evaporation through pores in the shell. The longer incubation required, the more compensatory reduction in shell porosity is needed. However, the pores are where oxygen enters and CO_2 exits, so that slowing also reduces rates of metabolism and growth as well. The tissues are adapted to that slower pace, and this persists throughout the life of chick and even the adult.

7.1 THE MALE BROWN KIWI

Apteryx australis incubates 1 to 2 eggs for 75 da., 1.67 times as long as predicted for its 400-g bird egg by Rahn and Ar's [58] equation (Figure 8). Thus the rate of embryonic development is $1/1.67 = 60\%$ of normal. Evaporation through shell pores is fully compensated, cut to 60% of standard porosity, so that water is conserved to last for the longer incubation. Because porosity also effects oxygen and carbon dioxide exchange, the metabolism of the embryo is proportionately slowed. This low metabolic rate persists throughout the life of juvenile and adult [8, 9, 12].

7.2 THE LIFE HISTORY OF THE BROAD-TAILED HUMMINGBIRD

Selasphorus platycercus is actually slower paced than it appears. Hummingbirds do have short wings which beat at high natural frequencies ($38\,\mathrm{s}^{-1}$). They dart around at high angular velocities. Mass-specific metabolic rates during hovering are among the highest known. Therefore they seem to live their lives at a rapid pace. However, incubation of the broad-tailed hummingbird's two eggs is 70% slower than predicted from the allometry of incubation for all birds [58]. Time between hatching and fledging is similarly prolonged (Figure 9). Longevity has also increased proportionately, as is indicated by the maximum longevity we recorded, 12.1 years [15, 16].

Hence we can add (a) internal consistencies in life histories, as seen in kiwis and hummingbirds, to the reciprocal scalings of (b) mammalian natality vs. life expectancy and (c) scaling of D_{AP} (upper boundary) vs. metabolic rates, as evidence that life history traits are indeed so strongly interrelated that they provide linkages for predictive purposes. If just one trait has been timed in the species of concern—or perhaps a near relative—the observed/expected ratio for that measured trait could be used to adjust an allometric prediction for an unmeasured one. Is extrapolation to whales too much for this practice?

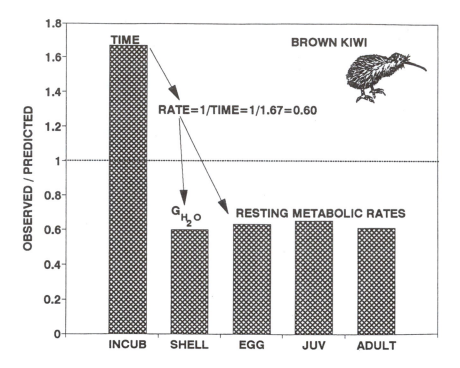

FIGURE 8 Incubation of the brown kiwi egg is retarded to 1.67 times "typical" predicted for egg size. Porosity of the shell has been reduced, restricting both evaporation (G_{H_2O}) and oxygen uptake. The resulting embryonic metabolic rate is paralleled by chick and adult metabolism.

7.3 THE BLUE WHALE

Clark's [23] harvest curve for blue whales gives us a sample test for pace-adjusted estimates of population growth and recovery from near extinction. Recovery plans and harvesting moratoriums for threatened and endangered species should be based upon as much knowledge of their reproductive biology as is possible. Compared to other marine mammals, whales were thought to have fairly late maturation, resulting in slower reproduction. Whales were contrasted with seals and sea otters, which were said to have the opposite, fairly early maturation and moderate to high reproductive rates The basis for this conclusion deserved reexamination. To compare "late" with "typical" we can take the mean of the masses, 141 T, listed for the blue whale by Silva and Downing [69] and compare with the mammalian scalings of gestation duration and reproductive maturity (Figure 10). The bovid ancestor of domestic cattle, the aurach, *Bos taurus*, (or other typical mammal scaled up to 141-T would be expected to have a gestation of 1446 da, but the blue whale does it in

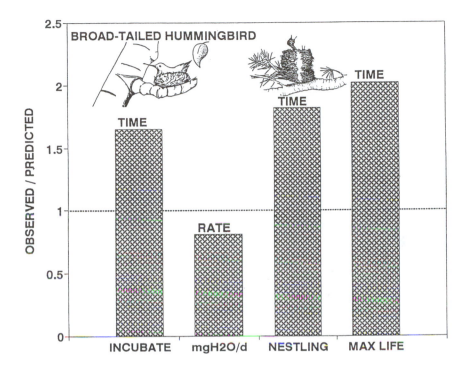

FIGURE 9 Incubation is slower in hummingbirds than predicted for size. Daily evaporative loss is reduced, prolonging water availability for the developing embryo. Slower incubation is followed by slower post-hatch development and aging.

335 d, taking only 23% as long as predicted (from the 66.3 $M^{0.26}$ of Blueweiss et al. [3]). Similarly, our 141-T ocean-going aurach would mature sexually in 23 y 5 mo, 19% of prediction from Economos [31]. Thus the reproduction of blue whales is actually accelerated, to take 19–23% of the time expected by extrapolation from terrestrial mammals (Figure 10). At the allometric r_{max}, a population of hypothetical 141 T cows could predictably double in 77 years. Corrected to the 21% average of O/P ratios, the blue whale population could, under ideal circumstances, double in 16.2 y. Using the logistic growth curve and calculating from figures in Clark [23], the blue whale population would take a minimum of 12.1 y to double, 15.7% of the unadjusted prediction or 75% of the adjusted scaling prediction.

7.4 THE GRAY WHALE

The good news is that great whales have a capacity for rapid recovery from low populations if human predation stops, a consequence of having an accelerated reproductive cycle that fits within an environmental one-year annual cycle—

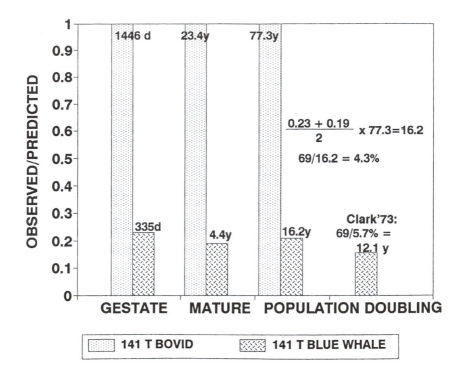

FIGURE 10   Gestation, maturation, and population doubling times of the 114 ton blue whale are each ca. one quarter of body-size prediction for mass ("114T BOVID").

witness the recovery of the gray whale (*Eschrichtius robustus*) and its removal from the endangered species list. The gray whale's gestation period is 395 d, or 41.6% of the 949 d predicted by the equation of Blueweiss et al. [3] for its body mass of 29,900 kg. It matures in 4.5 y, 30.8% of the Economos [31] prediction for its size. The mean of these O/P ratios is 36%. The minimum t_{2n} predicted for this size by equation is 20.2 y, which corrected by a factor of 0.36 O/P would be 7.3 y. The recovery t_{2n} of the gray whale was 18.3 y, which is 2.5 times the minimum (corrected) t_{2n} or 90.6% of the uncorrected prediction for minimum t_{2n}.

8   CONCLUSIONS

Lack of life history information about rare species hampers conservation efforts. Biological scaling (allometry) offers expedient surrogates for missing information needed for assessment and planning. Life history traits most likely evolved in suites of integrated adaptations, so if one trait is accelerated or retarded relative to a general size-based prediction, a similar quantitative adjustment can be anticipated in other traits. Coarse predictions from established regressions for a class of animals can be "fine tuned" on the basis of limited data. Kiwis and hummingbirds are outliers, which take longer to hatch than would be predicted from egg sizes. Internal consistency in life histories of these birds suggests that allometric predictions could be adjusted to life history pace if just one trait has been timed in the species of concern—or a near relative. The observed/expected ratio for that measured trait can be used to adjust an allometric prediction for an unmeasured one. The pace-adjusted allometric prediction for the blue whale (*Balaenoptera musculus*), used to test this, gave a minimum time for population doubling (at r_{max}) which exceeded the actual field estimate by a third, but both predicted and observed values indicated that whales are relatively fast in sexual maturation and gestation, especially compared to the previous conclusion that whales develop very slowly. This preliminary exploration suggests possibilities for more utilization of scaling in conservation. Size considerations are also applicable to census techniques and evaluation of abundance and biological diversity.

REFERENCES

[1] Allainé, D., D. Pointier, J. M. Gaillard, J. D. Lebreton, J. Trouvilliez, and J. Clobert. "The Relationship Between Fecundity and Adult Body Weight in Homeotherms." *Oecologia (Berlin)* **73** (1987): 478–480.

[2] Beuchat, C. A. "Allometric Scaling Laws in Biology." (letter) *Science* **278** (1997): 371.

[3] Blueweiss, L., H. Fox, V. Kudsma, D. Nakashima, R. Peters, and S. Sams. "Relationships Between Body Size and Some Life History Parameters." *Oecologia (Berlin)* **37** (1978): 257–272.

[4] Brockie, R. E., and A. Moeed. "Animal Biomass in a New Zealand Forest Compared with Other Parts of the World." *Oecologia (Berlin)* **70** (1986): 24–34.

[5] Brody, S. *Bioenergetics and Growth*. New York: Hafner, 1934. Reprint 1945.

[6] Brown, J. H. *Macroecology*. Chicago, IL: University of Chicago Press, 1995.

[7] Brown, J. H., and B. A. Maurer. "Evolution of Species Assemblages: Effects of Energetic Constraints and Species Dynamics on the Diversification of the North American Avifauna." *Amer. Natur.* **130** (1987): 1–17.

[8] Calder, W. A., III. "The Kiwi: A Case of Compensating Divergences from Allometric Predictions." In *Respiratory Functions in Birds, Adult and Embryo*, edited by J. Piiper, 239–242. Heidelberg: Springer-Verlag, 1978.

[9] Calder, W. A., III. "The Kiwi and Egg Design: Evolution as a Package Deal." *BioScience* **29** (1979): 461–467.

[10] Calder, W. A., III. "Ecological Scaling: Mammals and Birds." *Ann. Rev. Ecol. Syst.* **14** (1984): 213–230.

[11] Calder, W. A., III. "Size and Metabolism in Natural Systems." *Can. Bull. Fish & Aquat. Sci.* **213** (1985): 65–75.

[12] Calder, W. A., III. "The Kiwi and Its Egg." In *The Kiwi*, edited by E. Fuller and R. Harris-Ching, Ch. 10. Auckland, NZ: Seto Publishers, 1990.

[13] Calder, W. A., III. "The Scaling of Sound Output and Territory Size: Are They Matched?" *Ecology* **71** (1990): 1810–1822.

[14] Calder, W. A., III. *Size, Function, and Life History*. Cambridge, MA: Harvard University Press, 1984. Reprint. New York: Dover Publications, 1996.

[15] Calder, W. A., III. "Territorial Hummingbirds." *Natl. Cyeographic Resh. & Explor.* **7** (1991): 56–69.

[16] Calder, W. A., III, and L. L. Calder. "The Broad-Tailed Hummingbird." In *Birds of North America*, edited by A. Poole and P. Stettenheim, no. 16. Philadelphia, PA: American Ornithologists' Union and Academy of Natural Sciences, 1992.

[17] Calder, W. A., III, and L. L. Calder. "The Calliope Hummingbird." In *Birds of North America*, edited by A. Poole and P. Stettenheim, vol. 135. Philadelphia, PA: American Ornithologists Union and Academy of National Sciences, 1994.

[18] Calder, W. A., III, and L. L. Calder. "Size and Abundance: Breeding Population Density of the Calliope Hummingbird." *Auk* **112** (1995): 517–521.

[19] Carey, A. B., M. M. Hardt, S. P. Horton, and B. L. Biswell. "Spring Bird Communities in the Oregon Coast Range." In *Wildlife and Vegetation of Unmanaged Douglas-Fir Forests*, edited by L. F. Ruggiero, K. B. Aubry, A. B. Carey, and M. H. Brookes, 122–142. Report PNW-GTR-285. Portland, OR: U.S. Forest Service, Pacific Northwest Research Station, 1991.

[20] Carroll, R. *Vertebrate Paleontology and Evolution*. New York: W. H. Freeman, 1988.

[21] Caughley, G., and C. J. Krebs. "Are Big Mammals Simply Little Mammals Writ Large?" *Oecologia (Berlin)* **59** (1983): 7–17.

[22] Charnov, E. L. *Life History Invariants: Some Explorations of Symmetry in Evolutionary Ecology*. New York: Oxford University Press, 1993.

[23] Clark, C. W. "The Economics of Overexploitation." *Science* **181** (1973): 630–634.

[24] Cole, L. C. "The Population Consequences of Life History Phenomena." *Quart. Rev. Biol.* **29** (1954): 103–137.

[25] Congdon, J. D., and A. E. Dunham. "Contributions of Long-Term Life History Studies to Conservation Biology." In *Principles of Conservation*

Biology, edited by G. K. Meffe and C. R. Carroll, essay 7A, 181–182. Sunderland, MA: Sinauer Associates, 1997.

[26] Cotgreave, P., and P. H. Harvey. "Relationships Between Body Size, Abundance and Phylogeny in Bird Communities." *Functional Ecol.* **6** (1992): 248–256.

[27] Cutler, R. G. "Nature of Aging and Life Maintenance Processes." *Interdiscipl. Topics Geront.* **9** (1976): 83–113.

[28] Damuth, J. "Interspecific Allometry of Population Density in Mammals and Other Animals: The Independence of Body Mass and Population Energy-Use." *Biol. J. Linnean Soc.* **31** (1987): 193–246.

[29] Doak, D. F., and L. S. Mills. "A Useful Role for Theory in Conservation." *Ecology* **75** (1994): 615–626.

[30] Dunning, J. B. *CRC Handbook of Avian Body Masses.* Boca Raton, FL: CRC Press, 1992.

[31] Economos, A. C. "Beyond Rate of Living." *Gerontology* **27** (1981): 258–265.

[32] Fulghum, R. *All I Really Needed to Know I Learned in Kindergarten.* New York: Ivy Books, 1988.

[33] Gonzales, P. C., and R. S. Kennedy. "A New Species of *Crateromys* (Rodentia:Muridae) from Panay, Philippines." *J. Mamm.* **77** (1996): 25–40.

[34] Grosberg, R. K. "Competitive Ability Influences Habitat Choice in Marine Invertebrates." *Nature* **290** (1981): 703–706.

[35] Harris, J. "The Effect of Atmospheric Humidity on Attenuation Coefficients of Sound." *J. Acoust. Soc. Am.* **40** (1966): 148–153.

[36] Hayssen, V., and R. C. Lacy. "Basal Metabolic Rates in Mammals: Taxonomic Differences in the Allometry of BMR and Body Mass." *Comp. Biochem. Physiol.* **81A** (1985): 741–754.

[37] Hennemann, W. W., III. "Relationship Among Body Mass, Metabolic Rate, and the Intrinsic Rate of Natural Increase in Mammals." *Oecologia (Berlin)* **56** (1983): 104–108.

[38] Huff, M. H., and C. M. Raley. "Regional Patterns of Diurnal Breeding Bird Communities in Oregon and Washington." In *Wildlife and Vegetation of Unmanaged Douglas-Fir Forests*, edited L. F. Ruggiero, K. B. Aubry, A. B. Carey, and M. H. Brookes, 176–205. Report PNW-GTR-285. Portland, OR: U.S. Forest Service, Pacific Northwest Research Station, 1991.

[39] Juanes, F. "Population Density and Body Size in Birds." *Amer. Natur.* **128** (1986): 921–929.

[40] Kleiber, M. "Body Size and Metabolism." *Hilgardia* **6** (1932): 315–383.

[41] Kleiber, M. *The Fire of Life.* New York: Wiley, 1961.

[42] Linden, E. "Ancient Creatures in a Lost World." *Time* **June 20** (1994): 52–54.

[43] Martin, A. P., and S. R. Palumbi. "Body Size, Metabolic Rate, Generation Time, and the Molecular Clock." *Proc. Natl. Acad. Sci. (USA)* **90** (1993): 4087–4091.

[44] May, R. M., and D. I. Rubenstein. "1. Reproductive Strategies." In *Reproduction in Mammals*, edited by C. R. Austin and R. V. Short. 2nd ed. Book 4: Reproductive Fitness. Cambridge, MA: Cambridge University Press, 1984.

[45] McNab, B. K. "The Influence of Food Habits on the Energetics of Eutherian Mammals." *Ecol. Monogr.* **56** (1986): 1–19.

[46] Meffe, G. K., and C. R. Carroll. *Principles of Conservation Biology.* 2nd ed. Sunderland, MA: Sinauer Publishers, 1997.

[47] Millar, J. S. and R. M. Zammoto. "Life Histories of Mammals: An Analysis of Life Table." *Ecology* **64** (1983): 631–635.

[48] Morton, E. S. 1975. "Ecological Sources of Selection on Avian Sounds." *Amer. Natur.* **109** (1975): 17–34.

[49] Nagy, K. A. "Field Metabolic Rate and Food Requirement Scaling in Mammals and Birds." *Ecol. Monogr.* **57** (1987): 111–128.

[50] Odum, E. P. *Fundamentals of Ecology*, 79–81. Philadelphia, PA: Saunders, 1971.

[51] Onions, C. T., ed., W. Little, H. W. Fowler, and J. Coulson. *Oxford Universal Dictionary*, 2092, 3rd ed. UK: Oxford at the Clendenon Press, 1955.

[52] Parma, A. M., and NCEAS Working Group on Population Management. "What Can Adaptive Management Do for Our Fish, Forests, Food, and Biodiversity?" *Integrative Biol.* **1** (1998): 16–25.

[53] Pennisi, E. "Brazil's New Monkey." *Science* **277** (1997): 1207.

[54] Peters, R. H. *The Ecological Implications of Body Size.* Cambridge, MA: Cambridge University Press, 1983.

[55] Pimm, S. L. "Biodiversity Is Everything." *Quart. Rev. Biol.* **73** (1998): 51.

[56] Pyke, G. H., and H. F. Rector. "Estimated Forest Bird Densities by Variable Distance Point Counts." *Aust. Wildl. Res.* **12** (1985): 307–319.

[57] Rabinowitz, A. "Turning over a New Leaf: Muntjac Discovered in Myanmar." *Wildlife Conservation* **101(6)** (1998): 42–45.

[58] Rahn, H., and A. Ar. "The Avian Egg: Incubation Time and Water Loss." *Condor* **76** (1974): 147–152.

[59] Ray, G. C. "The Role of Large Organisms." In *Analysis of Marine Ecosystems*, edited A. R. Longhurst, 397–413. New York: Academic, 1981.

[60] Read, A. F., and P. H. Harvey. "Life History Differences Among the Eutherian Radiations." *J. Zool. Lond.* **219** (1989): 329–353.

[61] Rosenzweig, M. L. *Species Diversity in Space and Time.* Cambridge, MA: Cambridge University Press, 1995.

[62] Schieck, J. "Biased Detection of Bird Vocalizations Affects Comparisons of Bird Abundance Among Forested Habitats." *Condor* **99** (1997): 179–190.

[63] Schmidt-Nielsen, K. *Desert Animals: Physiological Problems of Heat and Water.* London: Clarendon, Oxford University Press, 1964.

[64] Schmidt-Nielsen, K. "The Unusual Animal, or to Expect the Unexpected." *Fed. Proc.* **26** (1967): 981–986.

[65] Schmidt-Nielsen, K. *Scaling: Why Is Animal Size so Important?* Cambridge, MA: Cambridge University Press, 1984.

[66] Schmitz, O. L., and D. M. Lavigne. "Intrinsic Rate of Increase, Body Size, and Specific Metabolic Rate in Marine Mammals." *Oecologia (Berlin)* **62** (1984): 305–309.

[67] Silva, M. "Allometric Scaling of Body Length: Elastic or Geometric Similarity in Mammalian Design." *J. Mammalogy* **79** (1998): 20–32.

[68] Silva, M., and J. A. Downing. "Allometric Scaling of Minimal Mammal Densities." *Conserv. Biol.* **8** (1994): 732–743.

[69] Silva, M., and J. A. Downing. *CRC Handbook of Mammalian Body Masses.* Boca Raton, FL: CRC Press, 1995.

[70] Silva, M., J. H. Brown, and J. A. Downing. "Differences in Population Density and Energy Use Between Birds and Mammals: A Macroecological Perspective." *J. Animal Ecology* **66** (1997): 327–340.

[71] Song, Y-L. "Population Viability Analysis for Two Isolated Populations of Hainan Eld's Deer." *Conserv. Biol.* **10** (1996): 1467–1472.

[72] Thompson, S. D. "Body Size, Duration of Parental Care, and the Intrinsic Rate of Natural Increase in Eutherian and Metatherian Mammals." *Oecologia (Berlin)* **71** (1987): 201–209.

[73] Van Dung, V., P. M. Giao, N. N. Chinh, D. Tuoc, P. Arctander, and J. MacKinnon. "A New Species of Living Bovid from Vietnam." *Nature* **363** (1993): 443–445.

[74] Verner, J. "Data Needs for Avian Conservation Biology: Have We Avoided Critical Research?" *Condor* **94** (1992): 301–303.

[75] Wallschlager, D. I. "Correlation of Song Frequency and Body Weight in Passerine Birds." *Experientia* **36** (1980): 412.

[76] West, G. B., J. H. Brown, and B. J. Enquist "A General Model for the Origin of Allometric Scaling Laws in Biology." *Science* **276** (1997): 122–126.

[77] Whitfield, J. "A Saola Poses for the Camera." *Nature* **396** (1998): 410.

Scaling and Self-Similarity in Species Distributions: Implications for Extinction, Species Richness, Abundance, and Range

John Harte

1 INTRODUCTION

Species-area relationships (SARs) describe the dependence of the number of species found in a censused patch of habitat on the area of that patch. The most commonly discussed form of an SAR is the power-law form:

$$S(A) = cA^z , \tag{1}$$

where S is the number of species found in an area A, and c and z (the SAR exponent) are constants. Because the parameters in this or other (nonpower-law) forms of the SAR generally differ across taxonomic categories, they are usually expressed for specific taxa such as plants or birds or butterflies. They may also differ from one type of habitat to another, from true islands of different area to different size patches arbitrarily delineated within a continental biome, and across widely disparate spatial scales.

Complementing the numerous studies demonstrating empirical support for Eq. (1) (see Rosenzweig [23], for a review), two major lines of effort characterize theoretical research on SARs. The first looks for the origin of the SAR, seeking to derive it from underlying assumptions about the spatial distribution of species and of individuals within species (see, for example, Preston [22], May [19], Coleman [3], Caswell and Cohen [1], Leitner and Rosenzweig [16]).

The second examines consequences of SARs for reserve design and for extinction rates under habitat loss (see, for example, MacArthur and Wilson [17], Diamond [6], Simberlof and Abele [24], May et al.[20], Pimm et al. [21]). Although the work I describe here may ultimately provide insight into the origin of the SAR, its immediate goal is to examine consequences, not causes. By casting the power-law form of the SAR in the form of a self-similar scaling relationship, I show that it is remarkably rich in implications for a host of fundamental and applied issues in ecology.

Among these implications are an endemics-area relationship (EAR) and an expression for the fraction of species in common to two patches as a function of patch size and interpatch distance (the "commonality" function). Both these results have implications in conservation biology: the EAR provides new insight into species extinction under habitat loss, while the commonality result provides an improved means of estimating species richness in biomes too large to census directly and completely. The dependence of species richness on the shape of a censused patch is also predicted from the self-similar SAR; plant census data are consistent with the predicted shape dependence. Moreover, the self-similar SAR implies unique species-abundance and species-range distributions. The derived abundance distribution deviates considerably from the widely assumed lognormal distribution (predicting far more rare species than does the lognormal), thereby invalidating the prevailing assumption that lognormality in the distribution of individuals is the basis for the power-law form of the SAR.

2 SELF-SIMILARITY AND THE SPECIES-AREA RELATIONSHIP (SAR)

My starting assumption is Eq. (1), which can be reexpressed in the form:

$$S(A) = \left(\frac{A}{A'}\right)^z S(A').$$ (2)

In general, this relation will not hold for all areas A and A', but rather only within some scale range. Thus, z may take on different values at very different spatial scales. Eqs. (1) or (2) are interesting and useful to the extent z is constant across a large range of areas; an enormous amount of experimental evidence [23] suggests this power-law form is indeed often valid over a range of at least one or two orders of magnitude of area.

Although the issue of incomplete sampling and imperfect censusing is certainly relevant to discussions of species distributions [3], *I restrict the entire discussion below to the limiting case in which censusing is complete and exact.* Many available plant censuses are sufficiently close to this limiting case to render it of practical interest.

Consider for the time being the limiting case in which z is constant over a range of areas stretching up to some largest area, A_0. Using the notation

A_i to refer to a patch of area $A_0/2^i$, and S_i to refer to $S(A_i)$, Eq. (2) can be rewritten as:

$$S_i = S(A_i) = S\left(\frac{A_0}{2^i}\right) = \left(\frac{1}{2^i}\right)^z S(A_0) = \mathbf{a}^i S_0 \qquad (3)$$

where

$$\mathbf{a} = 2^{-z}. \qquad (4)$$

For the typical case [22, 23] in which $z = 0.25$, $\mathbf{a} = 0.84$.

We note that Eq. (3) implies that

$$S_{i+1} = \mathbf{a}\, S_i. \qquad (5)$$

The constant $\mathbf{a}$ has a simple interpretation: *of the species found in an area A_i, $\mathbf{a}$ is the fraction found in at least the A_{i+1} patch that comprises the left (or equivalently, at least the right or top or bottom) half of A_i.*

Our statement of self-similarity is: $\mathbf{a}$ *is a constant that is independent of i.* Together, Eq. (5) and our statement of self-similarity provide a necessary and sufficient condition for the power-law form of the SAR (Eq. (2)).

One additional condition must be imposed, however. As patches are split in half, they may become increasingly long and skinny. This would be a problem because, as I show later, the average number of species found in a censused patch of area A depends on the shape of the patch. Thus we have to restrict the bisection process to one that preserves the shape of the resulting patches. One way to do this (suggested by Henry Horn at this SFI workshop) is to consider an A_0 patch with length-to-width ratio of $\sqrt{2}$. Bisection of such a patch perpendicular to the long dimension yields two patches of identical shape to each other and to the original patch. In what follows, I refer to patches that are roughly self-similar under bisection as "nicely shaped." *The following results apply only to such nicely shaped patches, unless stated otherwise.*

Finally I note that any real ecosystem will be heterogeneous with respect to habitat quality, so that $S(A)$ will depend on which patch of area A is censused; *thus the quantity $S(A)$ in this and similar expressions that follow refers to the average number of species found in all the nonoverlapping patches of area A that comprise the entire biome.*

3 THE ENDEMICS-AREA RELATIONSHIP (EAR)

Following custom, I define a species to be endemic to a patch of area A if it is found only in that patch. Letting $E(A)$ be the number of endemics in such a patch, then, as shown by Harte and Kinzig [9], there follows from our self-similarity condition (or its equivalent power-law form of the SAR) a simple endemics-area relationship (EAR):

$$E(A) = \left(\frac{A}{A'}\right)^{z'} E(A'), \qquad (6)$$

where

$$z' = \frac{-\ln(1 - 2^{-z})}{\ln 2}. \tag{7}$$

Equivalently, Eq. (6) can be expressed in a form similar to Eq. (1):

$$E(A) = c' A^{z'}. \tag{8}$$

In parallel with Eq. (3), Eq. (6) can, with a little algebra, be reexpressed as:

$$E(A_i) = (1 - \mathbf{a})^i S_0. \tag{9}$$

In parallel with the text following Eq. (5), we can state that: *of the species found in a patch of area A_i, the fraction found in the left, but not the right, half of that patch is $1 - \mathbf{a}$.*

Insert 1 provides a simple motivation of Eqs. (6)–(9); a more complete and general derivation is given in Harte and Kinzig [9].

Insert 1 Origin of the Endemics-Area Law

1000 species endemic to entire habitat ($Z = 0.20$)

$\boxed{1000(.5)^Z = 870 \mid 1000(.5)^Z = 870}$ $\Rightarrow 1000 - 870 = 130$ species confined to each half

$\boxed{130(.5)^Z = 113 \mid 130(.5)^Z = 113}$ $\Rightarrow 130 - 113 = 17$ species confined to each quarter

$\vdots$

Summarizing:

$130 = 1000(1 - .5^Z)$, which can be written as $1000(.5)^{Z'}$

$17 = 1000(1 - .5^Z)(1 - .5^Z) = 1000(1 - .5^Z)^2$, which can be written as $1000(.25)^{Z'}$

$\vdots$

where $z' = \frac{-\ln(1 - 2^{-Z})}{\ln 2}$

We note that because $z \le 1$, z' must be ≥ 1. For the typical case in which $z = 0.25$, $z' = 2.65$. Thus, Eq. (8) implies that the number of endemics grows faster than linearly with area. In other words, if two adjacent patches of area A each have E endemics, then the number of species endemic to both patches together must exceed $2E$. This, of course, must be the case: the total endemics list for the two patches combined will include the sum of those species endemic to each patch *plus* those species found in both patches (and thus not endemic to either patch alone) but found nowhere else in the biome.

A further consequence of self-similarity is an expression for the fraction of species found in a patch of area A_i and that are actually in both halves of the patch. In particular, it follows [9] that *of the species S_i found in A_i, a fraction $2\mathbf{a} - 1$ are in both of the A_{i+1} patches that comprise the two halves of A_i.*

Let me summarize results to this point. The power-law form of the SAR (Eq. (1)) is mathematically equivalent to the following self-similarity condition: let $\mathbf{a}$ be the fraction of species in a patch of habitat that is found in at least the left half of that patch (or equivalently at least the right half); then $\mathbf{a}$ is a constant independent of patch area. The fraction of species in that patch which is found in both halves is $2\mathbf{a} - 1$; the fraction found only in the left half is $1 - \mathbf{a}$, and the fraction found only in the right half is also $1 - \mathbf{a}$. Note that these relations imply that the odds the particular species is found either in both halves or only in one or the other half add up to 1, as they must.

4   A TEST OF THE EAR

Figure 1 plots the total number of plant species and the number of endemic plant species in the 48 contiguous states of the U.S. against state area on a log-log scale. The straight line fit to the total species data has a slope of 0.13 ($R^2 = 0.22, p < 0.01$); the straight line fit to the endemics data has a slope of 3.7 ($R^2 = 0.38, p < 0.001$). In comparison, Eq. (7), with $z = 0.13$, predicts $z' = 3.5$. Despite the wide range of habitat conditions spanned by the 48 contiguous states and our use of a data set that does not permit averaging over all patches of a given area, the agreement between the predicted and measured value of z' provides support for the theory.

Further tests of the EAR using gridded plant, butterfly, and bird census data sets from the U.K., Amazonia, South Africa, and the Colorado Rockies are in progress.

5   COMMONALITY

I use the term "commonality" to refer to the fraction of species in common to two patches; in other words, it is the number of species in common to the two patches divided by the average number in each of the patches. If the two patches each have an area A and are a distance D apart, then I denote that fraction by $\chi(A, A, D)$. Note that $\chi = 1 - T$, where T refers to the spatial turnover of species [2].

Intuitively, commonality should decrease with increasing interpatch distance, D. Self-similarity provides an explicit expression for the dependence of commonality on D. In particular, Harte and Kinzig [9] derived from the SAR (Eq. (1)) the following nearly exact formula for commonality:

$$\chi(A, A, D) \cong (2 - 2^z) \left(\frac{A}{D^2} \right)^z \tag{10}$$

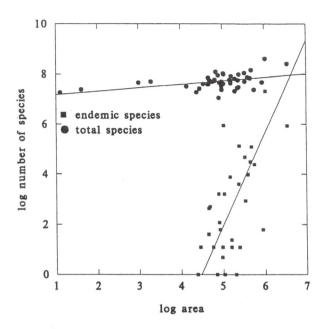

FIGURE 1 A log-log plot of total and endemic plant species in the 48 contiguous states of the U.S. Detailed description of the data set and fitting procedure are found in Harte and Kinzig [9].

where the "almost equal" sign takes account of a small (< 1.04) geometric correction factor that cannot be specified as a closed-form algebraic function of A, D, and z.

In the most general case, z is not constant across the entire spatial scale range from A to D^2 and Eq. (10) must be modified. For example, in the case in which there are two scales of z: $z(D)$ and $z(A)$, where $z(A)$ applies over the scale range characterizing the area of censused patches, and $z(D)$ applies over the scale range characterizing the distances between patches, this equation becomes:

$$\chi(A, A, D) = [2 - 2^{Z(D)}] \left[\frac{A_0^{Z(D)}}{D^{2Z(D)}} \right] \left[\left(\frac{A}{A_0} \right)^{Z(A)} \right]. \qquad (11)$$

To test Eqs. (10) and (11) across interpatch distance scales ranging from 0.5 m to 12 km, and censused areas ranging from 0.25 m^2 to 16 m^2, we censused vegetation on 1,920 quadrants, each of area 0.25 m^2 in mountain meadows in the Colorado Rockies. The quadrants are arrayed in a nested design (64 contiguous quadrants/plot, ten 4-m × 4-m plots per site with interplot distances within a site ranging from 5 to 70 m, and 3 sites located at distances from one another of 2 km to 12 km.

Our results indicate that the values of z obtained from fitting plot-scale data ($D \sim .5$ m $- 5$ m) for all 30 plots to Eq. (1) agree well with the values obtained at that same scale by fitting commonality data at plot scale to Eq. (11) [11]. Thus where estimates of z from both species richness data censused in ever-larger areas and from commonality data derived from censuses of separated patches are available, the two estimates are consistent. Using all data, with interpatch distances spanning the full range from m $D = 0.5$ m $-$ 12 km, a plot of log (commonality versus log(D) yields approximate straight line behavior (Figure 2), in support of the self-similarity assumption across the entire spatial scale explored. We note, however, that an intermediate distances ($D \sim 10$ m $- 70$ m) there is evidence for a flattening of the slope. The figure shows the data only for commonality between 0.5×0.5 m patches; a similar slope but different intercept is observed when commonality between 1 m^2 or 16 m^2 patches is similarly plotted; intercept values for different patch areas, A, are also consistent with Eq. (11) [11].

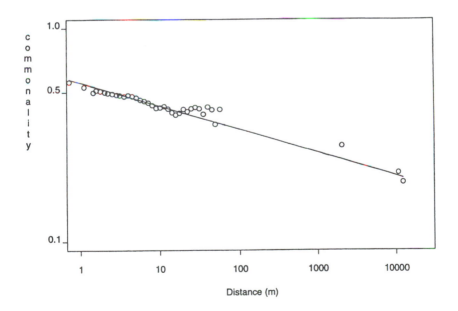

FIGURE 2   A log-log plot of commonality versus interpatch distance for patch areas of 0.5 m × 0.5 m and interpatch distances ranging from 0.5 m to 12 km.

6   DOES PATCH SHAPE AFFECT SPECIES RICHNESS?

Does the number of species found in a plot of area A depend upon the shape of the plot? To my knowledge, all published models that have been invoked to explain or derive SARs from underlying statistical assumptions such as species-abundance and/or species-range distributions (e.g., Preston [22], Coleman [3], Caswell and Cohen [1], Leitner and Rosenzweig [16]) implicitly or explicitly assume that species richness is independent of patch shape. Yet, as I show next (see also, Harte et al. [11]) our theory and existing data indicate that shape does indeed matter.

Consider two square patches, each of area A_i, and thus each with $2^{-iz}S_0$ species. If they share an edge, the interpatch distance is $D = A_i^{1/2}$ and so the fraction of species in common to the two patches is, by Eq. (10), $\chi = 2 - 2^z$. Hence the total number of species in the two patches is $2A_i - \chi A_i = [2 - (2 - 2^z)]A_i = 2^z A_i = 2^{-(i-1)z}S_0$. This, of course, is just the expected formula for $S(A_{i-1})$, as expected because the two adjacent A_i squares make a nicely shaped A_{i-1} rectangle. But now place the two 1×2 rectangles together so that they share a short edge and form a 1×4 rectangle. Now $D = 2A^{1/2}$. Using Eq. (10), it is straightforward to show that this patch, of area A_{i-2}, contains $[2 - (2 - 2^z)/2^z]A_{i-1} = [2 - (2 - 2^z)/2^z]2^{-(i-1)z}S_0$ species. For $0 < z < 1$, this expression is greater than the species richness expected from the SAR for a nicely shaped patch of area A_{i-2}: $S(A_{i-2}) = 2^{-(i-2)z}S_0$.

More generally, consider a rectangular patch of width one unit and length 2^N units. Then repeated application of Eq. (10) yields an approximate expression for the species richness of that patch:

$$S(1 \times 2^N) = S(1 \times 1) \prod_{i=0}^{N-1} \left[2 - \frac{(2 - 2^z)}{2^{iz}} \right]. \tag{12}$$

As $N \to \infty$, Eq. (12) predicts: $S(2^N) \to 2^N S(1)$, or $z_{\text{effective}} \to 1$. For $z = .25$, Eq. (1) predicts $S(\sqrt{8} \times \sqrt{8}) = 8^{0.25}S(1)$, whereas Eq. (12) predicts $S(1 \times 8) = 8^{0.39}S(1)$. Species richness is greater on long skinny patches than on square patches of the same area.

As shown in the Table below, our meadow plant census data confirm Eq. (12). The value of z in the first column is that obtained at each of the three sites by fitting census data for square or 1×2 rectangles on each of the ten plots to Eq. (1). The symbol $\langle \ \rangle$ refers to an average over all ten plots within a site. $S(1 \times 1)$ and $S(1 \times 8)$ are the average species richnesses for all quadrants and for all nonoverlapping 1×8 rectangles in the ten plots at each site. $z_{\text{eff,meas}}$ is $\log_8(\langle S(1 \times 8)\rangle/\langle S(1 \times 1)\rangle)$ and $z_{\text{eff,pred}}$ is obtained by evaluating:

$$z_{\text{eff,pred}} = \log_8 \prod_{i=0}^{2} \left[2 - \frac{(2 - 2^z)}{2^{iz}} \right] \tag{13}$$

using the value of z from column 1 in the table.

TABLE 1 Empirical test of Eq. (13); symbols are defined in the text.

	$\langle z \rangle$	$\langle S(1 \times 1) \rangle$	$\langle S(1 \times 8) \rangle$	$z_{\text{eff,meas}}$	$z_{\text{eff,pred}}$
Upper site	0.27	8.54	20.49	0.42	0.41
Middle site	0.29	7.37	17.35	0.41	0.43
Lower site	0.23	8.37	18.03	0.37	0.36

The data not only demonstrate shape-dependence of species richness, thus casting doubt on the assumptions that underlie many attempts to derive an SAR from underlying statistical assumptions, but also confirm the detailed prediction of our theory. We note that plant census data from tropical forests [4] and the UK [14] also indicate a positive effect of plot elongation on species richness.

7 APPLICATION OF THE EAR TO ESTIMATING EXTINCTION RATES UNDER HABITAT LOSS

How many species are expected to go extinct under habitat loss? The conventional application of the SAR to estimating species extinction under habitat loss consists of calculating, using Eq. (1), the number of species that can be sustained in the remaining habitat, and then subtracting this number from the number of species initially present. However, if a patch of habitat is removed from, say, the center of a large biome, the remaining habitat will not be nicely shaped and thus the SAR may not be applicable to the remaining habitat. Indeed, if the lost habitat is nicely shaped, a more reliable estimate of the number of lost species may be the number of species endemic to the lost habitat. Hitherto there was no theoretical basis for estimating the number of endemics in a lost patch, but with Eqs. (6)–(9) this approach can be used.

The two approaches can give very different results, in some cases differing by factors of 50 or more depending on the value of z and on the fraction of habitat lost [13]. The EAR approach results in very much lower estimates of extinction when small patches of habitat are lost, but it predicts a rapid increase in the rate of extinction as habitat loss progresses. Thus, in situations where extinctions have occurred at levels below those predicted by the SAR approach but consistent with the EAR approach, such as North American bird extinctions [13, 21], these seemingly low initial rates of extinction should be viewed not as grounds for complacency but as a forewarning of future extinctions that accelerate at a rate higher than anticipated by the conventional approach.

Neither the SAR nor the EAR approach is likely to be reliable if neither the lost nor remaining patch is nicely shaped. For the case in which multiple "islands" of nicely shaped habitat remain, Eq. (11) can be used to estimate the overlap of the species lists, and thus the total species list, in the multiple preserves. For the case in which multiple "islands" of nicely shaped habitat are lost, an expression for the commonality of endemics can be used to predict the total list of species unique to the collection of lost patches (see Kinzig and Harte [13]).

Of course any attempt to use biogeographic theory to estimate extinction under habitat loss must be accompanied by a "product-warning label," and the discussion above is no exception. Predictions should be considered only a rough guide to what may actually occur. In particular, even if the lost habitat is nicely shaped, the approach based on estimating endemism in the lost patch might overestimate extinction if individuals in a species endemic to the patch can leave the patch and establish residency elsewhere; on the other hand, it might underestimate extinction if the patch is a critical source area for species found elsewhere (i.e., species not endemic to the patch) or if edge effects from the lost patch threaten individuals found in the remaining area. Similar problems beset use of the SAR approach. Despite this, avian extinction data from North America are remarkably consistent with the EAR-based, but not the SAR-based, approach [13].

Finally, I note that the critique of the SAR-based approach to estimating extinction rates given above is quite different from that of Pimm et al. [21]; lacking an expression for $E(A)$, those authors advocated an expression for the expected number of lost species based on a formula for E that is only applicable when exactly half the area of the original habitat is lost (in which case the two approaches give the same answer when applied correctly).

8 APPLICATION OF THE COMMONALITY FORMULA TO ESTIMATING BIOME- OR LANDSCAPE-SCALE SPECIES RICHNESS

Our knowledge of the species richness of many biomes is quite incomplete. The number of insect species in tropical forests, for example, may not even be known to within a factor of ten [26]. If z is scale-independent, then knowledge of the SAR exponent, z, from small-scale census data can be used to estimate species richness at large scales; in other words, Eq. (1) would permit scaling up to large areas the number of species found in patches small enough to be censused in their entirety. If, however, z is spatial scale-dependent, as is likely, then direct use of Eq. (1) to estimate species richness at large spatial scales where z is not known, is impossible. Application of Eq. (1) to estimation of z is practical only for areas small enough to be censused, and thus hitherto there has been no practical means available for estimating z at large scales. Eq. (11) however, provides a way to estimate z at large spatial scales.

Insert 2 illustrates how these ideas can be converted to a strategy for estimating species richness in a large biome. The strategy is presented for a particular hypothetical case of beetles in Amazonia; it is based on the assumption (for the sake of illustration only) that z possesses two distinct scales, but the result can easily be generalized to a larger number or even a continuum of scale ranges for z. In the example, z is determined: (i) at small scale using Eq. (1) and complete census data in nested quadrants ($A = 1$ m^2 to 100 m^2), (ii) at large scale using Eq. (11) and commonality data between $\sim$ m^2 patches separated by large distances ($D = 1$ km to 3000 km).

Insert 2 How many beetle species in Amazonia?

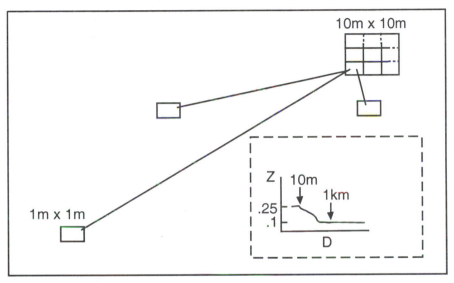

1. Fit of small scale data to $S = CA^Z$ (scale from $A = 1$ m^2 to 100 m^2 results in: $Z_{\text{small scale}} = 0.25$, $C_{\text{small scale}} = 3000$.
  [If naïvely applied to Amazonia, $S(10^{13}$ m$^2) = 5,335,000!$]

2. Suppose, however, that a fit of large-scale ($D = 1$ km to 3000 km) spatial commonality data to $\chi(1, 1, D) \sim (D^2)^{-Z}$ yields $Z_{\text{large scale}} = 0.1$.

3. Then, if both scales are joined at $A = 10^4$ m^2 (i.e., $D = 100$ m): $S(A = 10^4$ m$^2) = C_{\text{small}}(10^4)^{Z_{\text{small scale}}} = C_{\text{large}}(10^4)^{Z_{\text{large scale}}} \Rightarrow C_{\text{large scale}} = 12,000$.

4. And thus: $S(\text{Amazonia} = 10^{12}$ m$^2) = 12,000(10^{13})^{0.1} = 240,000$.
  (note: result is fairly insensitive to joining distance)

At the small scale, our hypothetical "two-scale" example assumes z is measured to be 0.25, while at the large scale, we assume $z = 0.1$. In reality, z

is likely to rise again at the largest spatial scales [23] as entirely new types of climatic conditions or habitat types are encountered, and so a three- or higher-scale calculation might be needed. The important point, however, is that the approach illustrated in Insert 2 is generalizable to whatever the number is of useful scale intervals into which it makes sense to divide the entire biome. Of course the effort needed to carry out the commonality measurements will increase as the number of distinct scale intervals increases.

9   THE SPECIES-ABUNDANCE DISTRIBUTION

The number of individuals in a species, i.e., the abundance, varies from species to species. A plot of the number of species versus abundance is the species-abundance distribution. Preston's [22] and then later May's [19] classic analyses assumed a particular form of that abundance distribution—the canonical lognormal—and led to the claim that the power-law form of the SAR resulted. Although an explicit abundance relation was assumed by these authors, no assumptions about range distributions were made. Further investigation, however, has raised some doubt about the validity of Preston's result (e.g., Connor and McCoy [5]) and in particular questioned the assumption that range doesn't matter to the origin of the SAR [16]. These doubts, combined with the concern that approaches such as Preston's and May's cannot predict the observed dependence of commonality on interpatch separation and of species richness on the shape of censused patches (see above), suggest the importance of revisiting the relationship between the SAR, on the one hand, and range and abundance distributions on the other.

Whereas Preston and May sought to derive an SAR from a particular abundance distribution, I will instead address the reverse question: what abundance distribution is implied by the SAR (or equivalently by self-similarity)? For a full derivation and more complete discussion of our result see Harte et al. [10]; here I simply sketch the approach and state the result.

I assume there is a smallest patch size, $A_m = 2^{-m} A_0$, such that the number of individuals in the biome $= 2^m$. In other words, each small patch A_m holds on average 1 individual (a notion that makes most sense, of course, for plants, not animals). Assume, first, that self-similarity holds across all scales from A_m to A_0, so that the number of species in A_0 is given by $S_0 = \mathbf{a}^{-m}$. In reality, a single scale-invariant value for $\mathbf{a}$ (or z) is not going to hold over the entire range of areas extending from A_m (area of single individual) up to A_0 (the entire biome). From a weaker form of these assumptions that do not require consideration of the patch with a single individual, the species-abundance distribution can still be derived [10], but only the species-abundance distribution resulting from the strong form of the assumptions is discussed here.

I define $P_i(n)$ to be the fraction of species found in a patch of area A_i that contains n individuals. Our interest ultimately is in $P_0(n)$, but to obtain that distribution, we derive it recursively from the $P_i(n')$ for $n' \leq n$ and $0 < i \leq m$. Note that $P_i(n) = 0$ for $n > 2^{m-i}$, that the sum over all n of $P_i(n) = 1$, and that $P_m(1) = 1$. Then the following double recursion relation results [10]:

$$P_i(n) = xP_{i+1}(n) + (1 - x) \sum_{k=1}^{n-1} P_{i+1}(n - k)P_{i+1}(k), \qquad (14)$$

where $x = 2(1 - \mathbf{a})$.

I have only obtained analytical solutions to this equation for the very smallest and largest values of n, but numerical solutions are revealing. Figures 3(a)–(b) show the numerical solution for the cases $i = 5 - 8$ and $x = 0.3$ (corresponding to $z = 0.23$) plotted on both a linear and a log scale for abundance.

Clearly, the species-abundance distribution is far from lognormal, being skewed more toward rarity (low n) than the lognormal. It thus appears that the self-similar SAR considered here is incompatible with a lognormal species-abundance distribution, and this feature of the solutions to Eq. (14) persists up to the largest values of i that we have explored numerically ($i = 28$). Available empirical species-abundance distributions tend to deviate from a lognormal [7, 12, 22, 25] with the lognormal underestimating the observed fraction of rare species. Sampling bias is, if anything, likely to result in an underestimate of rarity. It is noteworthy, then, that the deviation from lognormal of our predicted abundance distribution is in the direction of capturing this feature of the empirical distributions.

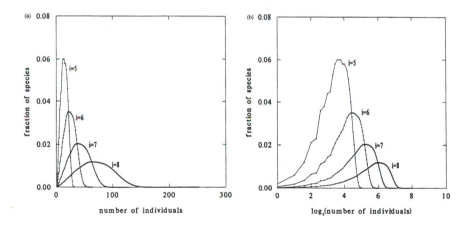

FIGURE 3 Theoretical species-abundance distributions from numerical solution to Eq. (11) for $x = 0.3(z = 0.23)$. The index i is the $\log_2$ of the area of the model biome, in units of minimum area containing a single individual. (a) linear abundance scale; (b) $\log_2$ abundance scale.

10   THE SPECIES-RANGE DISTRIBUTION

Analogous to the species-abundance distribution is the species-range distribution. Consider the species found only in a particular patch of area A_i. All those species have a range less than or equal to A_i. There are 2^i such disjoint patches in A_0, and no species with a range greater than A_i can be endemic to one of those 2^i disjoint patches. Thus, letting $S(R \leq A_i)$ be the number of species in A_0 with range less than or equal to A_i, we have

$$S(R \leq A_i) = 2^i E_i = 2^i \left(\frac{A_i}{A_0}\right)^{z'} S_0 = \left(\frac{A_i}{A_0}\right)^{z'-1} S_0 . \tag{15}$$

Dividing ranges into intervals that increase by factors of two, and defining $S(R = A)$ to be the number of species with range between $A/2$ and A, we have

$$S(R = A) = S(R \leq A) - S\left(R \leq \frac{A}{2}\right) = (2a - 1)\left(\frac{A}{A_0}\right)^{z'-1} S_0 . \tag{16}$$

For $z' = 0.25$, corresponding to $z' = 2.65$ and $a = .84$, we have:

Range	Fraction of Species
$A_0/2 < R < A_0$	0.68
$A_0/4 < R < A_0/2$	0.22
$A_0/8 < R < A_0/4$	0.07
$A_0/8 < R < A_0/16$	0.02

Thus the species-range distribution is a monotonically increasing function of range, with about 2/3 of the species having a range equal to or greater than half the area of the biome.

    This species-range distribution might, at first glance, seem to be overly weighted toward large ranges (that is, too many species with large range) but the result must be interpreted in the light of our manner of subdividing habitat and defining range accordingly. Consider, for example, a species that is found throughout a small patch that straddles the boundary between the upper left and lower left quadrants of the biome. By our definitions, this species would be considered to be endemic to an A_1 patch but not an A_2 patch. Its range would be in the interval $A_0/4 < R < A_0/2$ even though its actual region of occupancy might have an area many times, not just 2 to 4 times, smaller than A_0. Although this might seem like an artificiality of the approach, requiring (impossible to obtain) knowledge of exactly where the "center" of the biome is situated, that is not the case. The prediction can be readily tested from occupancy data by placing an appropriately gridded square over the biome and making a consistent set of bifurcations; the idea of a true "center" is meaningless and irrelevant.

Tests of Eq. (16) using digitized census data, along with a derivation of the distribution of a more conventionally defined range (for example, the circle or square containing all individuals in a species) are in progress.

11 THE ABUNDANCE-RANGE DISTRIBUTION

I turn, next, to the relation between abundance and range. In an interesting empirical study of the distribution of plant species in the United Kingdom, Kunin [15] showed that the amount of habitat occupied by a given plant species exhibits self-similar behavior when viewed at varying scales of resolution through a "censusing window" of a particular area. In particular, a plot of log[(area of the window or grid cell used to census)(number of such windows in which the species is observed within the landscape)] versus log(window area) yields a straight line over multidecade scale ranges.

Our theory predicts not only this straight-line behavior but provides an explicit relation between the slope of that line and the abundance of the given species. The slope is $1 - \ln(n)/\ln(N_0)$ where N_0 is the total number of individuals of all species in the biome and n is the abundance of the particular species. Kunin selected relatively rare species $(\ln(n)/\ln(N_0) \ll 1)$ and found slopes near 1 for each, but this expression can, in principle, be tested for more abundant species as well.

Moreover, self-similarity predicts another relation that can be tested across all the species in a biome: the shape of a plot of ln(number of occupied window squares of a particular area for each species) versus the ln(abundance of each species). Again a straight line is predicted, with a slope of $\ln(A_0/A_{\mathrm{window}})/\ln(N_0)$, where A_0 is the area containing the N_0 individuals. From this result, an explicit range-abundance relation follows, but for a definition of range that is dependent on the scale of observation. In particular, defining range to be (total number of window squares occupied)(area of window) yields the relation: $R = cn^b$, where R is range, n is species abundance, and b is the slope given above: $\ln(A_0/A_{\mathrm{window}})/\ln(N_0)$. For the UK breeding-bird census data, with a 10 km $\times$ 10 km window, the predicted slope is $b \sim 0.43$ in good agreement with the data [8].

12 ORIGINS OF SELF-SIMILARITY

In a more speculative vein, I turn briefly to the question: why is the distribution of species self-similar? One possibility is that species are distributed the way they are because they occupy niches in the physical landscape and that physical landscape is fractal or self-similar (see, e.g., Mandelbrot [18]). Perhaps, but I wish to suggest that the answer may instead lie in dispersal mechanisms in nature. Random walk processes yield self-similar behavior and random walk is a reasonable model for how a colonizer in a habitat disperses.

A possible generating model for the settlement of individuals within a species and species within a landscape is a Monte Carlo simulation in which at each time step in a random walk, a bifurcation (reproduction) occurs with individual settlers in a given species deposited at each node until the number of individuals for that species is "used up." The allotment of individuals for each species would be determined from the solution to Eq. (14). Lest this seem circular, since Eq. (14) was derived from self-similarity, I remind the reader that Eq. (14) is a consequence of self-similarity but not sufficient to produce it—in particular, one could have an abundance distribution that obeyed Eq. (14), for any value of x, and therefore any value of z, but in which each species occupied its own isolated subset of total landscape area.

Whether or not this propagule mechanism will result in a species distribution that at least in a statistical sense yields the SAR and all the other consequences of that relationship derived and discussed here may of course depend sensitively on details of the simulation. Such details include the distribution of time steps or the initial placement of first colonizers of each species. If that is the case, the model will not be as interesting as it will be if it generates observed patterns of self-similarity in a robust manner.

13 CONCLUSION

The widely obeyed power-law form of the species-area relationship (SAR) is equivalent to a self-similarity condition on the distribution of species within an ecosystem. Self-similarity leads to new and testable predictions. One is an endemics-area relationship (EAR), the application of which strongly alters predictions of species extinction under habitat loss. A second is a formula for the fraction of species in common to two censused patches separated in space. By measuring species commonality across a range of interpatch distances, the scale-dependence of the SAR exponent can be determined; this provides a means of estimating species richness in biomes too large to thoroughly census directly. The formula for species commonality also leads to a prediction for the dependence of species richness on the shape of censused patches; comparison with census data confirms this prediction. A species-abundance distribution and a range-abundance distribution are also shown to follow from the self-similar SAR. The self-similar species-abundance distribution differs from the traditionally assumed lognormal in that it is skewed toward rarity.

The SAR, cast in terms of self-similarity rules, is remarkably rich in implications. Limited data are consistent with the predicted consequences, but the number of untested implications across the many habitats and taxa far exceed the number tested. Moreover, the origin of self-similarity in species distributions remains an exciting research topic for the future.

ACKNOWLEDGMENTS

I am especially grateful to Ann Kinzig, my collaborator in much of this work, and to Jessica Green for her contributions to my understanding of the abundance distribution. Thanks as well to Sarah McCarthy, Kevin Taylor, Tim Blackburn, and Marc Fischer for advice, analysis, and data acquisition, and to the attendees of the SFI Workshop for stimulating discussions. Hospitality and support from the NERC Centre for Population Biology and support from NSF's REU program, the Winslow Foundation, and the Class of 1935 Endowed Chair at U.C. Berkeley were instrumental in pursuing this work.

REFERENCES

[1] Caswell, H., and J. E. Cohen. "Local and Regional Regulation of Species-Area Relations: A Patch-Occupancy Model." In *Species Diversity in Ecological Communities*, edited by R. Ricklefs and D. Schluter, 99–107. Chicago, IL: University of Chicago Press, 1993.

[2] Cody, M. L. "Bird Diversity Components Within and Between Habitats in Australia." In *Species Diversity in Ecological Communities*, edited by R. Ricklefs and D. Schluter, 147–158. Chicago, IL: University of Chicago Press, 1993.

[3] Coleman, B. "On Random Placement and Species-Area Relations." *J. Math. Biosci.* **54** (1981): 191–215.

[4] Condit, R., S. P. Hubbell, J. V. LaFrankie, R. Sukumar, N. Manokaran, R. Foster, and P. S. Ashton. "Species-Area and Species-Individual Relationships for Tropical Trees: A Comparison of Three 50 ha Plots." *J. Ecol.* **84** (1996): 549–562.

[5] Connor, E., and E. McCoy. "The Statistics and Biology of the Species-Area Relationship." *Amer. Natur.* **113** (1979): 791–833.

[6] Diamond, J. "The Island Dilemma: Lessons of Modern Biogeographic Studies for the Design of Nature Reserves." *Biol. Conservation* **7** (1975): 129–146.

[7] Gaston, K. J. *Rarity*. London: Chapman and Hall, 1994.

[8] Harte, J., and T. Blackburn. "Self-Similarity and Range-Abundance Relationships for British Birds and Mammals." *Amer. Natur.* (1999): in review.

[9] Harte, J., and A. Kinzig. "On the Implications of Species-Area Relationships for Endemism, Spatial Turnover, and Food Web Patterns." *Oikos* **80** (1997): 417–427.

[10] Harte, J., A. Kinzig, and J. Green. "Self-Similarity in the Distribution and Abundance of Species." *Science* **284** (1999): 334–336.

[11] Harte, J., S. McCarthy, K. Taylor, A. Kinzig, and M. Fischer. "Estimating Species-Area Relationships from Plot to Landscape Scale Using Species Spatial-Turnover Data." *Oikos* (1999): in press.

[12] Hubbell, S. P., and R. C. Foster. "Commonness and Rarity in a Tropical Forest: Implications for Tropical Tree Conservation." In *Conservation Biology: The Science of Scarity and Diversity*, edited by M. E. Soulé, 205–231. Sunderland, MA: Sinauer, 1986.

[13] Kinzig, A., and J. Harte. "Distributions of Endemic Species: Implications for Species Extinction Rates." *Ecology* (1999): submitted.

[14] Kunin, W. "Sample Shape, Spatial Scale, and Species Counts; Implications for Reserve Design." *Biol. Conservation* **82** (1997): 369–377.

[15] Kunin, W. "Extapolating Species Abundance Across Spatial Scales." *Science* **281** (1998): 1513–1515.

[16] Leitner, W. A., and M. L. Rosenzweig. "Nested Species Area Curves and Stochastic Sampling: A New Theory." *Oikos* **79** (1997): 503–512.

[17] MacArthur, R., and E. Wilson. "An Equilibrium Theory of Insular Zoogeography." *Evolution* **17** (1963): 373–387.

[18] Mandelbrot, B. *The Fractal Geometry of Nature*. San Francisco: W. H. Freeman, 1982.

[19] May, R. M. "Patterns of Species Abundance and Diversity." In *Ecology and Evolution of Communities*, edited by M. L. Cody and J. M. Diamond, 81–120. Cambridge, MA: Belknap Press, 1975.

[20] May, R., J. Lawton, and N. Stork. *Extinction Rates*, edited by J. Lawton and R. May, 1–24. Oxford: Oxford University Press, 1995.

[21] Pimm, S., G. Russell, J. Gittleman, and T. Brooks. "The Future of Biodiversity." *Science* **269** (1995): 347–354.

[22] Preston, F. W. "The Canonical Distribution of Commonness and Rarity: Part I." *Ecology* **43** (1962): 185–215.

[23] Rosenzweig, M. L. *Species Diversity in Space and Time*, 436. Cambridge, MA: Cambridge University Press, 1995.

[24] Simberlof, D., and L. Abele. "Island Biogeography Theory and Conservation Practice." *Science* **191** (1976): 285–286.

[25] Stone, B., J. Sears, P. Cranswick, R. Gregory, D. Gibbons, M. Rehfisch, N. Aebischer, and J. Reid. "Population Estimates of Birds in Britain and in the United Kingdom." *British Birds* **90(1–2)** (1997): 1–22.

[26] Wilson, E. O. *The Diversity of Life*. New York: W. W. Norton, 1992.

Index

A

abundance
  frequency distribution of, 15, 326
  range distribution, 339
acceleration reaction forces, 77, 79
adaptation, 222
adaptive management, 298, 301
aerobic capacity, 54
aerodynamic forces, 38–40
Alexander, R. McNeill, 37
algebra of complex scalings, 307
allocation models, 238–239
  comparison of, 247
  description of, 238
  See also Charnov's model,
    238–239, 241–242, 244, 247, 249,
    260
  See also Kozłowski and Weiner
    model, 238, 240, 247
allometric equation, 2–4, 6, 10,
    13–14, 39, 46, 114, 120, 169
  definition of, 19
  See also power laws, 2
allometric exponents, 48, 258
allometric predictions, 301, 319

allometric relationships, 4, 8, 34, 88,
    97, 117, 237, 259, 267, 311
  alteration of, 31
  definition of, 3
allometric scaling, 222, 224
allometric theory
  comparison methodology, 12–13
  interspecies comparisons in, 12
  intraspecies comparisons in, 12
allometric tradeoff, 232
allometry, 3, 8, 38, 199–200, 216,
    237, 318
  definition of, 114
  examples of, 4
  future research in, 14
  negative, 58, 64
  of plants , 169
  positive, 52, 55, 58, 64
  research history, 19
  simple, 27–28, 114
ammonites, 81
ammonoid shells, 74
anagenesis, 225
anagenic exponents, 224
anatomical models, 146

antennae, 68
architectural geometry, 167
area preserving, 203
    branching, 95–98
arterial bifurcation, 132, 136, 147,
    154–156
    area expansions in, 136–137
    asymmetry of, 133–134
    optimization principles, 130
    symmetry of, 133
arterial compliance, 124
arterial models, 145
arterial pressure, 116, 119
arterial pulse transmission
    characteristics, 121
arterial structures, self-similar, 162
arterial tree, 120, 122, 124, 130, 136,
    162
    delivering vs. distributing vessels,
    135
    optimal design, 125
    model, 146, 161
    See also binary branching trees,
    145
artificial selection, 26
Artiodactyla, 260
attenuation coefficient, 123
available resources, 238
average population density, 301
avian incubation, 307

B
bacteria, 221
basal ancestral group, 55
basilisk lizards, 71
bees, 38–39
bending, 75
Bessel function, 103
Biewener, Andrew A., 51
bifurcation index, 133
bifurcation law, 147
bifurcation symmetry, 151, 154
binary branching tree, 145, 147
biodiversity, 299
biological design
    principles of, 10
biological diversity, 1, 16, 19, 87
    See also species diversity, 2

biological network transport
    systems, 91
biological rate processes, 5
biological similarity criteria, 114–115
biological supply networks, 109
biological times, 5
biomechanical constraints, 169,
    180–181
biomechanics, 5, 8, 12, 54
birds, 261–262
    metabolic rates of, 8
bivalves, 81
blood flow, 117, 120, 122, 124, 154
    velocity, 151, 153
blood pressure, 10, 99–100, 114, 118,
    120, 151
    subsystems, 119
blood vessel, 76
    branching structure of, 129
blue whale, 316, 319
body mass, 3, 14, 19, 87, 313
body size, 14
    consequences of, 1–2
    optimization, 244, 250
    ranges, 299
    role in plant design, 169
bone strength, 51
Bonner, John Tyler, 25
boundary layer, 68, 74
    defined, 68
bovids, 55
brain size, 258–260
branch architecture, 168–169
branching
    angle, 149, 208
    area-preserving, 95–98
    asymmetry, 149
    geometric rules of, 12
    law, 132–137
    network, 92
    order, 206, 212
    rate, 155–156
    vascular, 122
    volume-servicing, 95, 98
branching structure, 58, 64
    hierarchy in, 109
Brody, Samuel, 2, 5
Brownian motion, 256, 258
bryozoans, 79

Buckingham's pi-theorem, 116
buds, 205

C
Calder, W. A., III, 2
capillaries, 94
carbon budget, 214
cardiac efficiency, 118
cardiac hypertrophy, 117
cardiac output, 114, 119, 121
cardiovascular system, 93
    allometry of, 117–120
    branching structure, 129
    function factors, 118
    human, 129–137
    invariant characteristics, 113
    of mammals, 10–11, 113–126
    variant characteristics, 113
carnivores, 55, 260
catapult mechanisms, 41, 43
CCO algorithm, 146, 149, 156, 161
CCO method, 160
CCO modalities, 150, 154
CCO models, 146, 149–150, 160, 162
    applications, 162
    developments, 162
    fractal aspects, 160
    geometry of, 159
    key features, 158–159
    of fascular trees, 160
    three dimensional, 160
    vascular tree, 148, 160
cell shape, 223
cell size, 222
census, 301, 310
    accuracy of methodology, 310–311
    techniques, 319
ceratomorphs, 55
character evolution
    independent events, 254–255
Charnov, E. L., 9, 239–240
Charnov's model, 238–239, 241–242,
    244, 247, 249, 260
    assumption, 239, 249
    comparison to Kozłowski and
        Weiner's model, 238, 240, 247
    productivity, 239
circulatory allometry, 91, 118
cladogenic exponents, 224–225

climate, 212
collateral vessels, 131
collector's curve, 18
colonial animals, 79
commonality, 329–332, 334
    application to species richness, 335
    definition of, 329
    field test for, 331–332
    formula for, 330
community, 79, 216
    composition, 281
    structure, 253
    patterns of, 261
comparative biology, 299
compartment models, 146
competition, 223
    direct, 261
competitive advantage explanation,
    261–262
complex systems, 10, 16
compound leaves, 206
computer-generated trees, 210
conducting tissue, 179
connective structure, 150
conservation biology, 298, 319
conservation planning, 299, 312
constrained constructive
    optimization (CCO) models,
    146, 149–150, 160, 162
    applications of, 159
copepods, 70, 78
corals, 77
coronary network, 129–132
    form of, 131
    role of function, 130
    strategic design of, 131
correlated response, 232
cost function, 100, 238
Cotgreave, P., 261–262
critical buckling, 201
cross-sectional area, 203
cross-species analysis (TIPS), 254,
    256, 258, 263
cross-species comparisons, 253–254

D
da Vinci, Leonardo, 168, 200, 203
Darwin, Charles, 26
deceleration parameter, 206

delivering vessels, 156
density, 301
  dependence, 240
  parameters, 302
  perfusion, 159
density-body size relationship, 267,
    270
detectability, 310
developmental trajectory, 212
digestion, 72
dimensionless numbers, 72
dissected tree, 200
distributing vessels, 154, 156
drag coefficient, 74
drag force, 79
dynamically scaled, 77

E
ecological allometries, 311
ecological communities, 190
ecological scaling, 15–16
economy of design, 9–10
ecosystems, 15
ectothermic, 5
ejection fraction, 119
elastic buckling, 180
elastic similarity, 27, 37, 58, 64, 203
elastically similar scaling, 55
elasticity, 33, 102, 122
embryologic development, 159
embryos, 80
Endangered Species Act (ESA), 298,
    301, 305, 308
endemics-area law
  origin of, 328
endemics-area relationship, 326–329,
    333–334, 340
  definition of, 328
endothermic, 5
energetic efficiency, 223
energetic equivalence hypothesis, 269
energetics, 76
energy
  acquisition, 47, 231
  conservation mechanism, 61
  conversion, 231
  dissipation, 109
  exchange of, 33
  investment, 238

energy (cont'd)
  management, 238
  metabolic transformation of, 9
  sinks, 238
energy use, 40, 186, 267
  equal, 8
  See also resource uptake, 8
Escherichia coli, 222
eutherian mammals, 55
evolution, 12, 14, 25, 37, 47, 67, 81
  of body size, 231
evolutionary adaptation, 80, 100, 222
  example of, 32
evolutionary dynamics, 222
exponential relationships, 75
extinction, 326
  comparison of methods, 333–334
  estimation of, 333
  rate, 256–258

F
fast lifestyle, 247
Felsenstein's independent contrasts
    method, 255–256, 261
Fibonacci series, 169
field exercises, 199, 214
fine tuning, 311
fish, 269
fitness, 10, 48, 80, 222, 239
  criterion, 211
flexible, 78–79
flexural stiffness, 77–78
flow velocity, 151, 153
fluid flow, 168
fluid motion, 73
flying, 68
form drag, 74
fossil fuel, 80, 214
fractal, 2, 8
  geometry, 169–170
  model, 146, 209
  parameters, 160
  recursion, 207
  self-similar, 96–97
  trees, 200, 205
frequency-magnitude relationships,
    15
frogs, 43–48
function changes, leakiness and, 70

function, transitions in, 81
functional shifts, 72

G
gametes, 79
genetic variation, 59
geometric constraints, 91
geometric model, 185–186
geometric optimization, 148, 153
geometric scaling, 4–5, 37, 43, 57
    at small size, 55–58
geometric similarity, 29, 37, 43, 45,
        52, 55–56, 61, 64
    See also isometry, 52
gestation, 316
    length, 259–260
gills, 68
gliding, 74
global warming, 200, 270
gorgonians, 77
gray whale, 317
growth
    deceleration, 206
    history, 202
    rate, 47, 223, 238, 249
Gutenberg-Richter law, 16

H
habitat, 67, 79–80, 262
    loss, 326, 333–334
hairs, 70
Harvey, Paul H., 253
heart
    efficiency of, 120–124
    geometric shape, 115–116, 122
    rate, 90, 99, 118, 123
        and life span, 119
        variance of, 125
    rhythms, 119
    size, 118–119
    See also cardiovascular system,
        113
heritability, 228
hierarchical branching network, 171
hierarchically scaled architectures, 11
Hill's equation, 44
Horn, Henry S., 25
hovering, 38–40
    aerodynamics of, 38

hovering (cont'd)
    helicopter theory of, 38
    scaling rules for, 39–40
hummingbirds, 38–39, 315, 319
Huxley, Julian, 2, 5, 30, 114, 169, 237
hydraulic architecture, 168, 183, 216
hydraulic function, 203
hydraulic segmentation hypothesis,
        183
hydrodynamic constraint, 175
hydrodynamic forces, 77, 79
hydrodynamic function
    example of locomotion, 68, 72
hydrodynamic resistance, 108

I
incubation, 315
    avian, 307
independent contrasts analysis
        (PICS), 256
inertial forces, 39
    ratio to aerodynamic forces, 39–40
interspecific allometry, 237, 240,
        242–243, 250
interspecific exponents, 250
interspecific slope, 242–244
intraspecific slope, 242–244
invariant quantities, 10, 19, 116–117
investment, 238
islands, 80
isometry, 3, 28–30, 47, 52, 200
    See also geometric similarity, 52

J
jumping
    catapult mechanisms, 41
    evolutionary considerations, 41
    model, 44–45, 48
    scaling rules for, 41–42

K
kinematics, 73, 75
kiwi, 319
Kleiber, Max, 5, 88
Kleiber's law, 89, 254
Korteweg-Moens velocity, 102–103
Kozłowski and Weiner's model, 238,
        240, 247

L

Lagrange multipliers, 100
Lamé relation, 116
land management, 301
landscape ecology, 301
Laplace's law, 115–116
leaf area, 210
leakiness, 70, 73
  definition of, 70
  See also hydrodynamic function,
    70
least squares regression, 4, 12, 53,
    258, 313
leaves
  dispersion of, 80, 210
Leonardo's rule, 214–215
Li, John K-J., 113
life expectancy, 8
  heart rate and, 119
life history, 8, 10, 67, 237, 247
  allometries, 239
  characteristics, 298
  information, 318
  variables, 259, 313
limb mechanical advantage, 60–64
limb posture, 53–54, 60
  changes in, 62, 64
lizard, 71
locomotion, 68
  energy cost, 54
  energy use, 61
  limb posture in, 53, 61
  See also hovering, 68
  See also hydrodynamic function,
    68
  See also jumping, 68
locomotor posture
  changes in, 53
locomotor stress, 53
locusts, 41–43
lognormal, 326, 336–337

M

Mach number, 117
mammalian arterial systems, 145
mammalian cardiovascular system,
    93
mammalian gestation, 307
mammalian respiratory system, 106

mammals, 4, 80
  cardiovascular system, 11, 113–126
  eutherian, 55
  metabolic rates in, 5
  See also terrestrial mammals, 52
mathematical models, 20
McMahon, T. A., xii, 203
mechanical advantage, 55
  limb, 60–64
  muscle, 61, 64
mechanical constraints, 55, 63, 173
mechanical design, 51
mechanical stress, 52
  definition of, 51
metabolic energy, 41, 52
metabolic heat, 90
metabolic rate, 8, 19, 47, 54, 67, 78,
    88–89, 94, 99–100, 119, 121, 132,
    177, 190, 224, 259, 300
  definition of, 9
  of frogs, 44
  of jumping insects, 41–42
  of mammals, 5
  of plants, 5, 15, 188
metabolic reserves, 231
metabolic support functions, 301
minimal viable population size, 301
model, 110, 247
  allocation, 238–239
  anatomical, 146
  arterial tree, 145–146, 161
  Charnov's, 238–239, 241–242, 244,
    247, 249, 260
  CCO, 146, 149–150, 160, 162
  fractal, 146, 209
  geometric, 185–186
  jumping, 44–45, 48
  Kozłowski and Weiner's, 238, 240,
    247
  optimization, 159, 162
  pipe, 93, 95, 167–168, 181, 203
  plant vascular systems, 93, 170,
    184
  population genetic, 258
  quarter-power scaling, 100, 110
  Shinozaki's, 203
  windkessel, 124
  Yoda's geometric, 14–15
molecular diffusion, 73

Monte Carlo simulation, 340
morphological allometries, 237
morphology, 67
mortality rate, 238–242, 244, 249
mortality, size-independence in, 239,
    241
Murray, Cecil, 2, 4
Murray's law, 99, 132, 134–135, 137
    vs. square law, 137
muscle
    effect of stiffness on, 56–57
    mechanical advantage, 53–54
    properties, 43–45, 49
    stress, 51, 54, 61
myocardial muscle, dual role of, 145
myocardial oxygen consumption, 120

N

natural selection, 9, 11, 19, 25–26,
    42, 80, 221, 237, 240, 314
    for efficiency, 27, 31, 34
    for size, 26–27
Navier-Stokes equation, 102
nonpulsatile flow, 99
numerical yield, 223

O

ontogeny, 12, 78, 81, 237
open tree structure, 131
optimal body size, 242, 244, 250
optimal design, 37–38, 48
optimal efficiency
    genetic basis of, 26
optimal life history, 241
optimal size, 76, 239–241, 244, 247
optimal tree, 211
optimization, 159, 162
    constrained, 162
    geometric, 148, 153
    models, 146
    of body size, 37–38, 249
    of global target function, 161
    structural, 148, 153
    targets, 162
oscillatory behavior, 103
oscillatory flow, 78
oxygen consumption, 94

P

Pareto distribution, 16
perfusion density, 159
perfusion volume, shape of, 150–151
peripheral resistance, 120
Peters, R. H., 2, 6–8, 13
phase constant, 123
phyllotaxic schemes, 169
phylogenetic effects, 313
phylogenetic framework, 254
phylogenetic information, 253–254,
    263
phylogeny, 12, 237
    reconstruction of, 256–258
physical factors, 72, 91
physical principles, 81, 91
phytoplankton, 269
phytosynthesis, 214
pi-numbers, 116–117
pipe model, 93, 95, 167–168, 181, 203
plant allometry, 169
plant architecture and form, 169
plant size, 169
plant vascular network, 107
plant vascular systems, model of, 93,
    170, 184
plants
    metabolic rates in, 5
    vascular, 15
    See also individual names, 15
Poiseuille formula, 103
Poiseuille impedance, 104
Poiseuille resistance, 101
Poisseuille's law, 146
population density, 8, 169, 188, 254,
    261–262, 267–268, 270–271
    average, 30
    variance in, 15
    See also abundance, 267
population doubling times, 318
population genetic models, 258
population growth rate, 226
population size, 301
posture
    adjustments in, 61
power functions, 2–3, 17–19
power law, 2, 87, 114, 133, 184
    See also allometric equations, 2
predator, 72, 75

pressure, 76
  gradient, 178
prey, 72, 75
primary productivity, 274
primates, 8
production rate, 238, 240–241,
    243–244, 247, 249
productivity, 187
propagation constant, 123
protowings, 72
pseudo-random number sequences
    (PRNSs), 149–150
pulsatile bellows pump, 93
pulsatile compression pump, 93
pulsatile flow, 101, 103
  impedances, 122–123
pulse transmission, 125
  invariant, 125
pulse wave velocity, 123, 125
pulsewave transmission, 151
Péclet number, 73

Q
quarter-power, 300
  allometric scaling laws, 91, 190
  exponents, 10, 19
  interspecific scaling, 242
  scaling, 52, 54, 110, 114, 120–121,
    242, 267
    common mechanisms theory,
      9–10
    elastic similarity theory, 5
    history of, 4–8
    in plants , 167
    metabolic rate in, 9, 54
    model of, 110
    theories of, 5–6

R
Raleigh indices, 115, 117
random walk model, 340
range-abundance relation, 339
recovery planning, 298, 309
recovery time estimates, 308
reflection coefficient, 123
regression models, 13
relative fitness, 222
reproduction, 76, 238
reproductive capacity, energy in, 238

reproductive capacity, size
    dependence of, 238
reproductive value, 211
resistance-capacitance model, 168
resource, 199
  allocation, 9, 238
  availability, 169, 238
  requirements, 169
  uptake, 9, 232
  use, 170, 186, 190
    in ecological communities, 14
respiration, 214
respiratory system, 91, 93, 97
restoration ecology, 301
Reynolds number, 31, 68, 117
ring-porous wood, 204
rotifers, 32

S
safety factors, 28, 30, 52–53, 55, 57,
    64, 203
  definition of, 51
scaling factors, 117
scaling invariance, 117
scaling law, 2, 9, 11, 90–91, 98, 105,
    113, 115, 133–136, 160, 216
  definition of, 19
  for vascular trees, 130
  modeling of, 20
  origin of, 170
  qualitative shifts, 254
  quarter power, 5
scaling predictions, 47–48
scaling relationships, 9, 19, 25, 58,
    109, 260
  among individuals, 12–13
  cross-species, 255–258
  levels of study, 11
  of populations, 14–18
  within organisms, 11–12
scaling relationships of allometry,
    299
scallops, 71
Schmidt-Nielsen, Knut, 2, 5, 169, 299
seaweeds, 77
self-similar
  fractal, 96–97, 160
  type of arterial structure, 162

self-similarity, 2, 4, 11, 18, 43,
    145–146, 326, 338, 341
  consequences of, 329
  definition of, 327
  origins of, 340
sessile organisms, 79
setulose appendages, 68
shade-adapted monolayer, 210
shape, 74, 150
shifts in hydrodynamic function, 68
Shinozaki's model, 203
shoot, 206
shrimp, 71
similarity principles, 117
sister taxon analysis, 256
size change, 72
  anatomical adaptation, 32
  effect on shape, 27–32
  transition function, 68, 73
size decrease
  implications of, 32, 55
size distributions, 244, 250
size increase
  convergent solutions to, 31
size refuge, 72
size-dependent production rate, 239
size-independent mortality, 241
skeletal support system, 51–52, 54,
    64
  stiffness in, 56–58
  stress on, 54
skeleton, 76
skin friction, 74
slime mold, 28
slow lifestyle, 247
spatial variability, in population
    densities, 280
spawning, 79
speciation rate, 256
speciation-extinction process,
    256–258
species diversity, 14, 18
  See also biological diversity, 3
species richness, 300, 326, 335–336
  effect of patch shape, 332–333
  estimation methods, 335–336
species-abundance distribution, 271,
    332, 336–338
  definition of, 336

species-abundance distribution
    (cont'd)
  lognormal, 338
  self-similar, 341
species-area relationship, 17, 333–334
  abundance distribution in,
    336–337
  definition of, 325
  mechanistic processes in, 17–18
  power-law form, 336, 340
  self-similar, 337, 341
species-range distribution, 332,
    338–339
species-time relationship, 17
square law, 135, 137
  vs. Murray's law, 137
stabilizing selection, 232
starfish, 72
Starling's law, 115
static stress similarity, 55
stationary, 212
stepwise optimization, 161
stiffness, 56–58, 60–61, 77–78
  importance of, 56–57
stochastic processes, 17
Strahler ordering system, 153, 206
stream invertebrates, 74
streamlined, 74
stroke volume, 118–119
Strouhal number, 38
structural optimization, 148, 153
sun-adapted multilayer, 210
surface texture, 74
surface-to-volume ratio, 227
surrogate values, 301
suspension feeding, 76, 79
  appendages, 68
swimming, 31, 68
symmetrical, 212
symmorphosis
  consequences of, 10
  definition of, 10

T
target of selection, 232
Taylor, L. R., 15
Taylor's power law, 15
temporal variability, 268
terminal dominance, 208

terminal units, invariance of, 109
terrestrial mammals, 52, 55, 60,
    63–64
  giant sizes of, 55
  limb posture in, 53
  mechanical advantage in, 53–54
  See also mammals, 52
thinning law, 14, 169, 186
  −3/2 geometric model, 15, 184
  −4/3 resource based, 15
thinning trajectories, 186
third variable problem, 259–261
Thompson, D'Arcy, 2, 4, 25, 32, 114
  theories of, 25–26
threated species, 308
transformations, 113
transition, 72, 81
  in function, 81
transport, 227
  metabolic cost of, 54
tree, 4, 12, 27, 64, 167, 255, 258
  differential scale constraints, 58–60
  max height of, 92
  trunk thickness of, 27–28
  See also branching structure, 58
tree ring data, 200–201
trophic role, 72
turkey-track module, 209

U
ursids, 55

V
value of biodiversity, 299
variance components, 228
vascular branching, 122
vascular plants, 167
  application to, 8
  architecture of, 12
  plants, description of, 167
  metabolic rates of, 15
  research themes, 168
vascular systems, 91
  effective length of, 125
vascular trees, 159
  scaling laws for, 130
ventricular function
    factors, 118
ventricular pressure, 116

ventricular tachycardia, 119
volume-servicing branching, 95, 98

W
walking on water, 71
water currents, 79
waves, 77, 79
windkessel model, 124
wings, 71, 74
Womersley number, 102
worms, 72

Y
Yoda's geometric model, 14–15
Young's modulus, 200

Z
Zamir, Mair, 129, 153–156
Zipf's law, 16
zooplankton, 269